Technology, Public Policy, and the Changing Structure of American Agriculture

CONGRESS OF THE UNITED STATES
Office of Technology Assessment
Washington, D. C. 20510

Recommended Citation:

U.S. Congress, Office of Technology Assessment, *Technology, Public Policy, and the Changing Structure of American Agriculture,* OTA-F-285 (Washington, DC: U.S. Government Printing Office, March 1986).

Library of Congress Catalog Card Number 85-600632

For sale by the Superintendent of Documents
U.S. Government Printing Office, Washington, DC 20402

Foreword

American agriculture is undergoing significant change and stress. Much of the recent change has been attributed to the financial farm crisis caused mainly by declining agricultural exports. However, underlying these financial difficulties are strong technological and structural forces which will cause further changes and adjustments in American agriculture for the remainder of this century.

Congress, concerned about the nature of these adjustments, requested the Office of Technology Assessment (OTA) to analyze the underlying technological, structural, and political forces which impact American agriculture and to determine the industry's probable future direction. Committees requesting the study include: the Senate Committee on Agriculture, the Senate Small Business Committee (the Subcommittee on the Family Farm), the Joint Economic Committee, the House Committee on Science and Technology, and the House Committee on Agriculture (the Subcommittee on Livestock, Dairy, and Poultry; the Subcommittee on Department Operations, Research, and Foreign Agriculture; and the Subcommittee on Forests, Family Farms, and Energy).

In the course of preparing this report, an interim report entitled *A Special Report for the 1985 Farm Bill* was transmitted to the requesting committees for their use during the debates and the writing of the Food Security Act of 1985 (1985 Farm Bill). The special report focused on assessment findings that were particularly relevant for issues debated in that legislation.

This report addresses the longer run issues that technology and certain other factors will have on American agriculture during the remainder of this century. It focuses on the relationship of technology to: agricultural production, structural change, rural communities, environment and natural resource base, finance and credit, research and extension, and public policy. The assessment identifies many benefits that new technologies will create, but these benefits will also exact substantial costs in potential adjustment problems. This report is a first step toward understanding these interrelated problems and identifying policies to ameliorate them.

OTA greatly appreciates the contribution of the advisory panel, workgroups, workshop participants, authors of the technical background papers, and the many other advisors and reviewers who assisted OTA from the public and private sector. Their guidance and comments helped develop a comprehensive report. As with all OTA studies, however, the content of this report is the sole responsibility of OTA.

JOHN H. GIBBONS
Director

Technology, Public Policy, and the Changing Structure of American Agriculture: Advisory Panel

Frank Baker
Director, International Stockmen's School
Winrock International Livestock Research and Training Center
Arkansas

James Bonnen
Professor
Department of Agricultural Economics
Michigan State University

William Brown
Chairman of the Board
Pioneer Hi-Bred International, Inc.
Iowa

Frederick Buttel
Associate Professor
Department of Rural Sociology
Cornell University

Willard Cochrane
Consultant
California

Jack Doyle
Director
Agricultural Resources Project
Environmental Policy Institute
Washington, DC

Marcia Dudden
Dudden Farms, Inc.
Iowa

Walter Ehrhardt
Ehrhardt Farms
Maryland

Dean Gillette
Professor
Harvey Mudd College
Claremont College

Roger Granados
Executive Director
La Coopertiva
California

Richard Harwood
Program Officer
Winrock International
Virginia

Charles Kidd
Dean
College of Engineering Science, Technology, and Agriculture
Florida A&M University

Robert Lanphier III
Chairman of the Board
DICKEY-john Corp.
Illinois

Edward Legates
Dean, College of the Agriculture and Life Sciences
North Carolina State University

John Marvel*
President and General Manager
Research Division
Monsanto Agriculture Products Co.
Missouri

Donella Meadows
Adjunct Professor
Resources Policy Center
Dartmouth College

Don Paarlberg
Consultant
Indiana

Don Reeves
Consultant, Interreligious Taskforce on U.S. Food Policy
Nebraska

Milo Schanzenbach
Schanzenbach Farms
South Dakota

*Resigned May 1985.

Technology, Public Policy, and the Changing Structure of American Agriculture: OTA Project Staff

Roger C. Herdman, *Assistant Director, OTA*
Health and Life Sciences Division

Walter E. Parham, *Food and Renewable Resources Program Manager*

Michael J. Phillips, *Project Director*

Yao-chi Lu, *Senior Analyst*

Robert C. Reining, *Analyst*

Juliette Linzer,[1] *Research Assistant*

Kathryn M. Van Wyk, *Editor and Writer*

Administrative Staff

Patricia Durana[2] and Beckie Erickson,[3] *Administrative Assistants*

Nellie Hammond, *Secretary*

Carolyn Swann, *Secretary*

---

[1]Through May 1984.
[2]Through July 1985.
[3]After July 1985.

Contents

Chapter 1
Summary

Contents

Tables

Chapter 1

Summary

Over the next 15 years, American farmers will be offered an extensive array of new biotechnologies and information technologies that could revolutionize animal and plant production. The adoption of these technologies will be critical for shoring up the United States' lagging ability to compete in the international marketplace. Indeed, 83 percent of the estimated 1.8-percent annual increase in agricultural production needed to meet world agricultural demand by year 2000 must come from increases in agricultural yields, yields that can only be possible through the development and adoption of emerging technologies.

Yet if current agricultural policies remain in force, this new biotechnology and information technology era will also generate marked changes in the structure of the agricultural sector and of the rural communities that support farming. Some of these changes are already evident: Farming is becoming more centralized, more vertically integrated. Large farms, though small in number, now produce most of this country's agricultural output. Operators of small and moderate-size farms, the so-called backbone of American agriculture, are becoming increasingly less able to compete, partly because they lack access to the information and finances necessary for adopting the new technologies effectively. Many such farmers must relocate, change to other kinds of farming, or give up farming altogether. The disappearance of these farm operations is causing repercussions for other businesses in the rural community and for the labor pool in general, which must absorb all those whose livelihood once depended on agricultural production.

This report is the first step toward understanding the social and economic costs, as well as the benefits, of the emerging technologies for U.S. agriculture. It analyzes the dynamic forces influencing change in the structure of agriculture. Although technology was found to be an important force in such change, it is only one of several such forces. Public policy, institutions, and economics have had and will continue to have important roles in shaping agriculture. OTA analyzed the relationships between all these factors, focusing on the 150 production technologies that are likely to be available commercially over the next 15 years. The study results are presented in this report in four parts.

Part I identifies and analyzes the productive capacity of those emerging technologies that will help shape and define American agriculture to the year 2000. Chapters 2 and 3 describe the emerging technologies, discuss how they will be used in agriculture, and analyze the impact these technologies will have on animal and plant agriculture.

Part II traces the historical changes in agricultural structure. It provides a perspective for analyzing technology's distributional impacts on agricultural structure by surveying the characteristics of that structure and the factors that affect it.

How the emerging technologies, the policies, and structural change relate to one another is the subject of chapters 6 through 12 in part III. The chapters analyze the results of this relationship on: 1) future structure, 2) agricultural finance and credit, 3) survivability of crop and dairy farms of various sizes, 4) environment, 5) rural communities, and 6) agricultural research and extension.

Part IV draws the implications of the analysis for policymakers. It shows the direction in which agriculture is headed and concludes with congressional policy options for improving the picture of U.S. agriculture.

AGRICULTURAL DEPENDENCY ON WORLD MARKETS

The financial condition of many American farmers in the 1980s has significantly deteriorated during a long period of surpluses. The decline in agricultural exports is largely responsible for this situation. And although exports are not this report's central focus, the future of U.S. agricultural exports loom large in the background of this report.

Agricultural exports have historically been responsible for lessening the negative trade balance caused primarily by the manufacturing and energy sectors. This importance of agriculture to the balance of trade has increased significantly over the past 30 years. However, the past several years have witnessed a drop both in the value of U.S. agricultural exports and in agriculture's share of total U.S. exports.

Several key factors are causally related to recent declines in U.S. agriculture:

1. a weak world economy,
2. the strong value of the dollar,
3. the enhanced competitiveness of other countries,
4. an increase in trade agreements, and
5. price support levels that permit other countries to undersell the United States.

Although all of the factors are important, agricultural experts are beginning to focus on the lower costs of production in other countries as the long-term primary factor in the decline of this country's competitiveness. The United States faces strong competition in wheat, corn, rice, soybeans, and cotton. Each of these major export commodities has been produced by at least one country at or below the U.S. average production costs since 1981. Estimates suggest that any historic cost advantage that the United States may have enjoyed in these commodities is now tenuous.

Future exports will depend on the ability of American farmers to use new technology to produce commodities more efficiently than competing countries can. If the United States cannot effectively compete with other countries in the export market, reduced exports will magnify the structural change and adjustment that U.S. farmers and the rural communities will face because of technological change.

EMERGING TECHNOLOGIES FOR AGRICULTURE

Technology has made U.S. agriculture one of the world's most productive and competitive industries. Americans have already witnessed the dramatic results of two major technological eras in agriculture. The mechanical era of 1920 to 1950 allowed farmers to make the transition from horsepower to mechanical power and greatly increased the productive capacity of U.S. agriculture. The chemical era of 1950 to 1980 further increased agricultural productivity by increasing the farmers' ability to control pests and disease and by increasing the use of chemical fertilizers. Now, in the 1980s, American agriculture is being propelled by a new major technological thrust—the biotechnology and information technology era. The effects of this new era on agricultural productivity may be more profound than those experienced from either the mechanical or chemical eras.

Below is a brief summary of the technologies examined for this study. A more complete description of the 150 technologies can be found in chapter 2.

Biotechnology

Biotechnology, broadly defined, includes any technique that uses living organisms or processes to make or modify products, to improve plants or animals, or to develop micro-organisms for specific uses. It focuses on two powerful molecular genetic techniques: recombinant deoxyribonucleic acid (rDNA) and cell fusion technologies. Using these techniques scientists can visualize the gene—to isolate, clone, and study the structure of the gene and the gene's relationships to the processes of living things. Such knowledge and skills will give scientists much greater control over biological systems, leading to significant improvements in the production of plants and animals.

Animal Agriculture

In animal agriculture, advances in protein production, gene insertion, and embryo transfer will play a major role in increasing efficiencies in animal production.

Production of Protein.—One major thrust of biotechnology in animals is the mass production in micro-organisms of protein-like pharmaceuticals, including a number of hormones, enzymes, activating factors, amino acids, and feed supplements. Previously, these biological products could be obtained only from animal and human organs and were either unavailable in sufficient amounts or were too costly.

Some of these biological products can be used for detection, prevention, and treatment of infectious and genetic diseases; some can be used to increase animal production efficiency. One of the applications of these new pharmaceuticals is the injection of growth hormones into animals to increase production efficiency. For example, several firms are developing a genetically engineered bovine growth hormone to stimulate lactation in cows. Trial results indicate that cows treated with the hormone increase milk production by 20 to 30 percent, with only a modest increase in feed intake. Commercial introduction of the new hormone awaits approval by the U.S. Food and Drug Administration, which is expected to approve the hormone within the next 3 years.

In the area of disease prevention and treatment, an immunological product currently exists on the market that prevents "scours" in calves. In addition, vaccines produced by rDNA methods are currently being tested for foot-and-mouth disease, swine dysentery and, most recently, coccidiosis in poultry.

Gene Insertion.—A new technique arising from the convergence of gene and embryo manipulations promises to permit genes for new traits to be inserted into the reproductive cells of livestock and poultry, providing major opportunities to improve animal health and productivity. Unlike the genetically engineered hormones discussed above, which cannot affect future generations, gene insertion will allow future animals to be endowed permanently with traits of other animals. In this technique, genes for a desired trait, such as disease resistance or growth, are injected directly into either of the two pronuclei of a fertilized egg. On fusion of the pronuclei, the guest genes become part of all the cells of the developing animal, and the traits they determine are transmitted to succeeding generations.

Embryo Transfer.—Embryo transfer, which is closely related to gene insertion, involves artificially inseminating a super-ovulated donor animal[1] and removing the resulting embryos nonsurgically for implantation in surrogate mothers which then carry them to term. Prior to implantation, the embryos can be treated in a number of special ways. They can be sexed, split (generally to make twins), fused with embryos of other animal species (to make chimeric animals or to permit the heterologous species to carry the embryo to term), or frozen in liquid nitrogen for storage. Freezing is of great practical importance because it allows embryos to be stored until the estrus of the intended farm animal is in synchrony with that of the donor. Embryos used for gene insertions must be in the single-cell stage, having pronuclei that can be injected with cloned foreign genes. The genes likely to be inserted into cattle may be those for growth hormones, prolactins (lactation stimulators), digestive enzymes, and interferons, thereby providing both growth and enhanced resistance to diseases.

Even though less than 1 percent of U.S. cattle are involved in embryo transfers, the obvious

[1]An animal that has been injected with a hormone to stimulate the production of more than the normal number of eggs per ovulation.

benefits of this technology will push this percentage upward rapidly, particularly as the costs of the procedure decrease. Recently, a genetically superior Holstein cow and her 14 embryos were purchased for $1.3 million.

Plant Agriculture

The application of biotechnologies in plant agriculture could modify crops so that they would make more nutritious protein, resist insects and disease, grow in harsh environments, and provide their own nitrogen fertilizer. While the immediate impacts will be greater for animal agriculture, the long-term impacts of biotechnology may be substantially greater for plant agriculture. The potential applications of biotechnology on plant agriculture include microbial inoculums, plant propagation, and genetic modification.

Microbial Inocula.—Rhizobium seed inocula already are used widely to improve the nitrogen fixation of certain legumes. Extensive study of the structure and regulation of the genes involved in bacterial nitrogen fixation will likely lead to development of improved inocula. Moreover, research on other plant-colonizing microbes has led to a clearer understanding of the role of these microbes in plant nutrition, growth stimulation, and disease prevention, and the possibility exists for the modification and use of these microbes as seed inocula.

Monsanto has announced plans to field test genetically engineered soil bacteria that produce a naturally occurring insecticide potentially capable of protecting plant roots against soil-dwelling insects. The company developed a genetic engineering technique that inserts into soil bacteria a gene from a micro-organism known as *Bacillus thuringiensis,* a micro-organism that has been registered as an insecticide for more than two decades. Plant seeds could be coated with these bacteria before planting. As the plants grow, the bacteria would remain in the soil near the plant roots, generating an insect toxin that protects the plants.

Plant Propagation.—Cell culture methods for regeneration of intact plants from single cells or tissue explants are now used routinely for propagation of several vegetable, ornamental, and tree species. These methods can provide large numbers of genetically identical, disease-free plants that often exhibit superior growth and more uniformity over plants conventionally seed-grown. Such technology holds promise for breeding in important forest species whose long sexual cycles reduce the impact of traditional breeding approaches. Somatic embryos[2] produced in large quantities by cell culture methods can be encapsulated to create artificial seeds that may enhance propagation of certain crop species.

Genetic Modification.—Plant genetic engineering is the least established of the various biotechnologies used in crop improvement, but the most likely to have a major impact. Using gene transfer techniques, it is possible to introduce DNA from one plant into another plant, regardless of normal species and sexual barriers. For example, it is possible to introduce storage-protein genes from French bean plants into tobacco plants and to introduce genes that encode photosynthetic proteins in pea plants into petunia plants.

Transformation technology also allows introduction of DNA coding sequences from virtually any source into plants, providing those sequences are engineered with the appropriate plant-gene regulatory signals. Several bacterial genes have now been modified and shown to function in plants. By eliminating sexual barriers to gene transfer, genetic engineering will greatly increase a plant's genetic diversity.

Information Technology

Animal Agriculture

Information technology is the use of computer- and electronic-based technologies for the automated collection, manipulation, and processing of information for control and management of agricultural production and marketing. The most significant changes in future livestock production resulting from information technology will come from the integration of computers

[2]Embryos produced from body cells rather than reproductive cells.

and electronics into modern livestock production systems that will help make the farmer a better manager. Animal identification, animal reproduction, and disease control and prevention are some promising areas for information technology in livestock production.

Electronic Animal Identification.—Positive identification of animals is necessary in all facets of management, including recordkeeping, individualized feed control, genetic improvement, and disease control. Research on identification systems for animals has been in progress for some years. Soon, all farm animals will be "tagged" shortly after birth by an electronic device, called a transponder, that lasts the life of the animal. For example, some dairy cows now wear a transponder in the ear or on a neck chain. A feed-dispensing device identifies the animal by the transponder's signal and provides an appropriate amount of feed for the animal.

Reproduction.—The largest potential use of electronic devices in livestock production will be in the area of reproduction and genetic improvement. An inexpensive estrus detection device will allow: 1) animals to be rebred faster after weaning; 2) animals that did not breed to be culled from the herd, saving on feeding and breeding space; 3) time to be saved because breeding can be done faster; and 4) easier embryo transplants because of improved estrus detection.

Disease Control and Prevention.—Herd recordkeeping systems for animal health are already being developed and refined in the dairy, swine, and poultry industries. These recordkeeping systems will eventually be linked with the animal identification systems discussed above. Examples of the types of information that can be recorded for each animal include production records, feed consumption, vaccination profiles, breeding records, conception dates, number of offspring, listing and dates of diseases, and costs of medicines for treatment or prevention of disease. Bringing all this information together will allow the veterinarian and a manager of the livestock enterprise to analyze quickly a health profile for each animal and to plan for improved efficiency in disease control programs.

Plant Agriculture

Pest Management.—Information technology is already being used in plant agriculture for the management of insects and mites. Design improvements and availability of computer hardware and software will produce marked changes in insect and mite management.

Availability at the farm level of microcomputers, equipped with appropriate software and having access to larger centralized databases, will accelerate transfer of information and facilitate pest management decisionmaking. The advantages, simply in terms of information storage and retrieval, will be of major importance. The ready storage of and access to current and historical information on pest biology, incidence, and abundance; pesticide use; cropping histories; weather; and the like at the regional, farm, and even field level will facilitate selection of the appropriate management unit and the design and implementation of pest management strategies for that unit.

Current software has already greatly improved the efficiency and accuracy with which pest management decisions can be made and implemented. Much effort is being devoted to the development of new software and the improvement of existing software. The resultant products, in conjunction with the rapid advances being made in computer hardware, will provide a powerful force that will lead to dramatic changes in the implementation of integrated pest management (IPM) and to increases in the level of sophistication of IPM.

Irrigation Control Systems.—Because irrigation decisions are complex and require relatively large amounts of information, a microcomputer-based irrigation monitoring and control system is especially useful in areas with soils having variable percolation and retention rates, where rainfall is especially variable, or where the salinity of irrigation water changes unpredictably. In this system, a network of sensors, with radio links to the central processor, is buried in irrigated fields. Additional sensors may include weather station sensors to estimate crop stress and evaporation rates, salinity sensors, and runoff sensors. The central processor uses such information to allocate water automatically according to crop needs in each field, subject to considerations of cost, leaching requirements, and availability of water.

Radar, Sensors, and Computers.—Through the use of radar, sensors, and computers the correct amount of fertilizer, pesticides, and plant growth regulators can be applied to plants by integrating tractor slippage and chemical flow. The correct rate of application of most agricultural chemicals is usually within a narrow range for a given crop and field. However, application rates are often variable from area to area within a field, owing to changes in the flow rate of chemical slurries and to changes in tractor wheel slip, grading, and drawbar tension. Economic and environmental costs are associated with applications of too little or too much chemicals. Control of application rate depends on the ability to estimate rate of flow through the chemical sprayer and on the vehicle's speed over the field. The speed indicated by sensors in the tractor drivetrain is usually greater than the actual speed over the ground, owing to slippage of the drive wheels. The amount of slippage can be monitored by a doppler radar device that compares actual speed to indicated speed in the drivetrain. When all this information is available, a computer can then adjust the spray line pressure to deliver the correct amount of chemicals at varying speeds and amounts of wheel slip.

THE CHANGING STRUCTURE OF AGRICULTURE

Agriculture is entering a new technological era at a time when the character of agriculture is changing rapidly. Emerging biotechnologies and information technologies will be introduced within a socioeconomic structure that has undergone considerable change in the last 50 years and that promises to continue to change throughout the remainder of this century.

One of the best ways to look at changes in the economic structure of U.S. agriculture is in terms of value of production as measured by gross sales per year. In this way farms can be usefully classified into five categories of gross sales, as shown in table 1-1.

Small and part-time farms generally do not provide a significant source of income to their operators. Most of these farmers obtain their primary net income from off-farm sources. However, this segment is highly diverse. This class of farms is operated either by subsistence farmers or by individuals who use the farm as either a tax shelter or a source of recreation.

Moderate-size farms cover the lower end of the range in which the farm is large enough to

Table 1-1.—Distribution of Farm Sizes, Percent of Cash Receipts, Percent of Farm Income, and Farm and Off-Farm Income per Farm by Sales Class, 1982

Sales class	Value of farm products sold	Number of farms	Percent of all farms	Percent of total cash receipts	Percent of net farm income	Average net farm income	Average off-farm income	Average total income
Small	<$20,000	1,355,344	60.6	5.5	−3.8	(615)	20,505	19,890
Part-time	$20,000-$99,000	581,576	25.9	21.8	5.4	998	13,220	14,218
Moderate	$100,000-$199,000	180,689	8.1	19.1	14.6	17,810	11,428	29,238
Large	$200,000-$499,000	93,891	4.2	21.0	20.4	48,095	12,834	60,929
Very large	≥$500,000	27,800	1.2	32.5	63.5	504,832	24,317	529,149
	All farms	2,239,300	100	100	100	$9,976	$17,601	$27,578

SOURCE: Compiled from *Economic Indicators of the Farm Sector: Income and Balance Sheet Statistics, 1983*, USDA Economic Research Service, 1984, table 59, using farm number and cash receipts distribution from the *1982 Census of Agriculture*, U.S. Department of Commerce, Bureau of the Census, 1984.

be the primary source of income. However, most families with farms in this range also rely on off-farm income.

Large and very large farms include a diverse range of farms. The great majority of these farms are family owned and operated. Most require one or more full-time operators, and many depend on hired labor full time. The degree of contracting (monitoring and controlling production to produce a specified quantity of homogeneous products for a buyer) and vertical integration is much higher in this class.

To appreciate how agriculture has changed just between 1969 and 1982, consider the following:

- The number of small farms declined 39 percent, while the number of very large farms increased by 100 percent.
- The share of cash receipts from very large farms increased slightly, from 29 to 33 percent, while cash receipts declined from 40 to 25 percent for small and part-time farms.
- The share of net farm income declined significantly (from 36 to 5 percent) for small and part-time farms, and increased from 36 to 64 percent for very large farms.

These trends indicate that small and part-time farms no longer can depend on the farm to provide an adequate income. Large-scale farms dominate agriculture. Moderate-size farms have a small share of the market and a stagnant share of net farm income. The agricultural sector can be described as a bipolar, or dual sector: As the moderate-size farm disappears, it leaves small and part-time farms clustered at one end of the farming spectrum and large farms clustered at the other, in terms of their importance to agriculture.

If present trends continue to the end of this century, the total number of farms will continue to decline from 2.2 million in 1982 to 1.2 million in 2000 (table 1-2). The number of small and part-time farms will continue to decline, but will still make up about 80 percent of total farms. The large and very large farms will increase substantially in number. Approximately 50,000 of these largest farms will account for 75 percent of the agricultural production by year 2000. The trend toward concentration of agricultural resources into fewer but larger farms will continue, although the degree of concentration will vary by region and commodity.

Moderate-size farms will decline in number and in proportion of total farms, have a small share of the market and a declining share of net farm income. These farms comprise most of the farms that depend on agriculture for the majority of their income. Traditionally, the moderate-size farm has been viewed as the backbone of American agriculture. These farms are failing in their efforts to compete for their historical share of farm income.

Table 1-2.—Most Likely Projection of Total Number of U.S. Farms in Year 2000, by Sales Class

	1982		2000	
Sales class	Number of farms (thousands)	Percent of all farms	Number of farms (thousands)	Percent of all farms
Small and part-time	1,936.9	86.0	1,000.2	80.0
Moderate	180.7	10.0	75.0	6.0
Large and very large	121.7	4.0	175.0	14.0
Total	2,239.3	100.0	1,250.2	100.0

SOURCE: Office of Technology Assessment.

MAJOR FINDINGS

Emerging Technologies and Future Agricultural Production

Like the eras that preceded it, the biotechnology and information technology era will bring technologies that can significantly increase agricultural yields. The immediate impacts of these technologies will be felt first in animal production. Through embryo transfers, gene insertion, growth hormones, and other genetic engineering techniques, dairy cows will produce more milk per cow, and cattle, swine, sheep, and poultry will produce more meat per pound of feed.

Impacts on plant production will take longer, almost the remainder of the century. By that time, however, technical advances will allow some major crops to be altered genetically for disease and insect resistance, higher production of protein, and self-production of fertilizer and herbicide.

In both plant and animal production, information technologies will be widely used on farms to increase management efficiency. Introducing to the marketplace these and the rest of the 150 emerging technologies forecasted in this study raises questions about the effects these technologies will have on crop yield, livestock feed efficiency, reproductive efficiency, and future food production.

Many people are concerned that the trends of major crop yields are leveling off and that the world may not be able to continue to produce enough food to meet the demand of a growing population. OTA analyses indicate that the emerging technologies, if fully adopted, will produce significant beneficial impacts on the performance of plant and animal agriculture. The most dramatic impacts will be felt first in the dairy industry, where new genetically engineered pharmaceuticals (such as bovine growth hormone and feed additives) and information management systems will soon be introduced commercially. New technologies adopted by the dairy industry will increase milk production far beyond the 2.6-percent annual growth rate of the past 20 years (table 1-3). Under OTA's most likely conditions, milk production per cow is expected to increase from the 12,000 pounds in 1982 to at least 24,000 pounds by 2000, an annual growth rate of 3.9 percent. Applications of new technologies also will increase the feed and reproductive efficiency of other farm animals.

Table 1-3.—Impact of Emerging Technology on Animal Production Efficiency in Year 2000

	Actual 1982	Most likely 2000	Annual growth rate[a] (percent)
Beef:			
Pounds meat per lb feed	0.07	0.072	0.2
Calves per cow	0.88	1.000	0.7
Dairy:			
Pounds milk per lb feed	0.99	1.03	0.2
Milk per cow per year (1,000 lb)................	12.30	24.70	3.9
Poultry:			
Pounds meat per lb feed	0.40	0.57	2.0
Eggs per layer per year	243.00	275.00	0.7
Swine:			
Pounds meat per lb feed	0.157	0.176	0.6
Pigs per sow per year	14.400	17.400	1.1

[a]Some of these figures differ from those in table 2-2 of the first report from this study, because actual 1982 figures were preliminary.

SOURCE: Office of Technology Assessment.

Because development of biotechnology for plant agriculture is lagging behind that for animal agriculture, equally significant impacts from biotechnology will not be felt in plant agriculture before the turn of the century. Development and adoption of the new technologies under the most likely conditions will, in the short run, increase the rates of growth of major crop yields at about the level of historical rates of growth (table 1-4). However, the impacts of these technologies will be substantially greater for plant agriculture after 2000.

Any conclusion about the balance of global supply and demand requires many assumptions about the quantity and quality of resources available to agriculture in the future. Land, water, and technology will be the limiting factors as far as agriculture's future productivity is concerned.

Agricultural land that does not require irrigation is becoming an increasingly limited re-

Table 1-4.—Impact of Emerging Technology on Crop Yields in Year 2000

	Actual 1982	Most likely 2000	Annual growth rate[a] (percent)
Corn—bu/acre	113	139	1.2
Cotton—lb/acre	481	554	0.7
Rice—bu/acre	105	124	0.9
Soybean—bu/acre	30	37	1.2
Wheat—bu/acre	36	45	1.3

[a]Some of these figures differ from those in table 2-2 of the first report from this study, because actual 1982 figures were preliminary.

SOURCE: Office of Technology Assessment.

source. In the next 20 years, out of a predicted 1.8 percent annual increase in production to meet world demand, only 0.3 percent will come from an increase in the quantity of land used in production. The other 1.5 percent will have to come from increases in yields—mainly from new technology. Thus, to a very large extent, research that produces new technologies will determine the future world supply/demand balance and the amount of pressure placed on the world's limited resources.

Table 1-5 shows the projections to year 2000 of increased production for some of the major U.S. commodities, based on the above yield projections, land availability, world demand, public policy, and other factors. OTA analyses indicate that with continuous inflow of new technologies into the agricultural production system, U.S. agriculture will be able not only to meet domestic demand, but also to contribute significantly to meeting world demand in the next 20 years. This does not necessarily mean that the United States will be competitive or have the economic incentive to produce. It means only that the United States will have the technology available to provide the production increases needed to export products for the rest of this century.

Under the most likely environment,[3] the aggregate growth rate in production of these commodities, which includes inputs of additional land resources and new technology, will be adequate to meet the 1.8 percent growth rate needed to balance world supply and demand in 2000. Under the more-new-technology environment,[4] production could increase at 2 percent per year, which would be more than enough to meet world demand. This increased production could, however, point to a future of surplus production. On the other hand, under the less new-technology environment[5] the production of major crops in 2000 would drop to 1.6 percent per year, a growth rate that would not allow the United States to meet world demand.

[3]Assumes to year 2000: 1) a real rate of growth in research and extension expenditures of 2 percent per year, and 2) the continuation of all other forces that have shaped past development and adoption of technology.

[4]Assumes to year 2000: 1) a real rate of growth in research and extension expenditures of 4 percent, and 2) all other factors more favorable than those of the most likely environment.

[5]Assumes to year 2000: 1) no real rate of growth in research and extension expenditures, and 2) all other factors less favorable than those of the most likely environment.

Table 1-5.—Projections of Major Crop Production[a]

			2000		
Crop	Unit	1984	No-new-technology environment	Most likely environment	More-new-technology environment
Corn:					
Production	Billion bu	7.7	8.6	9.3	9.7
Growth rate	Percent		0.7	1.2	1.5
Soybean:					
Production	Billion bu	1.9	3.0	3.2	3.3
Growth rate	Percent		3.1	3.4	3.6
Wheat:					
Production	Billion bu	2.6	3.3	3.5	3.5
Growth rate	Percent		1.5	1.9	2.0

[a]The projections shown in this table differ from those in table 2-3 of the first report from this study, because the previous figures were preliminary.

SOURCE: Office of Technology Assessment.

Emerging Technologies and the Future Structure of Agriculture

New technologies have historically had significant impacts on structural change. New disease control technologies gave poultry and livestock farmers unprecedented opportunities to specialize and vertically integrate. Improvements in farm machinery fostered large-scale, specialized farm units.

Like their predecessors, the emerging technologies examined in this study will make a considerable impact on farm structure, especially by 2000. Biotechnologies will have the greatest impact because they will enable agricultural production to become more centralized and vertically integrated. Although in the long run the use of new technologies will not increase the farmer's overall need for capital, there will be trade-offs: biotechnology will require less capital; information technology will require more.

The new technologies will allow increased control over end-product characteristics, for example less fat per unit of lean in meat animals or a specific color characteristic in corn. This implies that increased homogeneity within an agricultural product may result and that there will be a growing number of end products with engineered characteristics. This would require less sorting or grading to achieve increased homogeneity and a shift toward having more control over the production process so as to achieve homogeneity during production.

An anticipated economic consequence of this increased control over production is an increase in the practice of contracting. Contracting allows husbandry and cultural practices to be monitored and controlled closely during the production process. This greater process control leads to uniform product differentiation.

Biotechnologies will have relatively more important effects on resource concentration than will other technological developments. Even though mechanical technologies will continue to be important, they are not expected to have as important an impact on future structure. In particular, biotechnologies are expected to encourage closer coordination and greater process control in livestock production, permitting more contract livestock production. One example is the potential from these technologies for modifying milk at the farm rather than at the processing plant. This technology holds promise for producing more highly unsaturated fats in milk. If adopted, it would entail close coordination at the producer/first-handler markets and additional process control at the production level.

The biological technologies will encourage coordination in crop production, as well. However, the magnitude of change in this area is expected to be relatively less for crops than livestock. Part of the reason is that biotechnologies for livestock production are further advanced. The biotechnology era is expected to encourage closer vertical coordination, with a slight reduction in market access as a consequence. This situation would subsequently lead to fewer but larger farms.

The information technologies are expected to reduce barriers to entry and to increase market access without any significant change in vertical coordination or control at the producer/first-handler level—especially for crop agriculture. Information technologies hold the potential for significantly increasing the amount of information across markets. This impact would be attributable to improved communication of buyers' needs to production-level managers, which should result in more equality between buyers and sellers.

The largest farms are expected to adopt the greatest amount of the new technologies. Generally, 70 percent or more of the largest farms are expected to adopt some of the biotechnologies and information technologies. This contrasts with only 40 percent for moderate-size farms and about 10 percent for the small farms. The economic advantages from the technologies are expected to accrue to early adopters, a large proportion of which will probably be operators of large farms.

Impacts of Agricultural Finance and Credit

The severe financial stress of a large proportion of farmers and the recent regulatory and competitive changes in financial markets have combined to change significantly the financial framework of farming. The farm of the future will be treated financially like any other business—it will have to demonstrate profitability before a bank will finance its operation. Managing a farm efficiently and profitably, which will necessitate keeping up-to-date technologically, will be the key to access to credit.

The cost of credit, however, will be higher and more volatile. Interest on loans may be variable rather than fixed. Moreover, given the concentration in the banking industry, decisions about extending credit more likely will be made at large, centralized banking headquarters far removed from a loan applicant's farm. Loan decisions will thus be less influenced by the considerations of neighborly good will that frequently shaded decisions of local farm banks.

Congress will have to consider all these factors because the availability of capital will continue to be an important factor in agricultural production in general and in the adoption of agricultural technologies in particular. Readily available capital at reasonable rates and terms, plus technologies that aid profitability, provide a favorable environment for technology adoption. Emerging technologies, for the most part, will pass the test for economic feasibility.

The financing consequences of new technologies in agricultural production will probably depend on the relationships between three important factors: 1) the financing characteristics of the new technologies, 2) the creditworthiness of individual borrowers, and 3) the changing forces in financial markets that affect the cost and availability of financial capital. The financing characteristics suggest that most of the new technologies should be financed largely with short- and intermediate-term loans that are part of the normal financing procedures for agricultural businesses. However, the technical characteristics of the technologies, together with the factors constituting the creditworthiness of individual borrowers, suggest that increased emphasis in credit evaluations will be placed on the farmers' management capacity, on their ability to demonstrate appropriate technical competence in using the new technologies, and on building human capital, where appropriate. In some cases—particularly for Farmers Home Adminstration borrowers—significant investments in human capital, with related financing requirements, may accompany new technology adoption. This is consistent with the more conservative responses by lenders to the agricultural stress conditions of the early 1980s. Lending institutions themselves, in turn, must have sufficient technical knowledge and expertise to evaluate these management and credit factors along with other sources of business and financial risks in agriculture. Finally, some forms of new technology involving large investments and having long-run uncertain returns will probably rely more on equity capital for financing.

The changing regulatory and competitive forces in financial markets, including the preference for greater privatization of some credit institutions, means that the cost of borrowing for agricultural producers will likely remain higher and more volatile than before 1980 times and will follow market interest rates much more closely. Similarly, the continued geographic liberalization of banking and the emergence of more complex financial systems mean that the functions of marketing financial services, loan servicing, and credit decisions will become more distinct, with an increasing proportion of credit control and loan authority occurring subregionally and with regional money centers being located away from the rural areas. This will continue to fragment and dichotomize the farm-credit market so that commercial-scale agricultural borrowers will be treated as part of a financial institution's commercial lending activities and small, part-time farmers will be treated as part of consumer lending programs.

The competitive pressures on financial institutions and the risks involved will bring more emphasis on analyzing the profitability of various banking functions, including loan performance at the department level and individual customer level. Innovative lenders will strive more vigorously to differentiate their loan products and financial services, especially for more profitable borrowers, and will tailor financing programs more precisely to the specific needs of creditworthy borrowers. In turn, however, to compete for credit services these agricultural borrowers must be highly skilled in the technical aspects of agricultural production and marketing as well as in financial accounting, financial management, and risk analysis.

In general, most forms of new technology in agricultural production should meet the tests of both economic and financial feasibility, although the structural characteristics of the adopting farm units will continue to evolve in response to managerial, economic, and market factors. The structural consequences of these factors are severalfold:

1. a continuing push toward larger commercial-scale farm businesses, with greater skills in all aspects of business management;
2. continuing evolution in the methods of entry into agriculture by young or new farmers, with greater emphasis on management skills and resource control and less emphasis on land ownership;
3. the continuing development of a marketing systems approach toward financing agriculture, with more sophisticated skills in marketing analysis by farmers and higher degrees of coordination with commodity and resource markets;
4. more formal management of financial leverage and credit by farmers, with greater diversity of funding sources by farmers and better developed markets for obtaining outside equity capital;
5. further development in financial leasing and greater stability in leasing arrangements for real estate and other assets; and
6. more complex business arrangements in production agriculture that accommodate various ways to package effectively debt and equity financing, leasing, management, accounting, and legal services for the future farm business.

Emerging Technologies, Policy, and Survival of Various Size Farms

The size and, therefore, the survival of farms is affected by several factors. Clearly, there are economies of size in many commodity areas covered by farm policy. These economies motivate further concentration of resources. In addition, present farm policy, more than any other policy tool, makes major impacts on farm size and survival. Although very large farms can survive without these programs, moderate-size farms depend on them for their survival.

This study finds that substantial economies of size exist for several major commodities (table 1-6). The commodities include dairy, corn, cotton, wheat, and soybeans. With the exception of corn, economies of size do not exist uniformly in all the production areas studied for these commodities. Table 1-6 shows the areas in which economies of size do exist. It should be noted that the analysis considered only technical economies of size. If it had also included pecuniary economies, additional production areas would have been found to have economies of size.

Table 1-6 also shows commodities in which there will be significant gains in yield based on emerging technologies. All of the commodity areas except rice will experience substantial gains in yield *as well as* significant economies of size. (No economies of size were found for

Table 1-6.—Comparison of Commodities With Current Economies of Size and Future Technological Gains

Current economies of size (in descending order)	Greatest yield increases for the future (in descending order)
Dairy	Dairy
Arizona	
California	Wheat
New Mexico	
Corn	Soybeans
Illinois	
Indiana	Corn
Iowa	
Nebraska	Rice
Cotton	Cotton
Alabama	
Texas	
Wheat	
Kansas	
Montana	
Soybeans	
Iowa	

SOURCE: Office of Technology Assessment.

rice.) Dairy, in particular, leads all commodities in economies of size *and* production increases from new technologies. These forces will combine to shift over time the comparative advantage in dairy production from the smaller dairies in the Great Lake States and Northeast to the larger dairies in the Southwest and West.

Overall, the combination of future yield increases from new technology and current economies of size in these commodities means that there will be substantial incentives for farms to grow in size. These powerful forces will continue, and may even speed up resource concentration in U.S. agriculture.

This study finds that farm programs, which include Commodity Credit Corporation (CCC) purchases and price and income supports, have major impacts on rates of growth in farm size, wealth, and incomes of commercial farmers. Large farms increase their net worth significantly more than moderate-size farms under current farm programs and large farms account for a significantly large share of farm program payments. In particular, price supports provide most of the wealth and growth benefits to large farms.

Removing farm programs reduces the probability of survival more for moderate-size farms than for large farms. OTA's analyses find that large farms can survive and prosper without farm programs. And, because these farms account for the vast majority of farm program benefits, significant savings in Government expenditures could be realized if large farms were ineligible to receive program payments.

On the other hand, this study finds that moderate farms need farm programs to survive and be successful. Income supports, in particular, provide significant benefits to moderate farms, and the targeting of income supports to moderate farms is an effective policy tool for prolonging these farms' survival.

Those changes in tax policy that would be more restrictive have little impact on farm survival. Increasing the Federal tax burden on farmers reduces the average annual rate of growth in farm size uniformly for all farm sizes.

Currently the financial position of many farmers is under severe stress. The situation is serious and may not improve for some time. Two alternatives most discussed by policymakers are interest subsidy and debt restructuring programs. OTA finds that restructuring debt for highly leveraged farms does not appreciably increase their probability for survival. The interest rate subsidy substantially increases average net income more than debt restructuring. It is the more effective strategy to ease financial stress. In addition, large farms with high debts are not as dependent on these financial programs for survival as moderate farms are.

Impacts on the Environment and Natural Resources

In general, with a few notable exceptions, most emerging technologies are expected to reduce substantially the land and water requirements for meeting future agricultural needs. Consequently, these technologies are expected to reduce certain environmental problems associated with the use of land and water. The technologies are thought to have beneficial effects relative to soil erosion, to reduce threats to wildlife habitat, and to reduce dangers associated with the use of agricultural chemicals. New tillage technologies, however, may reduce erosion and threats to wildlife while increasing the dangers from the use of agricultural chemicals.

The new technologies are most likely to receive first adoption by farmers who are well financed and are capable of providing the sophisticated management required to make profitable use of the technologies. Most of these farmers will be associated with relatively large operations. Hence, the technologies will tend to give additional economic advantages to large farm firms relative to moderate and smaller farms, accentuating the trend toward a dual farm structure in the United States.

In addition, since many of the new technologies tend to be environmentally enhancing, public interest exists in research and education that can lead to the rapid development and widespread adoption of the technologies. That conclusion becomes even stronger if public policy is aimed at maintenance of the moderate-size farm. Larger farms, with their own access to research results and scientific expertise, may be able to advance the new technologies with relatively little publicly sponsored research. But moderate and small farms will have to depend on publicly sponsored research and extension education to gain access to the new technologies and to adapt them to their individual needs.

The new technologies will entail more stringent environmental regulations and stronger enforcement of regulations than at present. The complexities of some of the emerging technologies will pose significant challenges for those promulgating wise environmental regulations. The economic benefits of the technologies will be inviting, but users may have little incentive to use the technologies in ways that avoid unnecessary, adverse, third-party effects. Economic incentives or disincentives, including the use of excise taxes to discourage overuse of potentially threatening materials, represent a promising approach to the protection of environmental values than do direct regulation. Additional efforts to enforce existing regulations would hasten the adoption of the new technologies that seem less environmentally threatening. New regulations will be required, however, for dealing with some aspects of the emerging technologies.

Perhaps the most revolutionary of the new technologies are those associated with rDNA. While the specific applications of such technologies appear likely to reduce resource needs and threats to the environment that arise from agricultural activities, dangers may accompany the deliberate release of genetically altered micro-organisms. The revolutionary nature of the new biotechnologies and the lack of a scientifically accepted predictive ecology prevent specific evaluation of resource/environmental impacts associated with the deliberate release of new forms of life at this time.

Many scientists see little danger in the applications of rDNA technology in laboratory experiments. The proponents of biotechnology argue that genetic engineering has been used in plant breeding and animal husbandry for centuries and that genetically engineered micro-oganisms are no more dangerous than micro-organisms already in commercial use or that might be used in nature. However, the opponents of deliberate release argue that the new products of genetic engineering are different from the old ones. Scientists do not know how these new micro-organisms will behave in the environment and fear adverse consequences to the ecosystem. Both sides agree that more research should be conducted to assess the potential benefits and risks. Recently, the Environmental Protection Agency approved the first two field tests of genetically altered organisms.

Impacts on Rural Communities

The impacts of technological and structural change in agriculture do not end with the individuals who live and work on farms. A variety of additional consequences are expected at the level of rural communities, consequences that directly or indirectly affect farms and farmers. As with individual farmers, some communities are likely to benefit from change, while others are likely to be affected adversely. Much depends on the type of overall labor force in the community and on the opportunities for labor to move to other employment areas.

Hard-hit communities may need technical assistance to attract new businesses to their areas, to develop labor retraining programs, and to alter community infrastructure to attract new inhabitants. To accomplish these goals, Federal policy will have to be complemented by regional and local policies.

Those rural communities that benefit from changes in agricultural technology and structure may do so in several ways. For example, as agriculture becomes more concentrated, some communities will emerge as areawide centers for the provision of new, high-value technical services and products. Likewise, some communities will emerge as centers for high-volume food packaging, processing, and distribution. In both cases, the economic base of these communities is likely to expand. However, unless total demand for agricultural commodities increases substantially, centralization of services, marketing, and processing will be like a zero-sum game in many areas. The market centers will benefit at the expense of other communities. Many of the communities that are bypassed will decline as a result of the process of centralization.

Communities also may benefit in those parts of the country in which the number of small and part-time farms is increasing. This phenomenon results in an increase in population in many rural areas and an increase in total income and spending in some of these areas. The increase in small farms may sustain additional retail establishments than would otherwise be the case, since purchases by small farmers may tend to be more from local sources than those by larger farmers. The operators of these farms in many cases subsidize their own production from off-farm income.

A wide range of diversity is evident in the character, agricultural structure, patterns of change, and patterns of impact on rural communities in the five different regions of the United States studied for this report:

1. the CATF (California, Arizona, Texas, and Florida) region;
2. the South;
3. the Northeast;
4. the Midwest; and
5. the Great Plains and the West.

A clear picture of adverse relationships between agricultural structure and the welfare of rural communities is evident in the industrial-agricultural counties of the CATF region. Large-scale and very large-scale industrialized agriculture in these communities is strongly associated with high rates of poverty, substandard housing, and exploitative labor practices in the rural communities that provide hired labor for these farms. Very large-scale agriculture has been a strong source of employment in the CATF region for many years, although at very low wage rates. Emerging technologies may reduce the labor requirements throughout much of the CATF region by 2000. Increased unemployment will greatly increase the strain on these communities. A potential exists for the CATF region to increase its share of national agricultural production, which would mitigate the trend toward increasing unemployment. However, increased agricultural production in this region will tend to be constrained by the cost of irrigation water and the need to control environmental impacts.

The coastal zone of the South also has a substantial potential for structural change similar to that of the CATF region. Topography and climate favor large-scale, labor-intensive production of fruits, vegetables, and dairy products. The area also has a segmented, relatively unskilled labor force that could provide a source of low-cost labor similar to that of the CATF region. It is difficult to generalize about the rest of the South, owing to the diversity of agricultural structure and production. Evidence exists of a relatively strong association between rates of unemployment and agricultural structure. Unemployment rates tend to be lowest in counties with a predominance of moderate farms.

In the Northeast, dairy products are the single most important agricultural commodity group. Because dairy farms are likely to experience widespread failure as a consequence of the combination of technological change and public policies, the structure of agriculture in the Northeast is likely to change substantially during the next 10 to 15 years. However, rural communities in the Northeast have a low overall dependence on income from agriculture. Most productive agricultural counties in the Northeast are adjacent to metropolitan areas where greater employment opportunities and services are available. The most rural counties sometimes are not the most agricultural. Therefore, rural communities in the Northeast generally are not likely to experience adverse consequences from structural change, with the exception of a few localities with especially high dependence on dairy production.

No clear-cut evidence exists that rural communities in the Midwest were adversely affected by structural change during the 1970s. In general, alternative sources of employment in the manufacturing and service sectors were relatively prevalent and are expected to continue to be relatively good in the Midwest. Indicators of social welfare, in general, tended to improve as farm structure moved from small and part-time farms toward moderate to large farms during the 1970s. However, there was a tendency for population to decline in counties where the share of part-ownership of farms increased. As with the Northeast region, there is a reasonable expectation that technological change in the dairy industry will result in a mass exodus of small to moderate dairy farms during the next 5 to 15 years. Rural communities in dairy counties may not be adversely affected because off-farm employment is quite high in these counties. Those mixed agricultural counties on the western edge of the Midwest that are relatively dependent on agriculture are the most likely to suffer adverse consequences from structural change. If the percent of part-ownership increases as agriculture becomes more concentrated, population, median income, and retail sales may decline in these counties.

Strong potential exists for development of a high concentration of agricultural production in the Great Plains and the West, especially in terms of farm size, if not gross sales per farm. In turn, the number and percent of hired managers in this region is likely to increase. Unlike the South, there is a low potential for development of an industrialized agriculture with large numbers of hired field workers. The most likely adverse impact will be the loss of population and small retail firms in the region. In general, fewer alternate employment options will be likely in manufacturing and the service industries in this region than in the other regions of the country.

This study shows clearly that policies designed to prevent or ameliorate adverse impacts and promote beneficial impacts need to be crafted with consideration for regional structural/technological differences. Generalizing about the impacts of changing agricultural technology and structure on rural communities across regions of the United States is difficult.

Impacts on Agricultural Research and Extension

U.S. agriculture has been very successful to an important extent because of technological advances. However, agriculture's adoption of biotechnology and information technology raises several questions about the impact of technical advances on the performance of the research and extension system and about how that performance will ultimately affect the structure of agriculture.

Public research in the past was the driving force for agricultural production. Now, with the private sector becoming more involved in certain aspects of applied research, the public sector is emphasizing increased basic research. This situation leaves open the question of who will do applied research in the public sector. Although the public sector has allocated resources to research in biotechnology and information technology, extension has done little to make information about these technologies available to farmers. The extension service must thus decide what its mission will be, for extension policy will determine how effective moderate farm operators will be in gaining access to new technology. Without such access moderate-size farms will disappear even faster.

Consideration of specific changes in research and extension policy may be justified. The following areas have been identified as meriting consideration for policy changes:

- The social contract on which the agricultural research and extension system was created needs reevaluation. This issue should not be left for resolution by the courts. Specific guidelines must be developed that allow the system to compete while protecting the public interest and investment in the agricultural research and extension functions. Both Congress and the U.S. Department of Agriculture (USDA) should have a voice in this type of policy development.
- Some experts believe that increased private sector support for agricultural research sig-

nals less need for public support. Even though private sector support complements public support, basic biotechnology and information technology research is very costly. A reduced role for public research and extension would result in a slower rate of technological progress and a lower level of protection for the public. In addition, the public has a strong interest in maintaining an agricultural research component in each State to serve the problem-solving needs of that State's agriculture.

- Many agricultural problems are local or regional in scope. The applied nature of the system, having an agricultural experiment station and extension service in each State, has provided a unique capacity to identify and solve local or regional problems. Reality suggests that only certain universities have sufficient resources to compete for private sector support in biotechnology and information technology. The result is a confluence of forces that is creating a dichotomy of "have" and "have not" universities. There is, however, still an important role for even the smallest, poorest funded land-grant university. It plays an important part in a national system designed to deal with thousands of agro-ecosystems and to the existence of a decentralized system with nationwide capability. Because of these inequalities, there is concern that the traditional extension-research interaction and feedback mechanisms could break down, particularly in States that are not in a position to command a major biotechnology component.
- The role of extension is even more important than it has been in the past. New, more complex products require evaluation and explanation. In States where experiment stations have attracted substantial private sector support, the product testing function can be most objectively performed by extension. The recently passed 1985 farm bill gives explicit authority for extension to engage in applied research functions such as product testing and evaluation.
- While most agricultural research is not inherently biased toward large-scale farms, lags in adoption by small and moderate farms have the effect of such a bias. Unless special attention is given to technology generation and transfer to moderate farms, major structural changes could result, leading to the eventual demise of a decentralized structure that includes moderate farms. To the extent that preservation of these farms is a policy objective, special funding for and emphasis on the problems of technology generation and the transfer of that technology to moderate farms is warranted.
- Although the agricultural research system has received the benefits of increased funding from both private and public sources, extension funding has not materially increased. As a result, extension staff at the county and specialist levels are being caught up in a whirlwind of technological change. The result is a need for the injection of substantial staff development funding into the extension system.
- Basic organizational issues must be addressed by the Extension Service. The premise on which extension was developed was that of research scientists conveying the knowledge of discoveries to the extension specialist who, in turn, supplied information to the county agent who then taught the farmer. Over time, this concept has gradually but persistently broken down as agricultural technology has become more complex and insufficient resources have been devoted to staff development. Consequently, more emphasis has been placed on direct specialist-to-farmer education. More specialists have been placed in the field to be closer to their clientele, but at the cost of less contact with research scientists. As these changes have occurred, the role of the county agent has become increasingly unclear. Appreciation for and use of county agents as educators and technology transfer agents has declined. As a result of these changes, a basic structural reevaluation of the organization of the extension function of the agricultural research system is needed.

IMPLICATIONS AND POLICY OPTIONS FOR AGRICULTURE

The Issue of Farm Structure

This study indicates that the process of structural change in agriculture has already begun. Based on a continuation of current policies, past trends, and future technological expectations, the net result of this structural change could be the development of a farm structure composed of three agricultural classes:

1. The *large-scale farm segment* would be composed of a relatively small number of farms that produce the bulk of U.S. production. By year 2000 there could be as few as 50,000 large-scale farms producing as much as three-fourths of the agricultural production. This large-scale farm segment would be highly efficient in the performance of production, marketing, financial, and business management functions. Such farms would be run by full-time, highly educated business managers. Barring unforeseen acts of nature, farm operators would be able to predict their chances of making a profit before planting or breeding.
2. The struggling *moderate-size farm segment* would be trying to find a niche in the market and survive in an industrialized agricultural setting. The potential for the moderate farm finding that niche is rapidly becoming the center of the farm policy debate. Traditionally highly productive, efficient, moderate-size, full-time farms have been the backbone of American agriculture. It is still true that a moderate, technologically up-to-date, and well-managed farm with good yields is highly resilient. One key to the success of these farms clearly lies in the management factor. But more often than not, management has to be willing to accept a relatively low return on invested capital, time, and effort. With ever-increasing educational requirements associated with farming, there will likely be less willingness by successful managers of moderate farms to accept a lower return for their services and for invested capital. Another key to the survival of moderate farms lies in access to state-of-the-art technologies at competitive prices. Cooperatives traditionally have performed that role. But cooperatives by and large are not conducting or funding basic or applied research in biotechnology and information technology. Also, like their predominantly moderate-size farmer members, cooperatives, too, have encountered financial difficulty.
3. The *small, predominantly part-time farm segment* tends to obtain most of its net income from off-farm sources. However, this segment is highly diverse. It includes wealthy urban investors and professionals who use agriculture primarily as a tax shelter and/or country home. It also includes would-be moderate farm operators who are attempting to use off-farm income as a means of entering agriculture on a full-time basis. Finally, this segment includes a number of poor, essentially subsistence, farmers who are vestiges of the war on poverty in the 1960s. Such farmers remain a significant social concern that must be dealt with from a policy perspective, although traditional farm price and income policy hold no hope for solving their problems.

Contemporary farm programs have fostered this trend toward three farm-size classes. Payments to farmers on a per-unit-of-production basis concentrate most of the benefits in large farms that produce most of the output. Large farms have been in the best position to take advantage of new technologies arising out of the public sector agricultural research system.

Without substantial changes in the nature and objectives of farm policy, the three classes of farms will soon become two—the moderate-size farm will largely be eliminated as a viable force in American agriculture. In addition, the problems of the small subsistence farm will continue to fester as an unaddressed social concern.

This section sets forth the policy changes that would be required if it were decided by Congress that overt steps should be taken to foster a diverse, decentralized structure of farming where all sizes of farms had an opportunity to compete and survive in a time of rapidly changing technology. The objective of giving every farm the opportunity to compete and survive does not imply an unchanging and stagnant farm structure. It does imply a political and social sensitivity both to the impact of current farm programs on farm structure and to the different needs of large, moderate, and small farms for Government assistance. It can be expected that regardless of what Government does fewer commercial farms will exist in year 2000. However, Government can do much to ease the pain of adjustment.

Required Policy Adjustments

Substantive changes in policy direction are needed to address the structure issue. Specifically, separate policies and programs must be pursued with respect to each of the three farm segments—large farms, moderate farms, and small farms. The choice of any one set of policies to the exclusion of the other policy sets would imply that Congress desired to selectively enhance the status of one farm segment.

Policy for all farmers implies two basic policy goals:

- All farmers need to operate in a relatively stable economic environment where they have an opportunity to sell what they produce.
- All farmers need a base of public research and extension support whereby they can maintain their competitiveness in the markets in which they deal.

The needs of large farms can be met by addressing just these goals. The needs of moderate and small farms are more complex, however. Policy to address the needs of moderate and small farms must include the elements of large farm policy as well as additional elements.

Policy for Large Commercial Farms

A basic conclusion of this study is that large-scale farmers do not need direct Government payments and/or subsidies to compete and survive. However, this does not preclude the need for a commercial farm policy.

The criteria for determining what constitutes a large-scale farm is important but also somewhat arbitrary. The dividing line developed from this study is about $250,000 in sales for a crop or dairy farm unit under single ownership or control. This level of sales is generally required to achieve most of the economies of size found to exist in agricultural production.[6] Over time, this optimum size has had, and will continue to have, a tendency to increase. As this occurs, the farm size criteria for limiting program benefits would likewise have to increase.

Creating a Stable Economic Environment.—The policy goal of creating a relatively stable economic environment where farmers have an opportunity to sell what they produce implies the following major farm program initiatives:

- Direct Government payments to all farms having over $250,000 in sales would be eliminated. This implies the elimination of the target-price concept for this sales class. Elimination of payments to those farms would significantly reduce Government expenditures in agriculture.
- The nonrecourse loan would be converted to a recourse loan. The nonrecourse feature has resulted in the accumulation of large Government commodity stocks. The recourse feature would provide a continuing base of support for the orderly marketing of farm products.
- Aside from the recourse price support loan, Government credit to farms having over $250,000 in sales would not be available.
- An expanded international development assistance program would be established. Such a program would have to include an optimum balance of commodity aid and economic development aid. Its primary objective would be to help developing countries improve economic growth, thus becoming better future customers of American agriculture.
- A balanced macroeconomic policy that facilitates growth of export markets and

[6] The $250,000 figure is based on census data and the economies of size analysis discussed previously.

maintains a relatively low real rate of interest would have to be maintained.

Maintaining Technological Competitiveness.—The technological competitiveness of American farmers would be aided by continuing a policy that encourages public and private investment in agricultural research. The major thrust of the research and extension programs as they affect larger scale commercial farms would be as follows:

- The trend toward increased public sector emphasis on basic research would be continued. Increased reliance would be placed on the private sector for applied research in the development of new products.
- Even though public sector research would be aimed more toward basic research, an important problem-solving component would be maintained to adopt new technologies to various agro-ecosystems and to maintain newly achieved productivity from the evolution of pests and disease, decline in soil fertility, and other factors.
- Extension's role in direct education of, or consultation with, large-scale farmers would be deemphasized. Private consultants could play an increased role in technology transfer to the large-scale farm segment.

Policy for Moderate-Size Farms

Policy for moderate farms includes the aforementioned options as well as additional options tailored specifically to the needs of moderate farms. OTA finds, for example, that moderate farms having $100,000 to $250,000 in gross sales face major problems of competing and surviving in the biotechnology and information technology era. Some moderate farms will survive and some will not. This latter group should be assisted in their move to other occupations.

Policy for moderate farms requires the same stable economic environment and base of support for agricultural research and extension as for large farms. But, in addition, the following specific policy goals for moderate farms can be specified:

- The risk of moderate farmers operating in an open market environment would be reduced.
- New technologies that have the potential for adoption would be available to moderate farmers.
- Opportunities for employment outside agriculture would be created for those farmers who are unable to compete.

Diligent enforcement would be needed to assure that the benefits of programs established to favor moderate farms are limited to those farmers for whom they are intended.

Reducing Risks to Moderate-Size Farms.—The most difficult obstacle to survival facing the moderate farm is that of managing risk. Three options, that are not necessarily mutually exclusive, could reduce the risks confronting moderate farms.

1. Income protection could be provided through either a continuation of the current target-price concept for moderate farms only or through a device known as the marketing loan. Like the current nonrecourse loan, the marketing loan is a loan from the Government on commodities in storage. If the commodity is sold for less than the loan value, the farmer pays back only those receipts to the Government in full payment of the loan. The marketing loan, in essence, becomes a guaranteed price to the producer. The level of the marketing loan should be no greater than the average cost of production for moderate farmers.
2. The nonrecourse loan concept could be continued for moderate farms. However, the nonrecourse loan level should not be set any higher than the recourse loan suggested previously for large farms, or else the Government could end up acquiring most of the production from moderate farms.
3. Sharply increased assistance could be provided by the public sector to reduce the risk to moderate farms. Such assistance could be in the form of educational programs for example, on risk management, futures markets, contracting, and cooperative marketing.

Technology Availability and Transfer to Moderate-Size Farms.—OTA finds that agricultural research, as a general rule, is not inherently biased against moderate farms. Rather, moderate farms may be seriously disadvantaged either by lags in adoption or by lack of access to competitive markets for the products produced by new technology. The following initiatives could help curtail such problems of technology availability and transfer.

- Extension's evaluation of the increasing number of new products entering the market would be intensified. This increased effort would play the dual role of: 1) providing a check on the efficacy and efficiency of new products in biotechnology and information technology, and 2) eliminating the costs associated with individual farmer experimentation with those new products.
- Extension technology transfer services would be aimed specifically at moderate-size farms. The primary goal of such programs would be to ensure the same schedule of adoption of technologies for moderate-size as well as large farms.
- The development of cooperatives that emphasize technology supply and transfer services to moderate farms would have to be undertaken.
- Ample credit would have to be made available to moderate-size farms that have the potential to survive and grow. Government credit in concert with cooperative credit could be aimed specifically toward filling the needs of moderate-size farms. Emphasis should be placed on credit required to keep moderate farms technologically up-to-date.

Transition Policy to Other Agricultural Enterprises or Nonfarm Employment.—Regardless of the effectiveness of the initiatives discussed above, an accelerated need exists to assist farm families to either move to other agricultural enterprises or out of agriculture into other occupations. The need arises, therefore, for specific public action to facilitate the farmer's transition from the current farm operation into gainful, productive employment elsewhere. Specific initiatives to ease this process include the following:

- New opportunities for employment of displaced farmers need to be explored and developed within agriculture as the industry continues to evolve.
- To facilitate the transition to nonfarm jobs, special skills training programs aimed at those areas where significant employment opportunities exist must be considered. Jobs in rapidly growing service, health care, or care-for-the-aged industries provide contemporary examples.
- Financial assistance, similar to the famous G.I. bill, might be established to assist displaced farmers or rural residents during the period of transition while skills training is being received.
- In areas of severe financial stress, assistance may be provided in the form of Government purchase of land or production rights from displaced farmers at its "long-term fair market value." The returns from the land could be used by the displaced farmer for relocation and retraining. The Government could retain the land in conservation reserve status until it is needed for future production.

Policy for Small/Part-Time Farms

Policy for small/part-time farms includes several elements in addition to those mentioned under large farm policy.

With few exceptions, small farms, those having less than $100,000 in sales, are not viable economic entities in the mainstream of commercial agriculture—nor can they be made so. However, even a small increase in their farm income could have a significant multiplier effect on the local economy because of the large number of small farms. These farms survive because their operators have substantial outside income (part-time farmers), or because they have found themselves a niche in marketing a unique product with special services attached (often direct to consumers), and/or because they are willing to accept a very low return on resources contributed to the farming operation.

For the small farmers who have substantial outside income or who have found a niche in the market, Government's role would be severely restricted. They are as much able to take care of themselves as owners of large farms.

However, small subsistence farmers who have limited resources, and often limited revealed abilities, represent a genuine problem for which public concern is warranted—these indeed are the rural people left behind. Price and income support programs have done and can do little to solve their problems. These impoverished individuals are a social and economic problem. The following suggestions are made for dealing with the problems of subsistence farmers:

- Initiate a special study to identify those individuals and their specific statuses and needs. Develop social programs to meet those needs.
- USDA and the land-grant university bear a special burden of responsibility for serving the needs of these subsistence farmers. This responsibility has not generally been realized and, therefore has not been fulfilled. In the South, this responsibility falls particularly heavily on the 1890 land-grant universities in concert with the statewide extension education programs and the 1862 land-grant universities. In the North, the responsibility for serving the agricultural educational and research needs of subsistence farmers falls exclusively on the 1862 land-grant universities.
- USDA and these land-grant universities could be directed to develop jointly a plan for serving the agricultural research and educational needs of these farmers. Such a plan could include the delivery of farming, credit, and marketing systems designed to maximize the small farm's agricultural production and earning capacity.
- Specific farming systems must be developed to serve specifically the needs of small subsistence farms. Such systems should, to the extent practicable, encompass the use of new technologies.
- Credit delivery systems for small subsistence farmers could be developed specifically by USDA through the Farmers Home Administration. Such systems should consider the unique capital and cash flow-limiting factors associated with subsistence farmers who are often not in a position to take advantage of other farm programs such as price and income supports.
- Marketing programs geared to subsistence agriculture are essential for providing hope for this farm segment. The difficulty lies in the inability of these farmers to obtain access to the mass markets through which most agricultural production moves.

Policy for Rural Communities

The impact of adjustment in agriculture to changing technology will by no means be limited to the farm sector. Rural communities will be at least equally affected by increasing farm size, integration, and moderate farm displacement. Although, these effects will be felt initially by implement dealers, farm supply and marketing firms, or bankers, the reverberations will extend throughout the commmunity in terms of employment levels, tax receipts, and required services. Rural communities should assess these impacts and prepare to make needed adjustments. To ease the pain of adjustment the following actions are suggested:

- Comprehensive programs for community redevelopment and change need to be initiated throughout rural America. Such development plans should be fostered and facilitated by Federal and State government agencies.
- Increased employment opportunities in rural areas could be fostered by aggressively attracting new business activities in rural communities. Particular emphasis would be placed on attracting those businesses that develop technologies and serve the needs of high-technology agriculture in rural areas.
- Rural communities could be assisted in developing and modernizing the infrastructure needed to be a socially and economically attractive place to live. Some rural communities can serve as an attractive retirement residence for an aging population. But this would require that a higher level of social services be developed.
- Rural communities need to play a vital role in skills training for displaced farmers and

rural community employees. School and university outreach programs could be modified to serve this important role.

Policy for Technology and Environmental Resource Adjustment

One of the major reasons that American agriculture has been so productive is because technological change has been fostered by the public sector and nurtured by a profit-seeking private sector. As a result, American consumers have enjoyed a plentiful supply of low-cost food and natural fiber. In addition, agricultural exports have made a major contribution to the overall development of export markets, to the benefit of the general economy. Biotechnology and information technology promise to offer more of the same, with the added bonus of less chemicals used in the production of food—whether for the control of pests, disease, and weeds, or for the production of commercial fertilizer.

Maintaining the productivity and competitiveness of U.S. agriculture in the public interest requires a balance between public and private sector support for technological change. Yet it would be wrong to imply that there are no risks. The conferring of property rights on discoveries of the agricultural research system has shifted the agricultural research balance between the public and private sectors toward the private sector. While the effects of this shift appear to be positive, concerns exist that a substantial portion of the benefits of even public research could be captured by private firm interests. Distribution of these benefits may be so unequally distributed that competitive performance is impaired. In addition, no scientifically acceptable methodology exists for weighing the risks or hazards of biotechnology research. To deal with such issues, the following policy suggestions are made:

- Steps should be taken to secure the public interest on which the USDA and land-grant university agricultural research system has been based. Assurance must be provided that the benefits of publicly supported research and extension are not captured in the form of excess profits by the private sector based on research property rights and increased private sector funding of public research. The effect would be to stifle the process of discovery and the dissemination of new knowledge.
- Major investments must be made to foster the development of human capital that is in a position to cope with the process of rapidly changing agricultural technology. This need extends from the training and development of the most basic biological research scientists, through the extension specialist and county agent, to the farmer who adopts the new technology and the banker who supplies the loan for its purchase.
- Little is known about the adverse impacts of potential biotechnology developments on the ecosystem. These risks must be carefully assessed, monitored, and where necessary, regulated. Care must be taken, however, not to overregulate and thereby stifle the potential competitiveness and productivity of U.S. agriculture.

SUMMARY CONCLUSIONS

The biotechnology and information technology revolution in agricultural production has the potential for creating a larger, safer, less expensive, more stable, and more nutritious food supply. Yet it will exact substantial costs in potential adjustment problems in the agricultural sector and in rural communities. Those costs can be minimized by careful analysis, planning, and implementation. This study is only the first step in that direction.

Part I
The Emerging Technologies

Chapter 2

Emerging Technologies for Agriculture

Contents

Table

Figures

Chapter 2

Emerging Technologies for Agriculture

American agriculture is on the threshold of the biotechnology and information technology era. Like the eras that preceded it—the mechanical era of 1930-50 and the chemical era of 1950-70—this era will bring technologies that can significantly increase agricultural yields.

The immediate impacts of the biotechnologies will be felt first in animal production. Through embryo transfers, gene insertion, growth hormones, and other genetic engineering techniques, dairy cows will produce more milk per cow; cattle, swine, sheep, and poultry will produce more meat per pound of feed. Impacts in plant production will take longer to occur, almost the remainder of the century. By that time, however, technical advances will allow major crops to be altered genetically for disease and insect resistance, higher production of protein, and self-production of fertilizer and herbicide. Until then, crop yields will increase through the use of traditional technologies, but at less than past rates.

Both plant production and animal production will benefit from advances in information technology. Computers, telecommunications, monitoring and control technology, and information management will be widely used on farms to increase management efficiency.

Some of these new technologies will emerge unexpectedly; however, most will undergo a long process of development, from initiation of ideas to commercial introduction. Since the development of a new technology takes years, often decades, it is often possible to forecast future technologies while they are still in the laboratory. One method is to obtain collective judgments from experts who have direct access to the latest available information, a method OTA chose. OTA collected information from three rounds of a mailed survey to about 300 leading public and private scientists and research administrators who had broad, cross-cutting perspectives about future technologies (Lu, 1983). Based on these surveys and on subsequent interviews with scientists in various disciplines around the country, OTA thus identified the 28 areas of emerging technologies that are likely (with at least a 50-50 chance) to emerge before 2000 and to have major impacts on the agricultural sector. Many of the technologies examined for this study, such as growth hormones, monoclonal antibodies, superovulation, and embryo transfers, are already in the marketplace, while others are still in the laboratory and will not become available for commercial introduction until 2000.

This chapter presents an overview of the major advances in biotechnology and information technology and then describes in more detail the 28 areas of technologies that were assessed for this study. It should be noted that some of the emerging technologies assessed will be in neither the biotechnology nor information technology categories.

BIOTECHNOLOGY

Biotechnology, broadly defined, includes any technique that uses living organisms to make or modify products, to improve plants or animals, or to develop micro-organisms for specific uses. It focuses on two powerful molecular genetic techniques, recombinant deoxyribonucleic acid (rDNA) and cell fusion technologies. With these techniques scientists can visualize the gene—to isolate, clone, and study the structure of the gene and the gene's relationships to the processes of living things. Such knowledge and skills will give scientists much greater control over biological systems, leading to significant improvements in the production of plants and animals.

Animal Agriculture

One of the major thrusts of biotechnology in animal agriculture is the mass production in micro-organisms of proteinaceous pharmaceuticals,[1] including a number of hormones, enzymes, activating factors, amino acids, and feed supplements (Bachrach, 1985). Previously obtained only from animal and human organs, these biologicals either were unavailable in practical amounts or were in short supply and costly. Some of these biologicals can be used for the detection, prevention, and treatment of infectious and genetic diseases; some can be used to increase production efficiency.

Another technique, embryo transfer in cows, involves artificially inseminating a superovulated donor animal[2] and removing the resulting embryos nonsurgically for implantation in and carrying to term by surrogate mothers. Prior to implantation, the embryos can be treated in a number of ways. They can be sexed, split (generally to make twins), fused with embryos of other animal species (to make chimeric animals or to permit the heterologous species to carry the embryo to term), or frozen in liquid nitrogen.

[1]Pharmaceuticals that are proteins.

[2]An animal that has been injected with a hormone to stimulate the production of more than the normal number of eggs per ovulation.

These and other genetic engineering techniques are explained more fully under "Animal Genetic Engineering," later in this chapter.

Plant Agriculture

The application of biotechnologies in plant agriculture could modify crops so that they would make more nutritious protein, resist insects and disease, grow in harsh environments, and provide their own nitrogen fertilizer. While the immediate impacts of biotechnology will be greater for animal agriculture, the long-term impacts may be substantially greater for plant agriculture. The potential applications of biotechnology on plant agriculture include microbial inocula, plant propagation, and genetic modification (Fraley, 1985). All are explained later in this chapter under "Plant Genetic Engineering."

INFORMATION TECHNOLOGY

Agricultural information technologies can be classified as: 1) communication and information management, 2) monitoring and control technologies, or 3) telecommunications. The relationships of these classifications are shown in figure 2-1.

Communication and information management consists of onfarm digital communication systems, known generically as local area networks (LANs), combined with the microcomputer-based information processing technologies used by the farm operator as the central information processing and management system. This central computer system may include remote terminals with keyboards, display screens, and printers used for onsite data entry and readout by the farm operator. The computer terminals are indicated on figure 2-1 by the small boxes labeled "T."

Monitoring and control technologies automatically monitor and control certain aspects of a wide variety of production processes. These technologies, generally considered to be subsystems, are located at the site of production activities, such as livestock confinement systems, storage facilities, and irrigation pumping and control stations, and on mobile equipment such as tractors and combines. Monitoring and control systems can function autonomously, although they are increasingly being connected to the central onfarm information processing system through fixed links and low-power radio links to the onfarm LAN. The LAN connections between the central information management system and the onsite monitoring and control technologies are indicated by the boxes on figure 2-1 labeled "N," for network node. Several different kinds of local configurations of the LAN and the components of the onfarm computer system are possible. The arrangement shown here is just one of many possibilities.

Telecommunication technologies comprise the hardware and software that connect the onfarm systems with the rest of the world so that the farmer can communicate with people and with computer systems in other firms and institutions. Telecommunication systems may combine both voice and data communications. Three types of telecommunication technologies are shown on figure 2-1: satellite ground stations, low-power radio links, and telephone lines.

Figure 2-1.—General Configuration of Information Technologies in Production Agriculture

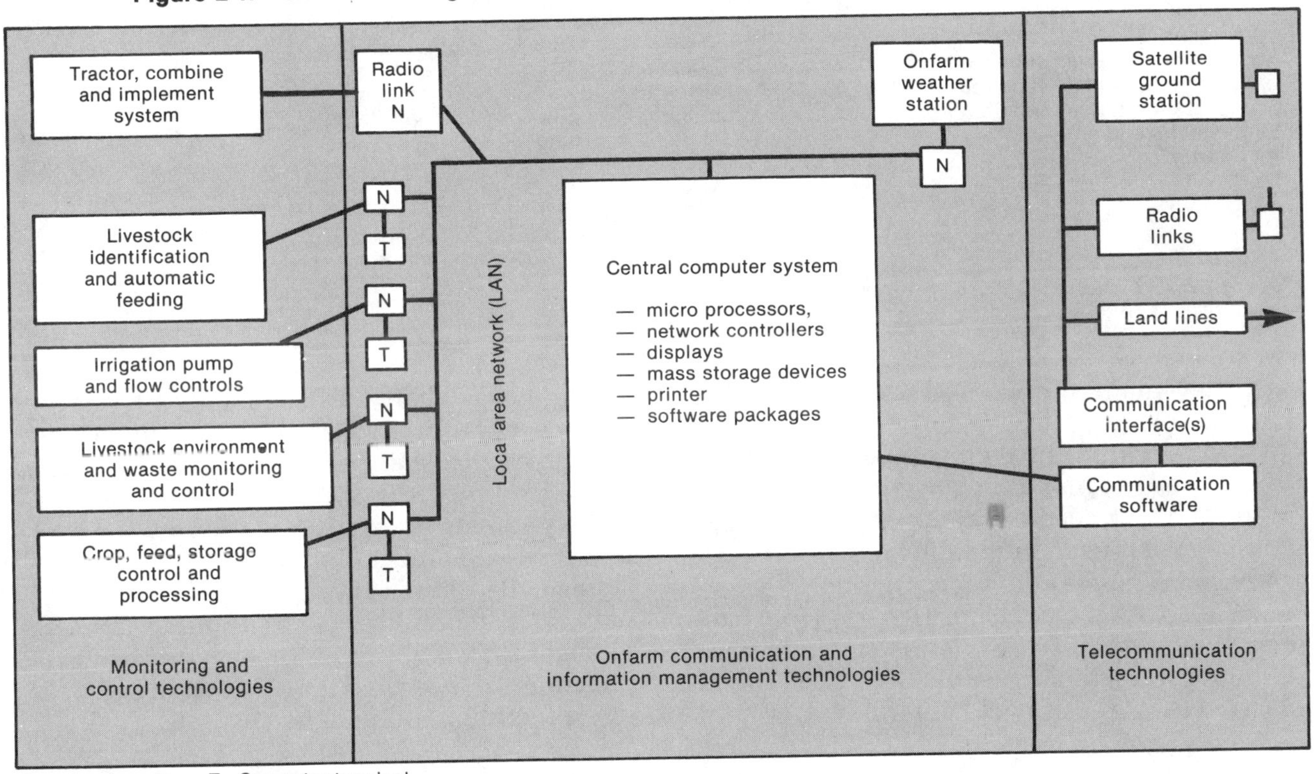

N - network node T - Computer terminal

SOURCE: Office of Technology Assessment.

SURVEY OF EMERGING TECHNOLOGIES

The 28 areas of technologies are shown in table 2-1. OTA commissioned papers by leading scientists in each of these technological areas. A summary of each paper is presented in this section.[3]

Animal Genetic Engineering

Genetic engineering includes a number of procedures by which genes can be manipulated for improving the health and productivity of plants, animals, and humans (Bachrach, 1985). Three important genetic engineering procedures are: 1) recombinant DNA (rDNA) techniques, also called gene splicing; 2) monoclonal antibody production; and 3) embryo transfer.

Recombinant DNA Techniques

Because of its power to alter life forms, rDNA technology is considered to be one of the great-

[3]The papers prepared by those scientists are referenced at the end of this chapter and are available in *Technology, Public Policy, and the Changing Structure of American Agriculture, Volume II—Background Papers* through the National Technical Information Service, U.S. Department of Commerce.

Table 2-1.—Emerging Agricultural Production Technology Areas

Animal	Plant, soil, and water
Animal genetic engineering	Plant genetic engineering
Animal reproduction	Enhancement of photosynthetic efficiency
Regulation of growth and development	Plant growth regulators
Animal nutrition	Plant disease and nematode control
Disease control	Management of insects and mites
Pest control	Weed control
Environment of animal behavior	Biological nitrogen fixation
Crop residues and animal wastes use	Chemical fertilizers
Monitoring and control in animals	Water and soil-water-plant relations
Communication and information management[a]	Soil erosion, productivity, and tillage
Telecommunications[a]	Multiple cropping
Labor saving[a]	Organic farming
	Monitoring and control in plants
	Engine and fuels
	Land management
	Crop separation, cleaning, and processing

[a]These technologies also apply to plant, soil, and water.

SOURCE: Office of Technology Assessment.

est achievements of biological science. Through this technology DNA fragments from two different species can be fused together to form new units called recombinant plasmids (figure 2-2). Such rDNA molecules might contain, for example, a gene from human insulin fused with DNA that regulates the reproduction of bacteria. When such molecules are inserted into bacteria, they instruct that bacteria to manufacture human insulin. Molecules of rDNA can now be inserted into a variety of bacteria, yeasts, and animal cells, where they replicate and produce many useful proteins, such as insulin, growth hormones, prolactin, prolaxin, enzymes, toxins, blood proteins, subunit protein vaccines, immunity enhancers (such as interferons and interleukins), and nutrients like amino acids and single-cell protein feed supplements. Recombinant DNA technology also produces DNA sequences for use as probes in detecting bacterial poisoning of foods and for diagnosing and treating infectious and genetic diseases.

Figure 2-2.—Recombinant DNA Procedure

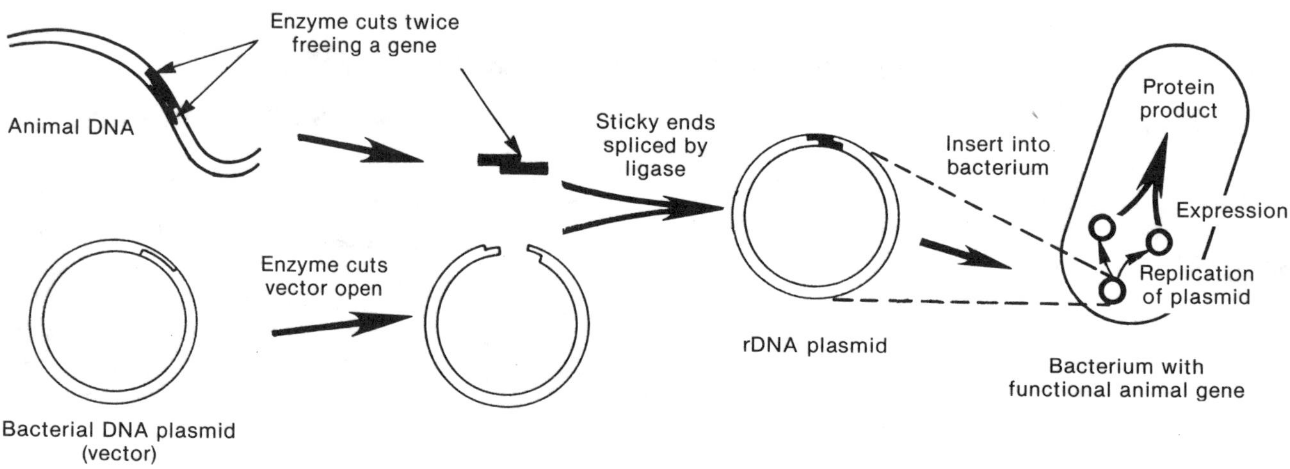

An animal gene is spliced into a carrier DNA (called a vector) for insertion into a micro-organism (a bacterium is shown) or alternate animal host cell, and is made to replicate and express its protein product.

SOURCE: Office of Technology Assessment.

One of the applications of the new pharmaceuticals is the manufacture of growth hormones that can be injected into animals to increase production efficiency. Monsanto, Eli Lilly, and other firms are developing genetically engineered bovine growth hormone (bGH) to stimulate lactation in cows. This hormone, produced naturally by a cow's pituitary gland, was synthesized by Genentech for Monsanto. It has been reported that daily injections of bGH into dairy cows at the rate of 44 milligrams per cow per day have resulted in an increase of 10 to 40 percent in milk yield. The response to injections is rapid (2 to 3 days) and persists as long as treatment is continued (Kalter, et al., 1984). More recently, it was reported that the bGH treatments have increased milk yield 25 to 30 percent in the laboratory and could increase milk yield 20 percent on the farm (Kalter, 1985). The new hormone now awaits approval by the U.S. Food and Drug Administration and is expected to be introduced commercially in 1988 (Bachrach, 1985; Hansel, 1985; *Chem. and Eng. News,* 1984).

Another new technique arising from the convergence of gene and embryo manipulations promises to permit genes for new traits to be inserted into the reproductive cells of livestock and poultry, opening a new world of improvement in animal health and production efficiency. Unlike the genetically engineered growth hormone, which increases an animal's milk production or body weight but does not affect future generations, this technique will allow future animals to be permanently endowed with traits of other animals and humans, and probably also of plants. In this technique, genes for a desired trait, such as disease resistance and growth, are injected directly into either of the two pronuclei of a fertilized ovum (egg). Upon fusion of the pronuclei, the guest genes become a part of all of the cells of the developing animal, and the traits they determine are transmitted to succeeding generations.

In 1983, scientists at the University of Pennsylvania and University of Washington successfully inserted a human growth hormone gene, a gene that produces growth hormone in human beings, into the embryo of a mouse to produce a supermouse that was more than twice the size of a normal mouse (Palmiter, et al., 1983). In another experiment, scientists at Ohio University inserted rabbit genes into the embryos of mice. The genetically engineered mice were 2.5 times larger than normal mice (Wagner, 1985).

Encouraged by the success of the supermouse experiments, U.S. Department of Agriculture (USDA) scientists at the Beltsville Agricultural Research Center and the University of Pennsylvania are conducting experiments to produce better sheep and pigs by injecting the human growth hormone gene into the reproductive cells of sheep and pigs (Hammer, 1985). USDA scientists provide scientists at the University of Pennsylvania with fertilized embryos from sheep and pigs at their Beltsville farms. After being injected with the human growth hormone genes, the embryos are returned to Beltsville for insertion into surrogate mothers.

The experiments of crossing the genetic materials of different species in general and of using the human growth hormone in particular have prompted lawsuits from two scientific watchdog groups: the Foundation of Economic Trends and the Humane Society of the United States. Both groups charge that such experiments are a violation of "the moral and ethical canons of civilization," and have sought to halt the experiments. The researchers argued that they are continuing the experiments cautiously and countered that the potential scientific and practical benefits far outweigh the theoretical problems raised by the critics. While the lawsuit is pending, the experiments are continuing.

Monoclonal Antibody Techniques

Antibodies are proteins produced by white blood cells in response to the presence of a foreign substance in the body, such as viruses and bacteria. Each antibody can bind to and inactivate a cell of the foreign substance but will not harm other kinds of cells. Until recently, the primary source of antibodies used for immunization and other purposes was blood serum from many animal species. However, such serum also contains antibodies to hundreds of other substances, and each antibody type was limited in quantity.

To produce large quantities of a single antibody, scientists now use a technique called monoclonal antibody production (figure 2-3). By fusing a myeloma cell[4] with a cell that produces an antibody, scientists create a hybridoma, which produces (theoretically in perpetuity) large quantities of identical (i.e., monoclonal) antibodies in a pure, highly concentrated form. An array of monoclonal antibodies can now be produced to fight major virus, bacteria, fungi, and parasites and to diagnose the presence of a specific agent in body fluid. The many important uses of monoclonal antibodies in agriculture include: the purification of proteins made by rDNA; the passive immunization of calves against scours; the detection of food poisoning; substitutions for vaccines, antitoxins, and antivenoms; sexing of livestock embryos; post-coital contraception and pregnancy testing; the imaging, targeting, and killing of cancer cells; the monitoring of levels of hormones and drugs; and the prevention of rejection of organ transplants.

Embryo Transfer

Embryo transfer is used for the rapid upgrading of the quality and productive efficiency of livestock, particularly cattle. In the process a superovulated donor animal is artificially inseminated, and the resulting embryos are removed nonsurgically for implantation in and carrying to term by surrogate mothers (figure 2-4). Before implantation, the embryos can be sexed with monoclonal antibody, split to make twins, fused with embryos of other animal species, or frozen in liquid nitrogen for storage until the estrus of the surrogate mother is in synchrony with that of the donor.

For gene insertions, the embryo must be in the single-cell stage, having pronuclei that can be injected with cloned foreign genes. The genes likely to be inserted into cattle may be those for growth hormones, prolactins (lactation stimulator), digestive enzymes, and interferons, collectively providing both growth and enhanced resistance to disease.

[4]Myelomas are cancerous, antibody-producing cells.

Figure 2-3.—Monoclonal Antibody Production

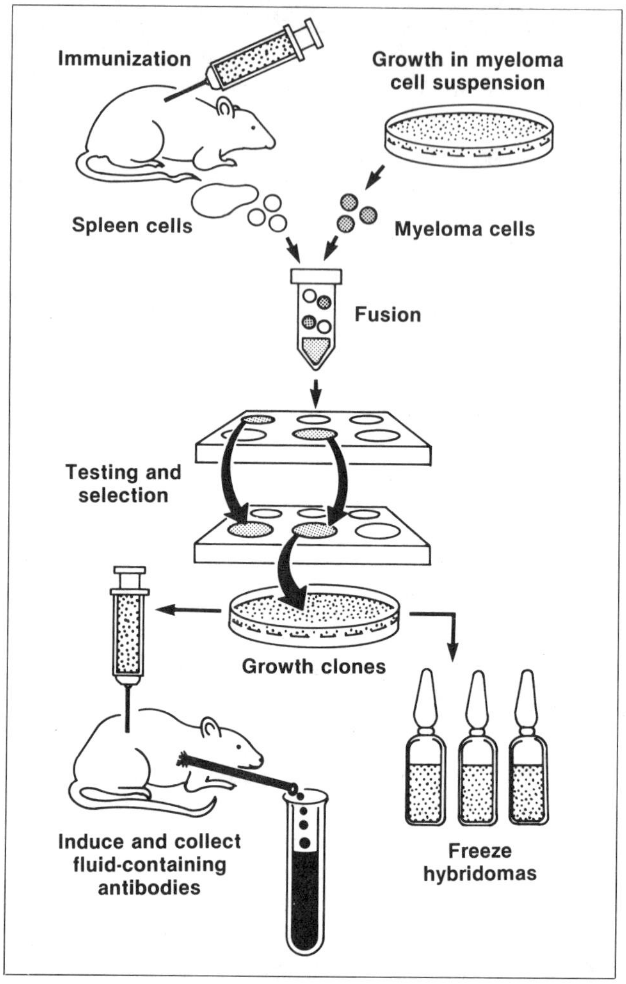

To produce monoclonal antibodies, spleen cells from a mouse immunized against a specific disease are fused with mouse tumor (myeloma) cells to create hybrid cells (hybridoma) that grow in culture. The hybridoma cells are then screened for the production of antibodies. Hybridomas that test positive are injected into a mouse, and the mouse becomes a living factory for the production of antibodies against the same disease. Other positive hybridomas are frozen for future use.

SOURCE: U.S. Department of Agriculture, Agricultural Research Service.

While less than 1 percent of U.S. cattle are involved in embryo transfers, the obvious benefits will cause this percentage to increase rapidly, particularly as the costs of the procedure decrease (Brotman, 1983). One company, Genetic Engineering Inc., already markets frozen cattle embryos domestically and abroad and provides an embryo sexing service for cattle breeders (*Genetic Engineering News,* 1983).

Figure 2-4.—Schematic Presentation of Cow Embryo Transfer Procedures

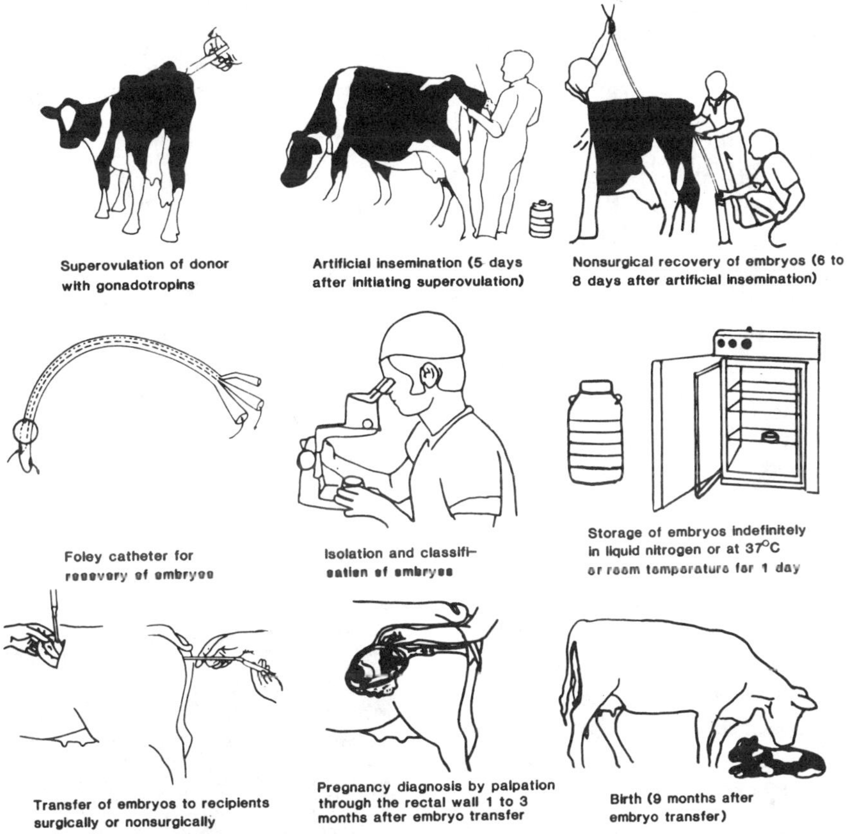

SOURCE: Adapted from G.E. Seidel, Jr., "Super Ovulation and Embryo Transfer in Cattle," *Science*, vol. 211, Jan. 23, 1981, p. 353.

Because of intense competition between hundreds of firms in the United States and abroad, a great many useful genetically engineered products and processes will be introduced during the 1980s.

Animal Reproduction

The field of animal reproduction is undergoing a scientific revolution that could scarcely have been visualized a decade ago (Hansel, 1985). Indeed, if all of the technology now available were used, a new kind of animal breeding system could be put into operation within 10 years.

By year 2000, artificial insemination may be replaced by a system best characterized as "artificial embryonation." In this system highly trained technicians will place embryos into the uteri of groups of outstanding female animals whose estrous cycles have been regulated by artificial means, such as hormone injections, ear implants, or intravaginal devices. The ova from this "superovulation" will be culled surgically or nonsurgically (by flushing) and then fertilized in the laboratory by spermatozoa from outstanding males. The fertilized ova can then be cultured, frozen, and stored until needed. Finally, the embryos will be placed in foster mothers nonsurgically.

Ultimately, it may be possible to sex the embryos by separating the X- and Y-bearing spermatozoa or by identifying the male embryos by immunological techniques so that recipient beef cows will receive primarily male embryos and dairy cows will receive primarily female embryos. Techniques for reducing early embryonic deaths, the major cause of infertility in all farm animals, are also likely to be developed within this time frame.

Achieving these goals will entail the funding of research in three major areas: 1) the development of improved estrous cycle regulation techniques; 2) the development of improved techniques for superovulation and embryo collection, storage, sexing, and transfer; and 3) the development of methods for reducing embryo mortality and improving fertility in all classes of farm animals.

Vigorous pursuit of research in these areas could result, by year 2000, in the marketing of large numbers of genetically engineered embryos containing genes that will improve fertility and fecundity and will result in improved rates of gain, improved carcass characteristics, increased milk production, and increased resistance to diseases in offspring. Despite recent spectacular breakthroughs in introducing human genes into laboratory animals, a great deal remains to be learned about the factors that control chromosomal integration of foreign DNA, the retention of that DNA during embryonic development, and ultimately the expression of DNA, without disruption of the formation and development of the embryo. These developments will affect the major drug companies, genetic engineering companies, equipment manufacturers, veterinarians, inseminators, and extension workers, as well as the Nation's farmers.

The ultimate goal of this research is to increase the efficiency of production so that fewer animals, and less input of labor will be needed to produce the needed animal products.

Regulation of Livestock Growth and Development

The rate and composition of growth is a critical factor in determining the cost of producing livestock products (Allen, 1985). While much is known about genetic and nutritional variables that influence animal growth, much less is known about the hormonal, cellular, and metabolic mechanisms that determine how and at what rate nutrients are partitioned into the growth of muscle, fat, bone, and the tissues of major concern. An understanding of these fundamental mechanisms is needed to provide a foundation for applying new technologies to the development of products to improve the rate, efficiency, and composition of animal growth.

The potential applications of genetic engineering, cloning, and immunology for the improvement of growth in food-producing animals are many. For example, recombinant DNA technology is responsible for providing sufficient quantities of bovine and porcine growth hormone so that scientists can now determine their role, mode of action, and potential use when administered to animals used for producing meat and milk. In the future, this kind of research may also lower the cost of beef production by permitting small cows, which have lower maintenance costs, to produce large market cattle of desirable composition. It also seems likely

Photo credit: U.S. Department of Agriculture, Agricultural Research Service

Ultrasonic techniques to measure backfat may someday provide data for evaluating fat and lean composition in live animals during various stages of growth.

that biotechnology will give rise to new products that can alter the inherent mechanisms of muscle protein and adipose (fat) tissue accretion so that the efficiency of meat production will be improved by the conversion of more nutrients into lean meat and less nutrients into fat. Such a development would be in keeping with the consumer demand for lean, but highly palatable, meat at a reasonable cost, and with the medical recommendations that the U.S. consumer reduce the intake of calories from dietary fat.

Other opportunities for advances involve the physical sciences. These include the need for more rapid, accurate, and economical ways of maintaining the identity of animals through the time of slaughter, and for determining the composition of the living animal and its carcass. Improved methods of identifying mammalian meat animals would be a basis for a national record system. This system would benefit producers, packers, regulatory agencies, and consumers, since it could provide information concerned with marketing, carcass merit, disease, and residue-monitoring programs.

A quick and accurate assessment of body composition not only would improve livestock production data and marketing procedures, but would be an example of new technology that could also be used to address human concerns about body weight and obesity. Current procedures used for determining body composition in livestock are too slow, inaccurate, or expensive for adoption by the industry. As a result, the real value differences between animals of low and high carcass merit, as affected by fat content, are normally not fully realized in the market when animals are sold alive.

The implications of applying these kinds of technologies for improving the production efficiency, composition, and consumer cost of animal products are numerous. They include the more efficient use of livestock feeds, possible changes in crop production priorities, improved composition of animal food products, improved production practices from more complete animal records, and implications related to human health. The application of these technologies will depend on understanding the fundamental principles or mechanisms involved in each major research area.

Animal Nutrition

The U.S. food animal industry is immense. Food animals provide 70 percent of the protein, 35 percent of the energy, 80 percent of the calcium, 60 percent of the phosphorus, and significant proportions of the vitamins and mineral elements in the average human diet in the United States (Pond, 1985).

The future of this industry will depend not only on profitability, but also on the industry's adoption of new technology and on the industry's response to consumer concerns about cost, esthetics, convenience, and health. Areas of nutrition research that may result in major advances in animal food production and use in the next 20 years include: 1) the relation of animal product consumption to human health, 2) alimentary tract microbiology and digestive physiology, 3) voluntary feed intake control, 4) maternal nutrition and progeny development, and 5) aquaculture.

Many consumers are concerned about the effect on human health of consuming animal food products because of the amount and composition of fat in those products as well as the amount of sodium, nitrates, and potentially harmful bacteria or chemical residues. Studies have suggested strong links between some of these factors and human cancer, osteoporosis, and cardiovascular disease. Research on-line is addressing these concerns by applying nutritional and genetic principles to the improvement of animal food products. For example, changes in animal fatty acid composition will be possible by using "protective" feed additives in specific animal diets. Changes in total animal fat content will probably occur through energy restriction, nutrient partitioning, and genetic selection. Sodium content of animal products can be reduced at the processing stage.

The direct impact of advances in this area will be animal food products that are safer for human health. The indirect impacts may be greater, however: to produce such products, producers may have to switch to more pasture, forage, and nonconventional feed resources. Such adjustments could change the total profile of agriculture.

Research into factors controlling voluntary feed intake and nutrient partitioning will result in the diversion of the use of nutrients from body maintenance to lean tissue growth and other productive functions. Such methods will save feed and provide opportunities for alternative uses of feed resources.

More complete knowledge of maternal nutrition in relation to fetal survival and prenatal and postnatal development may lead to significant increases in the amount of edible product per breeding unit. This outcome will be translated into savings in labor and resource use.

Finally, aquaculture has emerged as an important new field of animal agriculture in the United States. Research into specific nutrient requirements for different species of fish during all phases of the life cycle, and interactions between nutritional requirements and water environment, will provide new technology that will make the industry more competitive in animal agriculture. Future growth of private aquaculture will provide an additional supply of edible fish and shellfish for consumption by the U.S. population, whose per capita appetite for animal products may be saturated.

Animal Disease Control

Diseases of livestock are the greatest single deterrent to the efficiency of animal production (Osburn, 1985). Together, animal health-related problems and the resulting inefficiencies in reproduction limit the productive capacity of livestock enterprises to 65 to 70 percent of their potential. Although major epidemic diseases such as foot-and-mouth disease and tuberculosis have been eradicated or controlled, an estimated $17 billion or more annually is lost in production because of a variety of infectious diseases, parasites, toxins, and metabolic disorders.

Some of these losses result from a lack of understanding of animal health problems, such as reproductive inefficiency, neonatal death losses, or mastitis. Other losses relate to the change in structure of livestock enterprises to a system that has both fewer farms and a greater concentration of animals per farm. For example, dairy operations of up to 5,000 milking cows, and poultry operations of 100,000 or more birds, are now relatively common. In these large production units the introduction of an infectious disease can have devastating consequences.

The technologies that show the greatest promise for improving management schemes and controlling disease are: 1) data management and systems analysis, 2) rapid diagnostic tests, 3) selection for disease-resistant strains of livestock, 4) genetic engineering of micro-organisms and embryos, and 5) immunobiology.

Computers and computer programs already allow the farm manager to assess the well-being of each animal in large production units. Data on feed consumption, vaccination records, and conception dates, for instance, can be stored in the computer and retrieved quickly by the manager or veterinarian. Such systems can be coordinated with radiotransmitters used to identify each animal. Within 5 to 10 years such systems will be widely used by progressive animal producers.

Advances in biotechnology will include further development of animal-side test kits for rapid assessment of animal health. One of these tests, the enzyme-linked immunosorbent assay, can test for hormones (to determine pregnancy), detect drug residues in milk or feed, and diagnose disease (through antibody detection). If economical tests can be developed, their use will be widespread and immediate (5 to 10 years).

For certain intractable health problems, like parasites and mastitis, efforts are being made to breed disease-resistant strains of livestock. Advances in embryo transfer, gene insertion into embryos, and amplification of gene products will increase the number of more desirable offspring by year 2000.

Recombinant DNA technology is already being used to alter vaccines genetically so that pathogens in the vaccines cannot replicate in the inoculant and cause a mild infection that could spread to other animals. The development of vaccines for several viral diseases, such as bluetongue, should be possible in the next 15 years.

Finally, knowledge gained in the past two decades is being used to improve that system's efficiency. Ingredients (adjuvants) in vaccines are being used to pace the release of antigens into the body or to manipulate or favor certain immune responses. In addition, monoclonal antibodies are being used to detect and prevent disease. The major constraints to the use of these technologies include: 1) funding of field studies, 2) commercialization of products by the biological and pharmaceutical industries, and 3) cumbersome and expensive processes for assuring quality. The benefits of controlling disease will be a decrease in the cost of production for the farm operator and a decrease in food cost for the consumer.

Livestock Pest Control

Major insect pests cause losses to livestock and poultry of more than $2.5 billion (Campbell, 1985). Some insects, primarily the blood feeders, are pests of all warm-blooded animals. Others are host-specific, although related species may prey on several classes of livestock. Losses may be direct, in terms of decreased livestock products; or indirect, in the form of insect-transmitted disease, secondary infections, predisposition to other diseases, irritation that causes unthriftiness, and costs of insect control.

New technology, particularly for livestock insects that are difficult to control, will be more expensive and will have a lower cost-benefit ratio than that of current technology. Progress in new technology in the science of veterinary entomology is relatively slow for the same reason that adaptation of existing technology is slow—there are few scientists (60) doing research. Several technologies show promise for controlling insect pests of livestock, however.

Although animal producers will continue to use insecticides for the immediate future, progress is being made in such areas as habitat management (pasture rotation and brush control for ticks); integrated pest management (biocontrol, sanitation, and waste management for fly control at feedlots and dairies); and use of pest-resistant breeds in cross-breeding programs (Indian crossed with European cattle).

For blood-feeding insects research is directed at developing slow-release technology, whereby a chemical ingredient is formulated into a matrix that slowly erodes or vaporizes to release insecticide. For example, insecticide boluses are used in the stomach of animals, where they slowly release insecticide that destroys manure-developing fly larvae. Insecticide can also be implanted in an animal's body. Eartags impregnated with a slow-release insecticide have been very effective for horn fly control and have improved face fly control in cattle. As the insecticide vaporizes, it spreads over the haircoat of the animal, destroying insects that rest or feed on the animal. (However, horn fly resistance to the pyrethroid insecticides used in eartags has become widespread.) The newest of these technologies are implants that directly release insecticide into the bloodstream, destroying blood-feeding insects. However, implants and boluses will have a limited effect for migratory, blood-feeding insects unless many producers join the control effort.

Recombinant DNA technologies will be used for the molecular cloning of desired antigens, toxins, enzymes, or other biologically important molecules for use as research tools or in the development of vaccines for bluetongue, anaplasmosis, and other diseases for which insects are vectors. In addition, this technology will enhance the study of molecular genetics and metabolic control in *Bacillus thuringiensis,* a bacterium pathogenic to some insects.

Advances in genetics will allow scientists to manipulate the reproductive capabilities of pest species. These advances include the sterile insect release method and chromosomal translocation, among others.

If technology already available were used on a wider scale, livestock losses from insects could be reduced by one-third ($700 million). This outcome would entail at least a doubling of current extension efforts in livestock entomology. The new methodology discussed might reduce losses by another 15 to 25 percent, but at a lower cost-benefit ratio.

Environment and Animal Behavior

The effects of environment on animal well-being have become ever more important because of the trend toward production systems that confine a large number of animals together in a more artificial environment (Curtis, 1985). Confinement simplifies the environment, reducing an animal's opportunities to alter its surroundings to advantage. While such intensive systems increase production per unit of labor input or space, they can be detrimental to animal function and performance.

The advent of intensive production systems changed the relative importance of various environmental factors as well as the strategies for improving animal production through the application of technology. New technologies likely to emerge by 2000 as a result of current research lie in the areas of energy conservation, optimization of total stress, stress-altered disease resistance, and photoregulation of physiological phenomena.

Feed and fuel—sources of energy—account for much of the cost of animal production. Although the trade-offs between feed and fuel have been quantified for most species, the integration of additional research will result in further energy savings. For example, environmental temperature management schemes developed in an era of cheaper fuel are too luxurious today. Animal producers tend to maintain constant environmental temperatures for their stock, even though the animals evolved in the cyclical thermal environment of nature. In one experiment, when young pigs were allowed to regulate their own environmental temperatures, they inserted a daily 20° F fluctuation of warm afternoons and cool nights, resulting in unchanged pig performance but a 50-percent reduction in fuel use during cold weather. Lowering thermostat settings to parallel age-dependent changes in thermal requirements has also been found to save fuel. In some cases cooler surroundings spur appetites, so performance actually increases. Cost-effective, low-maintenance designs of heat exchangers and solar heating systems will affect further energy savings.

Either too much or too little environmental stimulation can have deleterious effects on the performance, health, and well-being of agricultural animals. To optimize total stress, more must be learned about how stress acts on and is perceived by animals. Devices that animals can use to regulate certain environmental factors are already being recommended to farmers. Computerized sensing devices and control equipment will make biofeedback-linked automation of environmental regulation a reality in animal agriculture.

Researchers are also investigating how the environment influences specific mechanisms of immunity to disease. A variety of common environmental stressors—temperature, crowding, mixing, weaning, limit-feeding, noise, and movement restraint—are known to alter animals' defenses against infectious agents. New techniques in basic science, coupled with more traditional neurobiological, endocrinological, and immunological approaches, can yield a better understanding of how stressors influence regulatory signals among lymphoid cell subpopulations.

The regulation of light is of particular interest in animal production. The advent of photoperiod management revolutionized the poultry industry 40 years ago. Light is managed in poultry confinement operations so that it stimulates poultry growth. In the last two decades the effects of photoperiod management have also been characterized for sheep reproduction. Although the results of similar studies on cattle and swine have been less definitive, some results have been encouraging: under controlled lighting, sows weaned heavier piglets, cows yielded more milk, and lambs grew faster. Experiments now in progress will produce information immediately applicable to animal production.

Crop Residues and Animal Wastes

Improved use of crop residues and animal wastes represents a tremendous potential for more efficient use of resources (Fischer, 1985).

Livestock on U.S. farms produce about 55 million tons of recoverable manure. Approximately 363 million tons of crop residues are produced annually in the United States. Several technologies and major lines of research and development exist in this area: 1) energy from manure, 2) animal feed from manure, 3) chemicals from crop residues, and 4) animal feeds from crop residues.

The high volume of manure production that occurs at many large feedlots and dairies is an opportunity in disguise. Manure has value both as a soil additive and as a source of energy for heat and electricity. Traditionally, manure has been either applied to the soil surface in an unprocessed form or disposed of in a sewage lagoon. Application of manure to the soil surface creates environmental problems in many areas and results in a loss of up to 90 percent of the useful nitrogen value of the manure. Technology is available to inject the manure below the soil surface, resulting in only a 5-percent loss of nitrogen (Suttan, et al., 1975).

Large farms may benefit from installing anaerobic digesters to produce methane from manure, for use as a heating fuel or as a substitute for propane in electric generators. The slurry that remains after digestion contains most of the original nutrient value and may be applied to cropland as fertilizer. Injection of the slurry is preferred, since most of the nitrogen after digestion is in the form of ammonia.

In many farm operations, it is profitable to process manure and use it as a source of nonprotein nitrogen and fiber in cattle and dairy cow rations. Manure is a low-cost source of nutrients, and reusing it as feed reduces the volume of animal wastes that must be processed or disposed of. If used for cattle feed, manure must first be concentrated, then processed by heat treatment or by ensiling.

Using crop residues as a source of chemical feedstocks and animal feed involves some complex trade-offs in most areas because crop residues are becoming widely valued for their ability to reduce soil erosion in combination with conservation tillage practices. Even when crop residues are completely tilled into the soil, they have significant value in maintaining soil structure and nutrient content. However, useful amounts of residues may be removed from fields in many parts of the United States where cropland slopes are gentle and residue density is high. The cost of transporting bulky crop residues generally constrains the area over which collection is economically feasible.

Several technologies under development have promise in areas where residue collection is economically feasible. Residues may be broken down into their component parts by mechanical, chemical, or biological processing, or a combination of all three. The principal components of crop residues are lignin, hemicellulose, and cellulose. Lignin can be used to produce solvents such as benzene, toluene, and xylene. Hemicellulose is readily converted into furfural, which is, in turn, a feedstock for the production of numerous chemicals. Plastic films and fibers and the simple sugar, glucose, can be produced from cellulose. Production of these chemicals is likely to require moderately large-scale technology based on industrial processes and equipment. Transportation costs reduce the likelihood that crop residues will be used as feedstock for industrial processing. Some farms may adopt direct combustion of crop residues for use as a source of heat for grain drying.

In the near term, the most likely process for conversion of crop residues is biological: ruminant animals. Most crop residues can be fed directly to ruminant animals as a source of roughage. A substantial potential exists for developing technologies to increase the palatability and digestibility of crop residues. Numerous efforts have been made to develop simple mechanical and chemical pretreatments, with some success. The problem is difficult, owing to the degree with which the digestible hemicellulose and cellulose are bound to the nondigestible lignin component in the residues of mature cash grain crops. Additional research and development leading to economic and effective pretreatments would have substantial benefits because the size of this resource is so large.

Plant Genetic Engineering

Biotechnology is not new to plant agriculture (Fraley, 1985). Plant breeding, agrichemicals, and microbial seed inocula have made major contributions to the remarkable development of American agriculture. Within the last decade, major advancements have been made in the understanding of gene function and architecture, and powerful methods have been developed for identifying, isolating, and modifying specific DNA segments.

The further application of biotechnologies in plant agriculture could modify crops so that they would make more nutritious protein, resist insects and disease, grow in harsh environments, and provide their own nitrogen fertilizer. While the immediate impacts of biotechnology will be greater for animal agriculture, the long-term impacts may be substantially greater for plant agriculture. The potential applications of biotechnology on plant agriculture will include microbial inocula, in vitro plant propagation methods, and genetic modification.

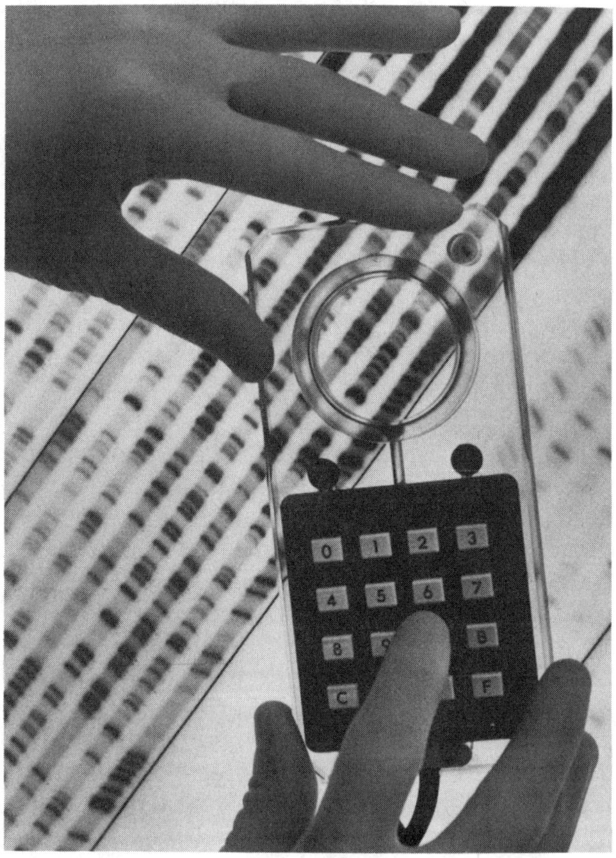

Photo credit: U.S. Department of Agriculture, Agricultural Research Service

Plant geneticist is determining the structure of a soybean DNA segment that resembles the movable genetic elements first discovered in corn. Each band represents a "letter" or nucleotide, in the genetic code.

Microbial Inocula

Research on plant-colonizing microbes has led to a much clearer understanding of their role in plant nutrition, growth stimulation, and disease prevention, and the possibility exists for their modification and use as seed inocula. Rhizobium seed inocula are already widely used to improve nitrogen fixation by certain plants (legumes). Extensive study of the structure and regulation of the genes involved in bacterial nitrogen fixation will likely lead to the development of more efficient inocula.

Two years ago, scientists at the University of California, Berkeley, genetically engineered ice-nucleation bacteria that inhibit frost formation in potato plants. To form ice, there must be nucleation sites around which the water molecules can form the regular ice structure. In the ecosphere, this role is performed by specialized bacteria called *Pseudomonas syringae,* which contain specific proteins that act as the nucleation centers for the growth of ice crystals. By colonizing plants in the manner of epiphytes,[5] these bacteria induce ice formation and thus cause frost damage to plants as the temperature drops below freezing (Feldberg, 1985).

Scientists constructed a new strain of bacteria in which the nucleation protein is absent or altered so that the bacteria can no longer play the role of nucleation centers. Having successfully constructed a new strain of bacterium, these researchers were ready to field test this new organism to see if it would outcompete the normal strains. If so, the new bacterium would protect crops from frost damage, and millions of dollars in lost crops would be saved. As the

[5]Plants that derive their moisture and nutrients from the air and rain and that usually grow on another plant. Spanish moss is an epiphyte.

novel bacteria were scheduled for release to the field, a coalition of public interest groups filed a lawsuit to postpone the field trials (see chapter 10 for more detailed discussion about this controversy).

Recently, Monsanto announced plans to field test genetically engineered soil bacteria that produce naturally occurring insecticide capable of protecting plant roots against soil-dwelling insects (House Committee on Science and Technology, 1985). The company developed a genetic engineering technique that inserts into soil bacteria a gene from a micro-organism known as *Bacillus thuringiensis,* which has been registered as an insecticide for more than two decades. Plant seeds can be coated with these bacteria before planting. As the plants from these buds grow, the bacteria remain in the soil near the plant roots, generating insecticide that protects the plants.

Plant Propagation

Cell culture methods for regenerating intact plants from single cells or tissue explants are being used routinely for the propagation of several vegetable, ornamental, and tree species (Murashige, 1974; Vasil, et al., 1979). These methods have been used to provide large numbers of genetically identical, disease-free plants that often exhibit superior growth and more uniformity over plants conventionally seed-grown (figure 2-5). Such technology holds promise for important forest species whose long sexual cycles reduce the impact of traditional breeding approaches. Somatic embryos[6] produced in large quantities by cell culture methods can be encapsulated to create artificial seeds that may enhance propagation of certain crop species.

[6]Embryos reproduced asexually from body cells.

Figure 2-5.—Plant Propagation—From Single Cells to Whole Plants

The process of plant regeneration from single cells in culture

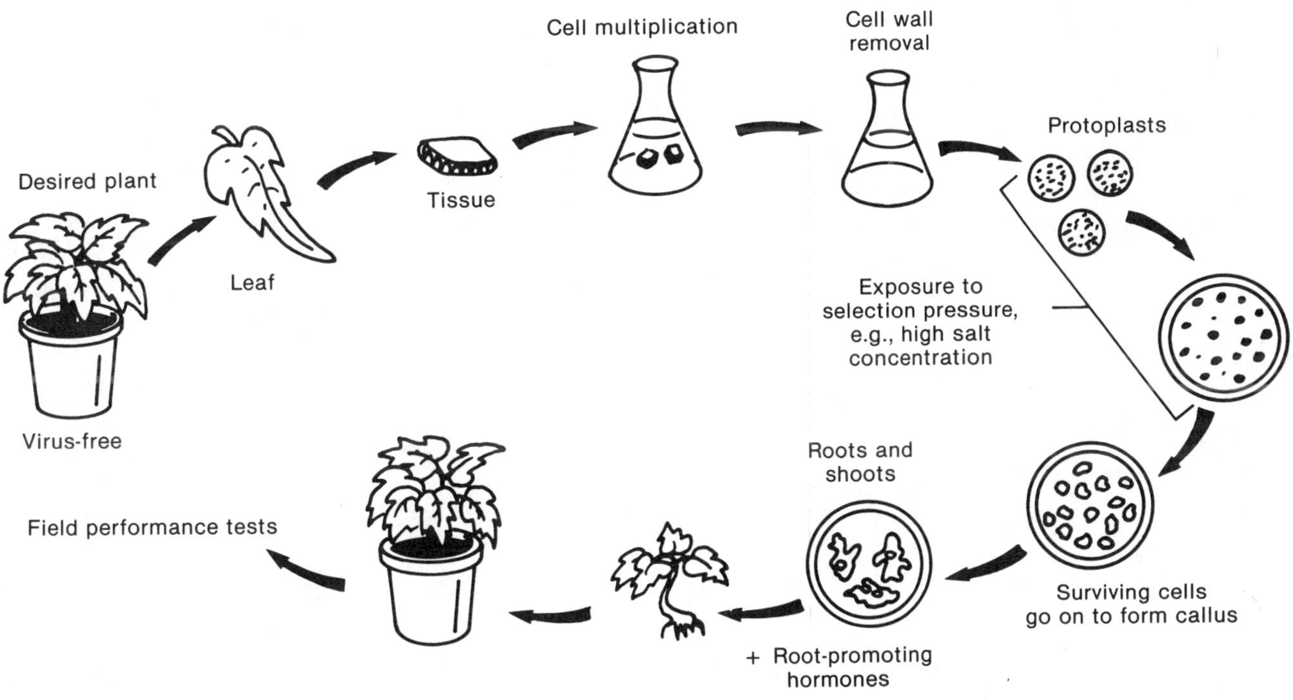

SOURCE: Office of Technology Assessment.

Genetic Modification

Three major biotechnological approaches—cell culture selection, plant breeding, and genetic engineering—are likely to have a major impact on the production of new plant varieties. The targets of crop improvement via biotechnology manipulations are essentially the same as those of traditional breeding approaches: increased yield, improved qualitative traits, and reduced labor and production costs. However, the newer technology offers the potential to accelerate the rate and type of improvements beyond that possible by traditional breeding.

Of the various biotechnological methods that are being used in crop improvement, plant genetic engineering is the least established but the most likely to have a major impact. Using gene transfer techniques, it is possible to introduce DNA from one living organism into another, regardless of normal species and sexual barriers (figure 2-6). For example, it has been possible to introduce storage protein genes from French bean plants into tobacco plants (Murai, et al., 1983) and to introduce genes encoding photosynthetic proteins from pea plants into petunia plants (Broglie, et al., 1984).

Transformation technology also allows the introduction of DNA coding sequences from virtually any source into plants, providing those sequences are engineered with the appropriate plant gene regulatory signals. Several bacterial genes have now been modified and shown to function in plants (Fraley, et al., 1983; Herrera-Estrella, et al., 1983). By eliminating sexual barriers to gene transfer, genetic engineering will greatly increase the genetic diversity of plants. This technology will have a major impact on the seed and plant production industries as well

Figure 2-6.—Gene Modification—Insertion of a Desired Gene Into the Host Plant Through Vectors (or gene taxis)

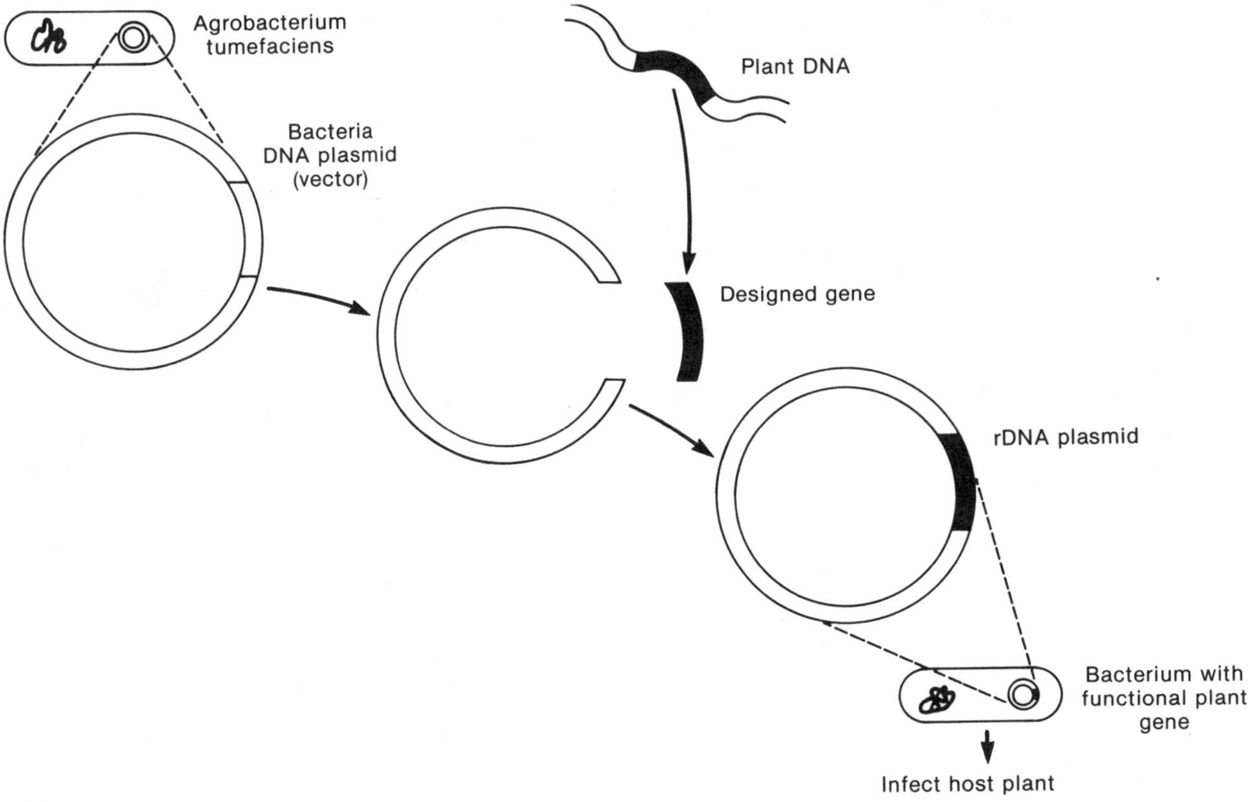

SOURCE: Office of Technology Assessment.

as on the chemical, food processing, and pharmaceutical industries.

The commercialization of plant biotechnology will require breakthroughs in several technical areas, including increased understanding of plant cell culture, plant transformation systems, plant gene structure and function, the identification of agronomically useful genes, and plant breeding. Increased research funding is needed in these specific areas and generally in the basic plant sciences and in molecular biology to accelerate technical development. Commercialization of plant biotechnology will also depend on other factors, including environmental regulation, university-industry relations, economic incentives, and consumer acceptance.

Improved plants produced by gene transfer methods should be commercially available in 7 to 10 years. The introduction of plants produced and selected using cell culture manipulations and certain biotechnology-derived microbial seed inocula or products could occur earlier.

Plant genetic engineering methods will initially emphasize the same targets for crop improvement (increased yield, improved qualitative traits, and reduced labor and production costs) as traditional breeding programs do. Ultimately, the technology will lead to improvements not even imagined in American agriculture.

Enhancement of Photosynthetic Efficiency

Photosynthesis is the fundamental basis for plant growth (Berry, 1985). Through photosynthesis, energy from sunlight is absorbed by chlorophyll-containing tissues of the plant and used to assimilate carbon dioxide into organic molecules. The photochemical reactions in the process are intrinsically very efficient. However, several factors inhibit photosynthetic efficiency in plants: 1) certain mechanisms of photosynthesis itself, 2) the efficiency of water and nutrient use, and 3) environmental stress. Research is ongoing in each of these areas.

Plants vary in their efficiency of photosynthesis. Higher plants have an enzyme (RuBP carboxylase) that causes oxygen to react in a side reaction during photosynthesis, diverting energy that would otherwise be used to fixate carbon dioxide. This oxygenase reaction, which appears to result from a metabolic defect in plants, is encouraged by the high-oxygen, low-carbon dioxide concentration of normal air. Artificially increasing the content of carbon dioxide in the air partially suppresses this mechanism and generally results in increased crop yields. This suggests that improvements in the mechanism of photosynthesis could result in increased yields, all else being equal.

Plants known as C_4 plants have developed a biological and morphological modification that reduces the impact of the oxygenase reaction. As a result, they waste less energy during photosynthesis. C_4 plants include corn, sorghum, sugarcane, and millet. Plants that cannot suppress the oxygenase reaction are called C_3 plants. They include wheat, soybeans, cotton, and rice.

C_4 plants have an advantage over C_3 plants when leaf temperatures are high and a disadvantage when they are low. Moreover, C_4 plants use nitrogen and water more efficiently in photosynthesis. Thus water use efficiency could be increased in warm, arid regions if more C_4 plants could be used.

A long-term prospect for improving photosynthetic efficiency lies in research to understand the basis for the oxygenase reaction and efforts to inhibit the reaction chemically or to modify the enzyme by using rDNA technology. Success will depend on many breakthroughs in understanding the chemistry and molecular biology of chloroplasts and in manipulating chloroplast genes.

Molecular biology has already yielded the ability to modify the sequence of amino acids in RuBP carboxylase to produce modified versions of the protein. This provides experimental tools of unimagined power for investigating the mechanisms of enzyme-catalyzed reactions.

Other research is being directed at improving the efficiency of use of water and nutrients through: 1) better management techniques that use microcomputer-based plant growth models, and 2) new instrumentation to monitor crop performance. Improved weather forecasting will also be important. Breeding plants for efficient water and nutrient use and for stress resistance is possible and has already had some impact. These technologies have the greatest immediate prospect for improving the efficiency of photosynthesis in the next decade, although a strong research effort is needed to realize these potentials.

Plant Growth Regulators

Plant growth regulators are natural or synthetic compounds that are applied (usually directly) to a plant to alter its life processes or structure in order to improve quality, increase yields, facilitate harvesting, or any combination of these (Nickell, 1985). Used commercially since the 1920s, plant growth regulators have had a variety of impacts. One of their earliest was in rooting powders and solutions for the propagation of cuttings. Another was the use of maleic hydrazide to prevent sprouting in potatoes and onions during storage.

The biggest boost to plant growth regulation came with the discovery that phenoxyacetic acids kill broadleaved plants (such as weeds) but not grasses. Using such chemicals in herbicides has out distanced economically all other uses of plant regulators and, until recently, dwarfed their general importance.

Overall research on plant growth regulation is currently multipronged. Industrial research is particularly directed at two major U.S. crops—corn and soybeans.

An increasingly important research effort is that for antidotes to herbicides. Called protectants, or safeners, such compounds can be applied to the crop, usually to the seed, to make it resistant to an herbicide. When the herbicide is applied to the crop row, it kills only the weeds.

USDA has used plant growth regulators so successfully in the guayule bush that it may be theoretically possible to have a rubber industry within the boundaries of the United States.

Photo credit: John Gardner, Brigham Young University

Scanning electromicrograph of a developing wheat head reveals vertebrae-like spikelets branching from its axis. By unlocking the hormonal secrets locked in the tissue of the spikelets, researchers hope to increase the number of spikelets per head, and the number of kernel-producing florets on each spikelet—thus increasing yield.

Ethephon, which is used to prevent coagulation of latex flow in rubber trees, eliminates the need to tap the tree daily. Plant growth regulators of the triethylamine type are used to increase the total rubber content of the guayule bush. A similar use of growth regulators is the use of paraquat on pine trees. The result is a significant increase in oleoresin content and the possibility that the naval store industry may take on new life in the Southeast United States.

The success in the sugarcane industry in the control of flowering, in the use of gibberellic acid to increase the tonnage of both cane fiber and sugar, and in the use of ripeners to enhance sugar yields allows industry to turn its attention to developing dessicants for use as harvest aids.

In the grape industry the successful use of gibberellins on grapes is stimulating studies on the control of abscission (the shredding or separating of plant organs such as fruit or leaves) and the use of ripeners to increase sugar content. Abscission agents have been used successfully on cotton, oranges, cherries, and olives, where it reduces the tenacity of the fruit sufficiently to allow easy harvest by hand-picking, mechanical harvest, or shaking. Abscission agents have also been used to thin apple blossoms, changing the yield pattern from alternating light-fruiting and heavy-fruiting years to annual, successfully bearing years.

Plant growth regulators can reduce harvesting costs by changing the shape of the whole plant or just its fruit to allow easier mechanical harvesting. Apples, grapes, and wheat are examples. Gibberellic acid is used with grapes, for instance, to lengthen the pedicel to each berry. This reduces the rotting that normally occurs because grapes grow too close together. The size and shape of both apples and grapes can be changed by cytokinins and gibberellic acid.

Regulators can also be used to speed or delay the maturation of fruit. Success has already been notable with navel oranges and with pineapple, peppers, cherries, coffee, tomatoes, and tobacco. In addition, the tremendous losses of food crops following harvest almost guarantees an increase in research to develop preharvest and postharvest preservation through plant growth regulators.

Finally, preliminary indications with Cycocel and other chemicals suggest that overcoming environmental limitations via plant growth regulators should be a fertile field for investigation.

A substantial number of new products or new uses for existing products can be expected in the 1990s. Because of the difficulty in registering new compounds, many of the advances will be extensions of uses of existing products. Since so much of the chemistry, evaluation, and expensive toxicology has already been done on existing products, finding new uses for those products might well have a greater impact than researching new compounds.

Plant Disease and Nematode Control

Plant diseases are caused by viruses, fungi, bacteria, nematodes, and other micro-organisms (Browning, 1985). Collectively, these organisms cause considerable losses before and after harvest, an estimated $18.6 billion annually. Only a few of the thousands of species of pathogens and insects cause concern, however; the rest are controlled by natural immunity. Many organisms that do cause loss may theoretically be controlled by managing more wisely the mechanism of host-plant resistance. This area is a major one for research.

Some beneficial micro-organisms help protect plants from disease. In addition to their nutritional benefits, nodulating bacterial and mycorrhizal (root-extending) fungi render some plants more disease resistant. Micro-organisms also provide a vast gene pool for improving plants and other micro-organisms through rDNA technology. That technology is already available for synthesizing microbes of naturally occurring products for use as pesticides. Such genetic engineering should lead to new biocontrol agents; for example, modified plant viruses that will give cross protection. One success story is that of crown gall, a serious bacterial disease of many woody and herbaceous plants. Crown gall is now controlled biologically by the K84 strain of bacterium that is a close relative of the bacterium that causes the disease. Inoculating a seed or transplant with K84 produces a bacteriocin that protects against crown gall.

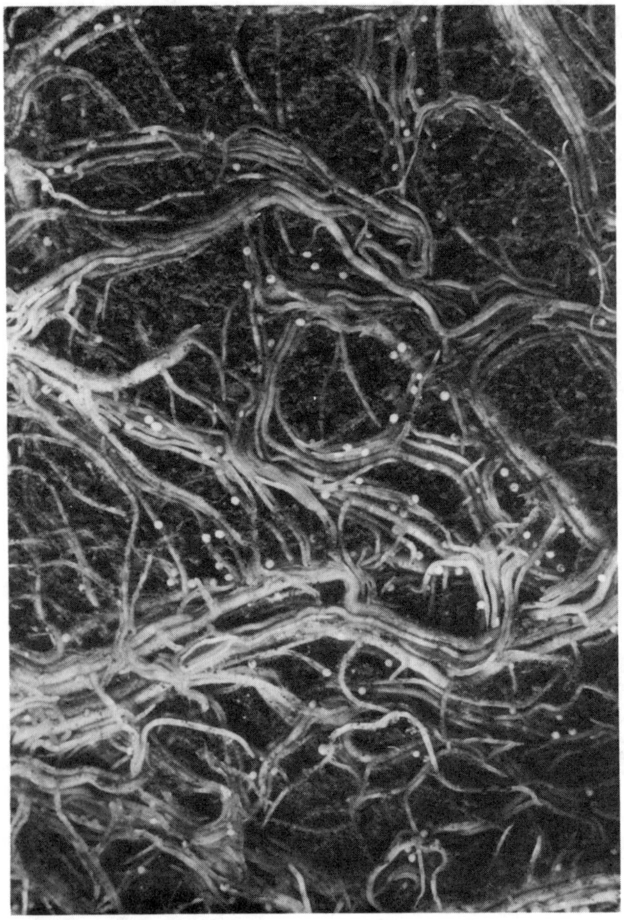

Photo credit: U.S. Department of Agriculture, Agricultural Research Service

Golden nematode cysts (about 0.5 mm long) on the roots of a potato plant.

Other examples of biocontrol include using disease-suppressive soils and pasteurizing the soil to kill pathogens but not thermophilic (growing at high temperatures), beneficial microbes. In addition, some cultural practices (fertilization, irrigation, and stubble and debris management) can be refined to effect biocontrol. For example, continuous cropping can be used to allow antagonists to pathogens to increase, as in the control of potato scab.

Disease-resistant cultivars can be bred and resistance-managed. Genetically, plant resistance is conditioned by major-effect genes and minor-effect genes. Although major-effect genes are easier to work with and give more dramatic results, their effectiveness in the field has frequently been disappointing. Thus researchers have turned to minor-effect genes, which are more difficult to work with but are the most successful way of controlling disease in the homogeneous cultivars demanded by mechanized Western agriculture. Major-effect genes show promise for controlling disease in heterogeneous cultivars, as occurred with multiline oat and wheat cultivars developed in Iowa and Washington. Even highly epidemic foliar pathogens can be controlled in this manner. A major line of research may result in using resistance genes to obtain diversity without sacrificing bona fide needs (as opposed to merely cosmetic needs) for uniformity. This may be one of the fastest ways simultaneously to control certain highly epidemic diseases and to reap the tremendous potential benefits from plant genetic engineering.

Additional work is needed at all levels of pesticide development, but is especially needed for completing the development of systemic pesticides that have two sites of activity on the molecule, thereby extending the pesticide's effective life. Research is also needed on more effective delivery systems for systemic pesticides.

Other research will be directed to developing naturally occurring chemicals that will stimulate the plant's defense mechanisms or enhance activity by biocontrol agents. Ultra-low-volume delivery systems will be needed for these and regular pesticides that are active at very low dosages.

A final important area for research is that of crop loss assessment. Although it is possible to assess plant loss from single pathogens, weeds, and arthropods (and a few combinations of these), such assessments are less precise when made for larger areas, several cultivars, and a wide variety of plant stresses. Research to improve crop loss assessment will help set research priorities and aid in making management decisions.

Management of Insects and Mites

Insects and mites are humankind's greatest competition for food and fiber (Kennedy, 1985).

Although less than 1 percent of all insect and mite species are considered agricultural pests, those pests cause average annual losses to agricultural production of 5 to 15 percent, despite the expenditure of millions of dollars each year for agricultural pest control. Thus, protecting crops from such losses will continue to be an important component of agricultural production.

Research on this problem is being conducted in the broad areas of: 1) chemical controls for insects and mites, 2) genetic manipulation of plants and insects and their natural enemies, and 3) information processing.

Because they are highly effective, economical, and fast acting, chemical insecticides and acaricides (for mites) are widely used for reducing insect and mite populations to subeconomic levels. Advances in insect physiology, toxicology, and analytical chemistry are leading to the discovery of new compounds that disrupt the normal growth and development processes of insects. Compounds with juvenile hormone activity that prevent an insect from molting to the adult stage, those with antijuvenile hormone activity that cause insects to molt prematurely to the adult stage, and those that interfere with the normal synthesis and deposition of exoskeleton all hold promise for the future. Similarly, advances in the chemistry of natural products and the study of plant defenses against insects and mites are leading to the identification of naturally occurring, insecticidal and acaricidal compounds with novel modes of action. Many such compounds are likely to be suitable for large-scale production via fermentation processes with genetically engineered micro-organisms.

With existing application technology only 25 to 50 percent of a pesticide is actually deposited on plant surfaces, and less than 1 percent actually reaches the plant. In addition to being wasteful, this situation greatly exacerbates undesirable effects to the environment. One factor is the incorrect mixing and calibration by pesticide applicators. Efforts are thus being made to design equipment that injects pesticides at the proper rate directly into the lines carrying water to the nozzle, eliminating the need for tank mixing. Other research will ensure more uniform droplet size, will control spray drift, and will improve adherence of the spray to the plant.

Photo credit: U.S. Department of Agriculture, Agricultural Research Service

A Mexican bean beetle larva—a devastating pest of snap and soybeans—becomes a meal for the spined soldier bug instead. The bug's pheromone may help farmers enlist its help in controlling many pest insects.

Advances in genetic engineering greatly increase the likelihood of new classes of insecticides and acaricides. Insect pathogens, including bacteria, fungi, protozoa, and viruses, are likely candidates for genetic engineering to enhance their utility as microbial insecticides. The pathogenic bacterium *Bacillus thuringiensis* is already commercially available and widely used to control caterpillars on certain crops. Genetic engineering holds great promise for expanding the spectrum of pests controlled by this bacterium.

Crop varieties resistant to insect pests have been used to manage insects with success in a number of important crops. Use of genetic engineering to transfer genes from resistant wild plants to crop cultivars holds great potential for insect and mite management, but requires very specific knowledge of the biochemical bases of the resistance crop to be transferred. In most cases, the requisite knowledge is not yet available.

Improvements in the design and availability of computer hardware and software will produce tremendous changes in insect and mite management at the research, extension, and farm levels. To contribute to crop profitability, insect and mite management entails the processing of tremendous amounts of information on the condition and the phenological stage of the crop, the status of insects and mites and their enemies in the crop, incidences of plant diseases and weeds and measures used in their control, weather conditions, crop production inputs, and insect and mite management options. Computers at the farm level, with access to centralized databases, will allow farm operators to design and implement pest management strategies for their farms. Some software systems are already in place and are continually being improved. In general, however, improvements in databases are awaiting advances in knowledge about pest dynamics and crop pest interactions.

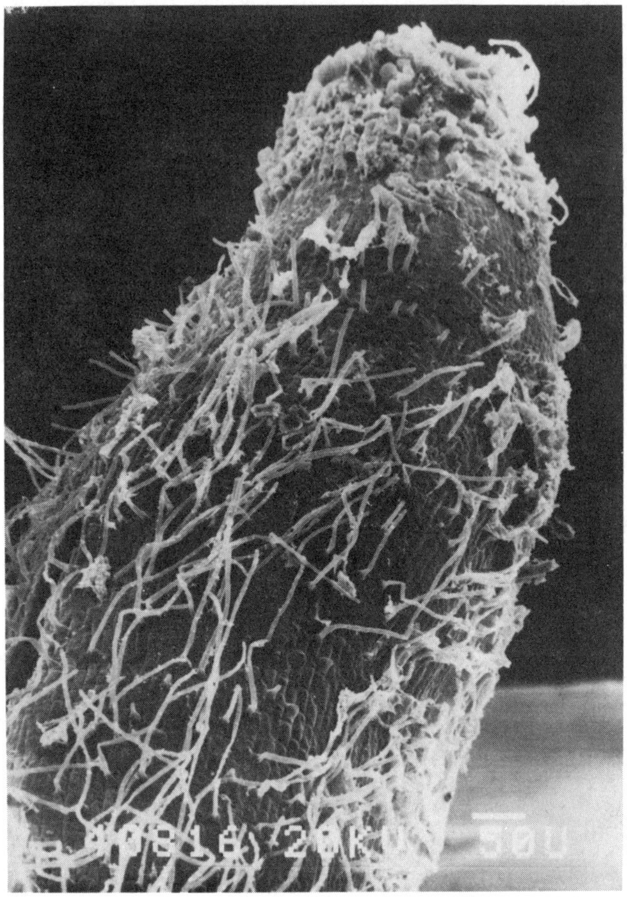

Photo credit: Howard Berg, University of Florida

A scanning electron micrograph of a root tip from a sorghum (*Sorghum bicolor*) plant with kidney bean shaped bacteria (*Azospirillum brasilense*) on its surface. Such nitrogen-fixing bacteria may live on the root surface or in the surrounding soil. The white, threadlike projections are root hairs.

Biological Nitrogen Fixation

Nitrogen is a critical nutrient for crop production (Alexander, 1985). Although abundantly available—either as atmospheric nitrogen (N_2) or in organic complexes in the soil—nitrogen in these forms cannot be used directly by plants. It must first be changed to ammonia (NH_3) or nitrate (NO_3). Thus the large supply of nitrogen needed to grow crops is most commonly provided by nitrogen fertilizers. However, such fertilizers are expensive, and their production consumes a nonrenewable resource, hydrocarbons.

Nitrogen can also be provided through biological nitrogen fixation, a process by which certain bacteria and blue-green algae use an enzyme, nitrogenase, to convert N_2 to NH_3. The most important of these bacteria agriculturally belong to the genus *Rhizobium*. These bacteria enter the roots of legumes and form nodules in which they "fix," or convert, nitrogen in the air to forms used by plants. A legume may receive all of its nitrogen needs this way, given the right *Rhizobium*. In turn, the rhizobia are somewhat protected from microbial competition and predation and from other detrimental effects in the soil environment.

Other kinds of nitrogen-fixing bacteria live near cereal crops and grasses, possibly providing small, beneficial amounts of nitrogen to the plants and receiving needed organic compounds but no protection from detrimental effects in return. This relationship is known as associative fixation.

If its magnitude can be increased, the process of biological nitrogen fixation offers an attractive way to supply the large nitrogen demand of crops without the extensive use of nitrogen fertilizers. To this end, considerable research has been done in the last decade on the biochemistry and genetics associated with the process, and much useful information has been gleaned from this basic research. Research is also under way to determine the possibility of developing cereal crops that fix their own nitrogen, and recent studies have provided needed approximations of the amount of nitrogen provided by associative fixation.

To provide enough nitrogen biologically to sustain high crop yields, however, the stresses affecting legumes and rhizobia must be better understood, and improved bacterial strains and other ways to overcome these constraints must be found. These developments will come from a combination of well-established techniques and agronomic practices as well as new technologies. For example, conventional strain selection and genetic manipulation may be used to produce strains of rhizobia that can compete with soil micro-organisms or that can resist abiotic stresses such as pesticides, drought, and high temperatures. Plant breeding will be used to develop legumes that are better acclimated to soil conditions, have greater photosynthetic activity and less photorespiration, can resist nodulation by less effective soil rhizobia in favor of inoculated rhizobia, and can prolong the duration of fixation. Less likely to come to fruition in this century, but of great importance, will be the development of cereals that can fix their own nitrogen in their tissues or root zones.

If funding is adequate, greater nitrogen fixation from legume-bacterial symbiosis will be realized in the next 10 years, and that from the associative fixation of cereal roots will be realized in 15 years. The benefits of these and future improvements will be the reduced use of hydrocarbons for fertilizer production, an increase in the availability of fertilizer worldwide, and less contamination of ground water.

Water and Soil-Water-Plant Relations

The distribution of vegetation over the Earth's surface is controlled more by the availability of water than by any other factor (Boersma, 1985). In the United States, agriculture accounts for over 80 percent of the water consumed; about 98 percent of that water is used for irrigation of crops, particularly in the more arid Western States. Several factors complicate the availability of water for irrigation: 1) cities, industry, and farming are in fierce competition for the water available; 2) ground-water sources are gradually being depleted; 3) the costs of pumping and distributing surface water are gradually increasing; and 4) many surface and groundwaters are being contaminated by a variety of pollutants. Thus techniques to conserve adequate supplies of fresh water have become important.

Many important contributions have been made by studying water requirements of crops. Although this information has helped in planning reservoir and canal sizes, the hope for breeding plants with lower requirements for water has not been realized, and no technologies have been advanced that would help realize this goal in the next 15 years. Nearly all improvements in water use efficiency have come from improved irrigation techniques, especially the timely application of the amount of water needed and application in a manner that minimizes evaporation. (At present, nearly all the water taken up by the plant is immediately passed through and evaporated at the leaf surfaces. Only a very small fraction becomes part of the plant's permanent structure.)

Progress in improving the water use efficiency of crops will hinge on gathering the information needed to develop a theoretical framework of the mechanisms that influence uptake, use, and loss of water—in humid regions as well as arid and semiarid regions. Dramatic progress in the development of instrumentation now permits researchers to measure many plant physiological responses in real time. It also allows the recent measurements of plant hormones and

Photo credit: U.S. Department of Agriculture, Agricultural Research Service

California cotton fields are the testing grounds for this laser-aligned traveling trickle irrigation system, which links traveling and trickle concepts to improve irrigation efficiency for row crops. Here, wheel towers—operating laterally from a concrete-lined irrigation canal—carry a water line across the field.

enzymes, which provide additional indications of water stress.

Once the mechanisms of water use efficiency have been identified and a better understanding of the plant as an integrated whole is gained, biotechnology may help in the development of more water-efficient plants. Already, recent experiments suggest that tissue culture may provide material less susceptible to water stress. For example, alfalfa and rice cell lines have been obtained that tolerate 2 percent sodium chloride, a salt concentration lethal to nonselected cells.

For the near term, however, traditional methods of plant breeding must be relied on, even though there is increasing evidence that for many crops the limits to improvement by this method are being approached. To break through this yield plateau, the breeder must work with the physiologist and biochemist to understand the stress response hierarchy and eventually to control enzymes, membrane characteristics, and mechanisms for communication in the plant.

The technologies available for immediate application are those that prevent losses in transport, particularly those for farm distribution of irrigation water. These include drip irrigation, below-ground distribution of water, deficit irrigation, water harvesting, time and frequency of application, and the forecasting of time and frequency of application.[7]

Land Management

Land is one agricultural resource that cannot be replaced. Thus a variety of methods and technologies have been developed to conserve soil while increasing yields. These land management technologies include conservation tillage, controlled traffic farming, custom-prescribed tillage, multicropping systems, and organic farming.

Conservation tillage is a tillage and planting system that leaves 30 percent of the crop residue on the soil after planting. The use of the various forms of this system has increased at over 13 percent annually from 1972 to 1982. The specific system used depends on local crops, soil type, moisture levels, and pest infestation, among other factors. Most conservation tillage methods eliminate the use of the moldboard plow, using instead chisel plows or heavy disks in conjunction with heavy-duty planting equipment to cut through soil residues. Mulch-till

[7]For more information on this area see the OTA study *Water-Related Technologies for Sustainable Agriculture in U.S. Arid/Semiarid Lands,* 1983.

Photo credit: U.S. Department of Agriculture, Agricultural Research Service

Facilities at the National Tillage Machinery Laboratory include nine outdoor and two indoor soil bins for evaluating tillage and traction machinery concepts in various soil conditions.

equipment, for example, entails using wide sweeps and blades up to 30 inches wide to cut horizontally several inches below the soil surface. The process loosens and aerates the soil, providing a good seedbed while leaving residue on the soil surface (Battelle, 1985).

Controlled traffic farming is a crop production system in which the crop zone and traffic lanes are distinctly and permanently separated, thus reducing the soil compaction that results from large, heavy machinery making several passes over a field. Although primarily a research concept, controlled traffic farming is practiced in the United States to the extent that current machinery systems allow.

Although it is also in the research stage of development, custom-prescribed tillage is used in some agricultural production systems. This approach to tillage integrates knowledge of soil dynamics, machinery, climate, and crop production economics, and entails making a prescription for various components of the tillage system. The specific machines to be used, as well as the sequence and time of their use, is defined in the prescription.

Multicropping is the practice of planting more than one crop on a field during the same growing season. Such crops can be grown sequentially (double cropping) or simultaneously (intercropping). For example, corn and soybeans can be grown in the same field in strips, reducing soil erosion, using nutrients more efficiently, and increasing crop yield. Currently available machinery and practices are used to perform the field operations needed in multicropping.

Organic farming reduces or eliminates chemical inputs in favor of more "natural," and supposedly safer, inputs. The products from this method are sold to markets willing to pay a premium for the assurance that chemical fertilizers and pesticides have not been used in production. Organic farmers generally prefer to use fewer technological inputs than do conventional farmers, including lower levels of mechanization. They also derive their nitrogen requirements from planting leguminous crops in rotation with nonleguminous crops and sometimes by adding animal manure. If this system were adopted on a large scale in the United States, the need for more mechanization technologies would be reduced, with the exception of the area of waste handling systems for livestock.

No new or unique machinery is needed to further implement conservation tillage, multicropping, or organic farming. However, considerable interdisciplinary research will be needed to implement controlled traffic farming and custom-prescribed tillage commercially. While these concepts have many perceived economic benefits, their true cost-benefit relationships must be evaluated for the wide variety of crops, terrain, soil types, and climate existing across the United States.

Soil Erosion, Productivity, and Tillage

The quantity and quality of harvested crops depend on the amount of land, the suitability of its soil for growing crops, the biology of crops, and the environment (Foster, 1985). Most crops are grown on clean, tilled soil, leaving the soil exposed and unprotected. Severe erosion can result, and over time so much soil is lost that crop yields decrease and some land may be forced from agricultural production. Excessive soil erosion is estimated to occur on about 30 percent of U.S. cropland, but its effects on productivity are thought to have been masked by new technological inputs like hybrids, fertilizers, and chemicals.

Soil erosion is the detachment of soil particles by the erosive effects of rain, surface runoff, and wind. When erosion removes soil more rapidly than it can be formed, soil becomes thinner with less rooting depth for crops. When the topsoil becomes thinner than the tillage depth, subsoil becomes mixed with topsoil during tillage, degrading the soil. Erosion also removes the fine silt, clay, and organic particles most important for good soil quality. The resultant increase in sand content of the soil reduces the soil's productive potential. Sediment from erosion can create off-site problems through deposits in road ditches, reservoirs, and river channels. Sediment or the chemicals it transports can also pollute off-site air and water.

Four major lines of research on erosion control are proposed: 1) improved conservation farming systems, 2) improved methods for assessing erosion's impacts, 3) evaluation of the potential for restoring productivity to severely eroded soils, and 4) improved understanding of how to use public policy to encourage soil conservation.

Of all factors affecting erosion, crop residue left on the soil surface is most effective in reducing erosion. Research on improved conservation systems will thus emphasize conservation tillage, including reduced tillage, minimum tillage, and no-tillage. These types of conservation tillage differ only in the amount of soil disturbance and in the amount of crop residue left on the soil. When matched to soil conditions, conservation tillage can potentially provide greater economic return and often equal or greater yield than that of conventional tillage. For example, no-tillage works well on well-

Photo credit: U.S. Department of Agriculture, Agricultural Research Service

Crop residue left on the soil surface is an effective way of reducing erosion. Here grain sorghum is growing in barley stubble.

drained, sloping soils but not on cool, poorly drained soils in the Corn Belt. Although conservation tillage has the fewest drawbacks of all erosion control practices, considerable development of the method is still required.

The degree of erosion's impact is a major issue that needs a conclusive answer. The principal tool used to estimate erosion by water is the Universal Soil Loss Equation. The tool for estimating wind erosion is the Wind Erosion Equation. Recent developments in erosion theory and the availability of powerful, portable computers make possible new methods that are more detailed and more accurate for estimating erosion over a varied landscape, erosion from individual storms, and average annual erosion. Remote sensing technology and special image processing equipment will aid in the collection of data. New field studies have been initiated and several mathematical modeling techniques have been developed to evaluate the effect of erosion on crop yield.

If eroded soils can reasonably be reclaimed, the problems of erosion may be less serious than presently thought. Current research in the Piedmont region shows that conservation tillage and multiple cropping (explained later) can be used to restore productivity. Much research must still be done in this area.

Although several practices are available for controlling erosion, many have drawbacks that hamper adoption by farmers. As a result, various policy alternatives are used and have been suggested to provide incentives to farmers to implement soil conservation. Improving the use of public policy will entail the incorporation of major analytical tools into an integrated package compatible with affordable computer resources. Such tools will include models for climate, erosion, water quality, crop yield, pests, and economics.

The major potential impact of this technology on agriculture will be significantly improved erosion control with little loss, if not gain, in crop yield, improved water quality, improved farmability, and increased profit. It is hoped that this technology will provide farming systems with enough positive benefits that erosion control becomes a side benefit.

Multiple Cropping

Multiple cropping is the intensive cultivation of more than one crop per year on the same land so as to use land, water, light, and nutrients efficiently (Francis, 1985). Double cropping, or the sequential planting of two crops, such as wheat in the winter and soybeans in the summer, is the only pattern commonly used in the United States. Intercropping, the simultaneous culture of two or more crops in the same field at the same time, is popular with low-resource farmers.

Although widely used in the lesser developed countries by farmers with limited land and resources, multiple cropping systems have not been extensively explored for their applications in this country. Yet, in addition to their efficient use of resources, intensive cropping systems offer several other benefits: vegetative cover through much of the year, which prevents erosion; the need for less fertilizer, owing to the contributions of legumes in these systems; and moderate to high potential yields that are sustainable over time.

Relatively little research attention has been paid to these systems in temperate agricultural regions. If such systems are to be widely adopted in the United States, major new technological advances may be necessary in four areas: breeding crops for intensive planting systems, understanding competition by plant species for growth factors, improving plant nutrition through fertilizers and microbiology, and developing mechanization for multiple cropping.

Crop breeding for multiple cropping systems can lead to the development of crops that can endure the stress conditions found in multiple-species crop combinations. Varieties and hybrids already exist that are well adapted to double cropping and reasonably well suited to relay cropping, the planting of two or more crops with an overlap of the significant part of the life cycle of each crop. Further refinement is needed in developing new hybrids and in further selecting for adaptation. Results could be available in 15 years.

The competition for growth factors by crops that are grown together or sequentially is not well understood. Such competition includes that between two plants of the same species, between two crops of different species, and between crops and weeds. Competition has been studied in grass/legume mixtures for pasture systems, and basic work on crop/weed competition gives insight on species interactions. Some of the results and much of the methodology can be applied to intercropping. Since existing varieties can be used for most preliminary work, results could be available in 6 to 10 years.

Multiple cropping entails a greater input of nutrients or an alternative approach to plant nutrition. Low-resource alternative cropping systems include rotations, minimum-tillage methods, and use of low levels of fertilizers that do not disturb the biological balance in the soil. Research on nitrogen fixation is an active area at present, but a basic understanding of plant nutrition could take 10 to 15 years to develop.

Machines already available can be used for planting and for most other cultural operations. Through modifications of existing tillage, planting, and cultivating equipment, the farmer can accomplish multiple cropping. However, the development of a combine that can harvest two crops simultaneously is necessary for intercropping to have widespread applications. This short-term objective could be achieved within

5 years, using expertise from the commercial sector.

The principal impacts from multiple cropping will be reduced production costs and increased output per year from a given unit of land. The greater sustainability of production and the reduction in energy use would lead to a more stable agricultural sector.

Weed Control

The cost of weeds to agricultural production is one of the most expensive factors in crop production, amounting to more than $20.2 billion annually (McWhorter and Shaw, 1985). Losses caused by weeds include not only direct competition of weeds to reduce crop yields, but also reduced quality of produce; livestock losses; weed control costs; and increased costs of fertilizer, irrigation, harvesting, grain drying, transportation, and storage.

Weeds can be defined as plants growing where they are not wanted. They range from trees and shrubs to grasses and even cultivated crop species. Volunteer corn, for example, is becoming an increasing problem in soybean production as more conservation tillage practices are being adopted.

In modern agriculture, weeds are controlled through integration of crop competition, crop rotation, hand labor, and biological, mechanical, and chemical methods into integrated weed management systems (IWMS). Since 1950, the use of mechanical power for weed control has increased 30 percent, and herbicide use has increased sevenfold. However, manual labor has decreased 40 percent. As a result of modern weed control technology, farming is now less physical and more technological.

Although significant progress has been made in developing new weed control technology, weeds continue to cause severe reductions in yield and quality. Weeds often limit expanded use of conservation tillage and multicropping. New difficult-to-control weed problems develop through ecological shifts and because more established weeds develop increased tolerance to herbicides.

New weed control technologies needed include: 1) improved chemical and biological methods, 2) allelopathic chemicals to bioregulate weeds, 3) crop cultivars with improved tolerance to herbicides and the discovery of the nature of weed resistance to herbicides, and 4) the development of improved IWMS for conservation tillage and for annual multicrop production.

Development of selective herbicides has spearheaded the advances in weed control technology during the last 30 years and will continue to be important in the foreseeable future. Major breakthroughs needed in this area include a nonselective chemical to control vegetation in fallow fields, more selective chemicals for control of broadleaved weeds in dicotyledonous crops (e.g., cotton, soybeans), and a chemical that can be applied postemergence for effective contol of perennial weeds. There is also interest in control of weeds by bioagents, particularly with native pathogens like fungi.

Photo credit: U.S. Department of Agriculture, Agricultural Research Service

Seed-killing methyl isothiocyanate kept crabgrass seeds (*Digitaria sanguinalis*) in flask on right from germinating. One week after the seeds were placed in flasks the untreated crabgrass seeds in flask on left have germinated. The chemical degrades rapidly in the soil, usually within a few days.

The effectiveness of many herbicides is limited by soil activity; for example, some microbial populations rapidly degrade certain herbicides, limiting the residual effects of the herbicides. Advances in controlled-release technology could aid in this and other problems by reducing volatility and rates of application, reducing herbicide movement through the soil profile, increasing crop selectivity, and reducing environmental exposure. Also helpful is a class of chemical protectant that slows the action of soil micro-organisms, permitting more cost-effective control.

Crops can be protected against the toxicity of certain herbicides through chemical antidotes called safeners, another class of plant protectant. When applied to seeds or soil, these chemicals make an otherwise susceptible plant species tolerant to an herbicide without affecting the weed control aspect of the herbicide.

Plants themselves release secondary chemicals during metabolism that can be toxic to other plants. Such allelopathic chemicals are being studied for their potential use in weed control.

Developments in genetic engineering may allow the availability of herbicide-tolerant crop cultivars in agronomic crops in the next 10 to 15 years. Many weeds have evolved a tolerance to herbicides. The availability of herbicide-tolerant crop cultivars would permit the use of herbicides at higher rates to reduce the evolution of tolerance and would permit the use of herbicides that were previously nonselective.

Finally, research efforts need to be increased to develop more effective IWMS. Basic ecological research is needed to understand weed population dynamics, weed threshold levels, and shifts in weed populations caused by control technology. Research is also needed on how to use rotational tillage to aid in controlling the weeds that develop through several years of conservation tillage. Perennial weeds become particularly troublesome after only 2 or 3 years and have forced many farmers to return to conventional tillage.

Improved weed control technology will result in a slow but steady decrease in production costs and an estimated 10 percent increase in the cost of weed control. Increased use of conservation tillage will necessitate increased herbicide use in the next two decades.

Commercial Fertilizers

The substantial use of commercial fertilizers—about 50 million tons per year—is generally credited with 30 to 50 percent of the cost of U.S. agricultural production (Davis, 1985). Corn and wheat are the most heavily fertilized crops.

Commercial fertilizers supply crops with one or more of the primary plant nutrients (nitrogen, phosphorus, and potassium) in forms usable by crops. Nitrogen and phosphorus are produced in the United States; most (about three-fourths) potassium must be imported from Canada. Nitrogen, phosphate, and potassium intermediates are produced in large plants and then shipped to small plants for combination into final products.

Although expenditures for research and development (R&D) in fertilizer technology are less than 10 percent of that for the entire chemical industry (as a percent of sales), the R&D that exists is aimed at maximizing fertilizer effectiveness, minimizing costs, and protecting the environment.

At present, one-half of the nitrogen applied to the soil is lost to the plants through a variety of inefficiencies, some of which are still not understood. Several new types of nitrogen products under development might improve efficiency of use. They include products with inhibitors to decrease undesirable transformations in the soil (nitrification and urease inhibitors), products coated for controlled release (e.g., the sulfur-coated urea sold for turf and horticultural uses), and acidified products that decrease the volatilization of ammonia (the reaction of urea with mineral acids in the soil). In addition, the use of urea phosphate, urea-nitric phosphate, and sulfur-coated urea may allow closer placement of fertilizer to seed without inhibiting germination. Urea phosphate may also aid in recovering phosphorus, 80 percent of which is unused by the plant and remains fixed in the soil in insoluble forms.

Other research is under way to decrease the energy required to produce, transport, and apply fertilizers. The escalation in oil prices following the oil embargo spurred efforts to design new energy-efficient plants and to retrofit existing plants. In addition, several new urea processes that have been announced will decrease production energy requirements by 25 to 50 percent. New phosphoric acid technology also promises energy savings. To avoid dependence on oil or natural gas (the raw material for ammonia for nitrogen fertilizers), technology for the production of ammonia from coal is in advanced stages of development. Finally, efforts are being made to increase the nutrient content of fertilizers so that the energy expended in transporting and handling will be decreased.

Another area under development is that of phosphate fertilizer production. Because reserves of high-quality phosphate ore are being depleted, researchers are attempting to use lower quality phosphate ore in fertilizers. Their efforts focus on removing the carbonate impurities in such ore and on determining what effect such impurities would have on the efficacy of phosphate fertilizers.

In reduced tillage agriculture, R&D efforts are directed at developing urea-nitric phosphate, urea with urease inhibitors, and urea phosphate and urea sulfate, all of which have the potential to decrease ammonia loss from surface-applied urea. Also, new types of equipment are being designed for precision placement of fertilizer and for simultaneous application of fertilizer with seed.

New developments in the industry evolve rather slowly because of the low level of R&D. Therefore, any new technology that is likely to be introduced by 1990-2000 would have to be under development now. No revolutionary or radically new products or processes appear to be near commercialization.

The future direction of energy prices will probably be the major factor affecting the commercialization of new technology. Because the production of nitrogen, particularly ammonia production, is the most energy-intensive operation for the industry, new nitrogen technology is especially geared to energy prices. The high cost of new facilities is also a deterrent to the adoption of technology. In many situations the industry will prefer to debottleneck or add to existing plants to conserve capital.

Organic Farming

Organic farming uses many conventional farming technologies but avoids, where possible, the use of synthetically compounded fertilizers, pesticides, growth regulators, and animal feed additives (Liebhardt and Harwood, 1985). It relies on crop rotations, crop residues, animal and green manures, legumes, off-farm organic wastes, mechanical cultivations, mineral-bearing rocks, and biological pest control. Organic farmers tend to integrate their farming techniques to a greater extent than conventional farmers do.

In the last 6 to 8 years, several studies have compared organic farming with conventional farming. Although final conclusions must await more rigorous studies and a wider sample of farms, preliminary conclusions show some interesting benefits of organic farming: first, yields per acre are generally equal to or only slightly less than those from conventional farming. Some organic farms have significantly higher-than-average yields. Second, production costs are lower by a high of 30 percent and an average of 12 percent, while energy inputs per unit produced are lower by 50 to 63 percent. Few or no insecticides, fungicides, and herbicides are used. Third, soil erosion is significantly reduced through various cultivation practices. Although organic farming maintains soil quality better and reduces contamination of air, water, soil, and the final food products, much research is needed to determine just why organic practices have this effect and to determine how to maximize the integration of organic practices.

One of the most significant factors in reducing production costs and energy inputs in organic farming is nitrogen self-reliance. Many organic farmers increase nitrogen fixation in their crops by seeding legumes between rows of grain crops during the growing season or after harvest. Research is under way to breed plants that fix nitrogen more efficiently or that fix nitrogen longer in the season. By 1990, research already on-line in this area should be well developed.

For weed control, cover crops are used in rotation; for example, sorghum crops are used to suppress nutsedge. In addition, crop residues are used in conservation tillage to suppress sensitive weed species. Much information about weed control should be available by 1990; however, the technology for wide application of crop rotations will not be available for at least 5 to 10 years.

Organic farms appear to cycle nutrients more efficiently than conventional systems do. One reason is the reduction in erosion that occurs, which allows better soil tillage and better maintenance of productivity. Furthermore, some organic practices enhance the soil's ability to suppress disease. Scientists hope to identify the helpful bacteria and bacterial byproducts involved in disease resistance and to harness them as biocontrol agents.

Biocontrol agents are also used to control insects. One example is the tansy, an insect-repellent plant that shows potential for controlling the Colorado potato beetle. Another example is the use of an antijuvenile hormone, extracted from a common bedding plant, that induces premature metamorphosis in insects, shortening their immature stages and rendering the adult females sterile. In 10 to 20 years, biological pest repellents will probably dominate the market because of their safety for users, consumers, and the environment.

Converting from conventional to organic farming takes about 3 to 5 years, during which yield may initially be reduced. Some of this problem relates to the nitrogen content of the soil and to weed pressure.[8] However, detailed studies of holistic systems are needed to understand better the extremely complex changes in nutrient flow in soil during organic and conventional farming. The potential impact of such studies on U.S. agriculture in the next 10 years could be considerable. If farmers shifted to organic production, farms would be more diverse biologically and economically, and the small farm could remain economically competitive and ensure diverse, competitive food production systems.

[8]This can be minimized by selecting the correct crops and structuring the production system to avoid nutrient deficiency or weed problems.

Communication and Information Management

Technology for communication and information management helps farm operators collect, process, store, and retrieve information that will enable them to manage their farm so as to minimize costs, maintain and improve product quality, and maximize returns. There are three basic components to such technology: 1) microcomputer-based hardware systems for information processing, storage, and retrieval; 2) high-speed LANs for onfarm communication of digital information; and 3) applications software. The computer allows farm operators to keep track of more detailed information, apply complex problem-solving techniques to this information, and thereby make better, more timely, decisions.

Microcomputers appropriate for onfarm use cover the range of business-class computers. Larger and more complex farm operations will generally benefit from larger, more complicated computer systems. Onfarm computers are likely to be subject to more adverse operating environments than those found in typical nonfarm businesses. Thus some additional equipment and adaptations are needed for onfarm operations (Battelle, 1985).

While LAN technology is rapidly becoming more mature and standardized, onfarm installations are likely to be more expensive per node than the typical business system. Farm nodes are generally much farther apart than nodes of the average office system. Farm installations placed among several separate buildings are also more susceptible to lightning-induced electrical problems. Photoelectric isolators at every node will enable use of copper wiring between nodes. Alternately, use of LANs with fiber optic cabling will eliminate problems from electromagnetic interference.

Photo credit: Dr. S.L. Spahr, University of Illinois

Example of microcomputer-based system for onfarm use. This system collects, processes, stores, and retrieves information to control computer feeders and electronic milk flow meters.

Many software packages sold for use on farm computers are general-purpose packages that are identical to those used in other businesses. Spreadsheet programs and database management systems fall into this category. Other packages have only minor modifications and upgrades. The most expensive class of software is generally that written for specialized applications. Few farms are large enough to afford custom programming for their own operations. The range of specialized applications programs that have been developed and are being developed by extension personnel at land grant colleges is quite large. Agricultural software from commercial sources and the land grant institutions is generally task-specific.

Another promising software concept is that of a fully integrated system that would allow the farm operator to simulate the outcome of small and large changes in production practices. The software could generate distributions of prices and weather impacts and simulated biological growth functions. It could produce detailed listings showing expected costs, returns, production schedules, cash flows, and net income streams, working within the constraints of those assets and productive potentials that the operator chooses to consider fixed. Such software would give operators much greater ability to maximize income and flexibility in planning for growth and in responding to changes in the economic and technical environment.

Monitoring and Control Technology

Many processes in plant and animal production may be monitored and controlled by new and emerging electronic technologies. In some cases these devices are designed simply to detect certain conditions and report the information to the farm operator. In other cases, the technology operates essentially autonomously, without operator attention. Devices of this nature are usually programmable, can operate continuously, can be designed to be very sensitive to changes in target variables, and can respond very quickly. These devices, therefore, offer improvements in speed, reliability, flexibility, and accuracy of control, and sometimes reduce

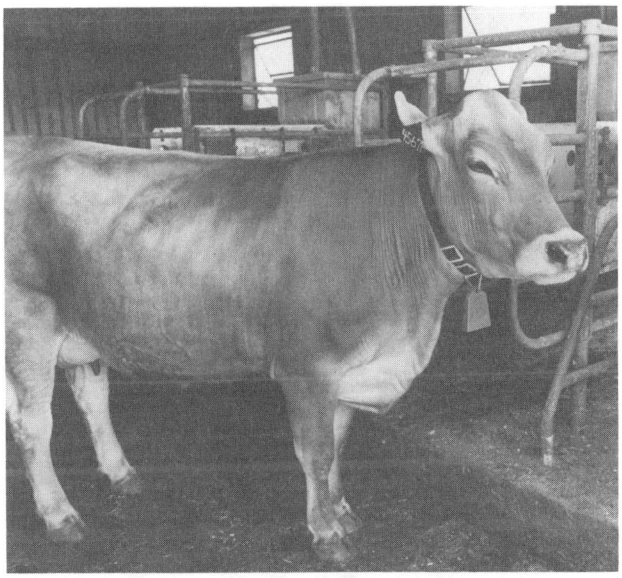

Photo credit: Dr. S.L. Spahr, University of Illinois

Electronic animal identification unit around cow's neck with automatic dispensing grain stall in background. Cow goes into stall, is identified electronically, and has grain dispensed to her automatically. Using computer controls, the feed dispensed is individualized to provide each cow a different amount of feed and a different protein percentage based on her nutrient needs

labor requirements (Battelle, 1985). Some applications of this technology include irrigation control, pest monitoring and control, and the automatic animal identification and feeding system in livestock operations.

Positive identification of animals is necessary in all facets of management, including recordkeeping, individualized feed control, genetic improvement, and disease control. All animals could be identified soon after birth with a device that would last the life of the animal. The device would be readable with accuracy and speed from 5 to 10 feet for animals in confinement and at much greater distances for animals in feedlots or on pasture. Research on identification systems for animals has been in progress for some years, especially for dairy cows. For example, an electronic device now used on dairy cows is a low-power radio transponder that is worn in the ear or on a neck chain. A feed-dispensing device identifies the animal by its transponder and feeds the animal for maximum efficiency, according to the lactation cycle and the life cycle of that animal. This technology also permits animals in different stages of production to be penned together yet still be fed properly.

The largest potential use of electronic devices in livestock production will be in the area of reproduction and genetic improvement. Estrus in dairy cows can be detected automatically by using sensors that remotely detect small changes in the body temperature of the cows. Such an estrus detection device could prove profitable in several ways:

- Animals could be rebred faster after weaning and could increase the number of litters per year.
- Animals that did not breed could be culled from the herd, saving on feeding and breeding space.
- Time would be saved because breeding would be done faster.
- Embryo transplants would be easier because of better estrus detection.

Environmental control of livestock facilities is another area where monitoring and control

Photo credit: U.S. Department of Agriculture, Agricultural Research Service

Fifteen center-pivot sprinklers, all operated by a master computer, "rain" water onto 150 to 210 acres of corn per pivot at the Condon Ranch near Sterling, Colorado.

devices can be used (figure 2-7). Microprocessors will be used to alleviate odorous gases and airborne dust in ventilation systems.

One of the applications of monitoring and control technology in plant agriculture is in the management of insects and mites (Kennedy, 1985). Improvements in the design and availability of computer hardware and software will produce significant changes in insect and mite management at all levels (research, extension, pest management, personnel, and farmer). Centralized, computer-based, data management systems for crop, pest, and environmental monitoring information have been developed and are being

Figure 2-7.—Configuration of Monitoring and Control Technologies in Agriculture

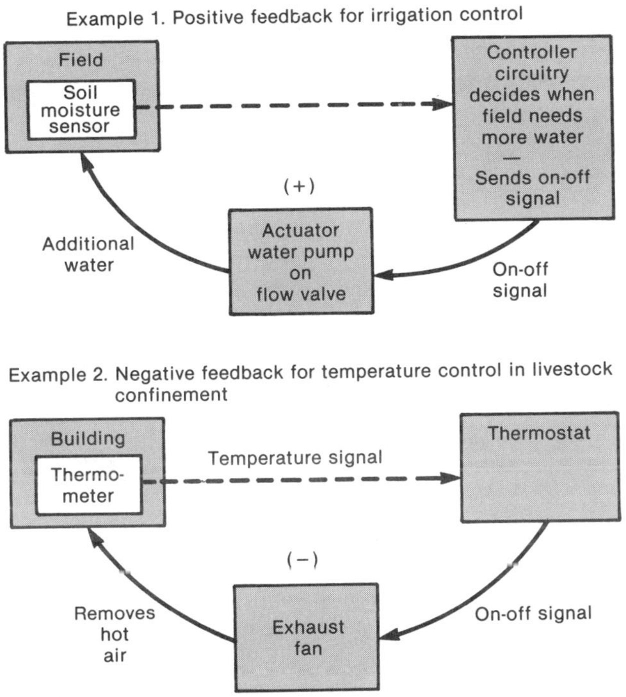

SOURCE: Office of Technology Assessment.

evaluated for use on a regional scale by a USDA Animal and Plant Health Inspection Service regional program. Such systems will provide rapid analysis, summarization and access to general crop summaries, observer reports, pesticide and field management information, reports of new or unknown pests, general pest survey information, and specified field locations with pest severities.

Other software systems designed to facilitate directly the implementation of pest management programs are in use and are continually being improved. The Prediction Extension Timing Estimator model (Welch, et al., 1978) is a generalized model for the prediction of arthropod phenological events but is sufficiently flexible to be used for management in many agricultural and nonagricultural systems. For example, it is used as a part of the broader biological monitoring scheduling system developed in Michigan by Gage and others (1982) for a large number of pests on a wide variety of crops (Croft and Knight, 1983).

An irrigation control system is another example of using monitoring and control technology (figure 2-2). Since irrigation decisions are complex and require relatively large amounts of information, microcomputer-based irrigation monitoring and control systems are especially useful in areas where soils have variable percolation and retention rates, where rainfall is especially variable, or where the salinity of irrigation water changes unpredictably. In this system, a network of sensors is buried in the irrigated fields, with radio links to the central processor. Additional sensors may include weather station sensors to estimate crop stress and evaporation rates, as well as salinity sensors and runoff sensors. The central processor can then automatically allocate water to each field according to the needs of the crops in each field, subject to considerations of cost, leaching requirements, and availability of water.

Telecommunications

Telecommunications technology provides links for voice communications and the transmission of digital data between farms and other firms and institutions. Through such technology, farms, firms, and institutions can be joined together in a large number of formal and informal networks. These networks enable farmers to have relatively rapid, inexpensive, and reliable access to central databases, centralized software packages, and information on weather, markets, and other subjects of interest. Virtually the same technology will be applied to both animal and plant agriculture. Telecommunications include high speed, low speed, and radio telecommunications, satellite base communications, and remote sensing technology (Battelle, 1985).

High-speed or high-bandwidth communications allow the farmer to send and receive much larger amounts of data at lower costs per bit of information. This capability is needed for videotext services, teleconferencing, and, in many cases, satisfactory real-time use of remote computer facilities.

High-speed telecommunications is still undergoing substantial amounts of development. New transmission capabilities or new technologies are needed for bringing high-speed telecommunications to most rural areas. High-bandwidth telecommunications can be provided by technologies that range from conventional high-capacity, coaxial cable, microwave relay systems to fiber optics systems and high-bandwidth direct transmit/broadcast satellite systems. High-bandwidth send-and-receive service for the average farm operation is not likely to be available for some time.

The existing telephone system is capable of handling the demand for slow-speed telecommunications services in many rural areas. The latest generation of microcomputers, modems, and communications software is capable of automatically accessing remote databases and quickly downloading and uploading information at regular intervals without operator attention. Rural areas that install fiber optic telecommunication systems will have enormous information capacity that will easily support very high data rates. In fact, the perennial dream of low-cost, two-way videoconferencing, education, and entertainment may well become a reality in these rural areas by 1990 or 2000.

A number of emerging radio telecommunication technologies will provide improved service in rural areas without the need to rewire the local telephone networks. These technologies can be put into two groups: ground-based, low-power radio repeater systems, such as cellular mobile phone systems; and satellite-based communication systems. In principle, the cellular radio technology being installed in major cities can be expanded to smaller cities, towns, and rural areas at higher power levels for use in voice and data communications. For applications where data transmissions are sufficient and instantaneous communications are not necessary, technology for packet radio messages may provide substantial savings. Packet radio systems use ground-based repeater stations to funnel messages with a standard, or "packaged," format from distributed users to one another or to a point where the messages can be inserted into a national telecommunication network. Messages are entered at each user station, then converted into encoded "packets" complete with addresses and distribution instructions. Each user station then transmits to the local repeater station when the transmission channel is free. This technology may enable cellular radio repeater technology to be extended to especially remote and sparsely populated areas and to areas where the basic telephone system is inadequate and is unlikely to be upgraded.

Satellite-based communication technologies may provide very high-capacity telecommunication channels for rural areas. These systems may be the only feasible high-capacity link for some especially isolated rural areas. Large farms may opt to establish their own ground stations for satellite-based telecommunication, but new generations of communication satellites may have the power to serve many small individual subscribers in remote rural locations.

Almost all commercial satellite communication systems employ satellites in geosynchronous orbit.[9] Alternately, the feasibility of using low-cost, low-Earth orbit satellites for the collection, storage, and rebroadcasting of message packets has been demonstrated by amateur radio groups. Commercial satellites using this design could enter service by 1990.

Remote sensing is a collection of technological systems used to detect, process, and analyze reflected and emitted electromagnetic radiation at a distance. This includes the National Oceanic and Atmospheric Administration weather satellites, land and ocean resource mapping satellites (the Landsat series), airborne camera and electronic sensor systems, and ground-based photogrametric and radiometric sensors. Information from remote sensing technology is used for a wide range of applications. Some examples are weather reporting and forecasting, land use planning, environmental monitoring, crop production estimates, soil mapping, range and forest management, mineral exploration, and watershed management.

Remote sensing technology in the form of weather forecasting has already made a great

[9]Traveling in orbit around the Equator at the same speed as the Earth rotates.

impact on agricultural production. Weather reports and forecasts help farmers decide when to plant and when to harvest. Fruit growers depend on local weather forecasts to help make frost protection decisions.

Farmers can also use remotely sensed information to make other management decisions. Soil moisture levels can be estimated accurately for large northern plains wheat farms that depend on stored soil moisture. Selection of fields for rotation, seeding, and fertilizer rates could then be planned for the available moisture on different parts of the farm to optimize net income.

Remote sensing technologies provide crucial and timely information for the process of estimating global crop production. These crop estimates can have large impacts on price levels and price variability. Estimates of crop production in different countries are an important factor in the administration of commodity and export policies.

Labor-Saving Technology

Labor-saving technologies have made a significant dent in the cost of labor for animal production and, to a lesser extent, for field crops. The change to large-scale confinement operations of livestock and poultry has dramatically reduced labor costs through the automation of feeding, waste disposal, and egg collecting. For field crops, reductions have come from using larger tractors, combines, and tillage equipment.

Opportunities still exist, however, for reducing labor costs, particularly through: 1) mechanization of fruit and vegetable operations, and 2) robotic farming. Researchers and growers are exploring ways to use these technologies with other technologies to change cultivars and cultural practices, rearrange work patterns, develop labor-aid equipment (e.g., conveyors and hoists), improve human relations, and develop labor replacement equipment (Battelle, 1985).

Mechanical harvesting is most applicable for fruits and vegetables that are to be processed or dehydrated, because such products will not show the effects of mechanical handling. Most fruits and vegetables targeted for the fresh market must still be harvested by hand.

The most economically important of the process vegetables are the potato and the tomato, both of which are mechanically harvested. The development of mechanized tomato harvesting is a particularly good example of technological success: the concurrent development of a mechanical tomato harvester and a new, high-yield process tomato, shaped for easy mechanical harvesting, gave California a production increase of 300 percent with only a 50-percent increase in land. Many other process vegetables are harvested mechanically, and research is still under way to automate the harvesting of cauliflower, lettuce, okra, and asparagus.

Of the fruit crops, citrus crops are the largest in total value. Although oranges would seem to be ideally suited to mechanical harvesting (80 percent of the crop is processed), the "bag and ladder" method of hand picking remains the most economical and widely used method. For mass removal of some crops, mechanical or oscillating-air tree shakers, usually in conjunction with abscission chemicals, are used. (Mechanical shakers are also used to harvest process grapes and process deciduous fruits, such as apples, pears, and peaches.) Technology trends in citrus production point to higher density plantings and the maintenance of trees at a height of 5 meters or less. If high fruit yields result, there is good potential for development of over-the-row equipment for production and harvesting.

The use of robotics in agriculture is likely to be centered on high-value, labor-intensive crops like oranges. Research is also being done on apple harvesters that will use ultrasonic sensors to detect tree trunks and steer around the trees. It is conceivable that by 1990, reductions in cost and increases in the speed of operation will make such robotic technology economically attractive. Robotics may also have applications in animal agriculture—for example, in checking calving and farrowing, identifying estrus, managing feeding, and handling manure.

Future labor replacement in agriculture will likely involve some aspect of electronics technology, much of which will be adapted from offshoots of military and aerospace technology. Such technology will have to be adapted to withstand the variety in agricultural environments and will have to have better cost-benefit ratios for widespread adoption. Many new electronics technologies may affect the quality more than the quantity of labor. People with higher level skills will be needed to operate and maintain the new, more complex equipment.

Engines and Fuels

Continued improvements in engines and fuels can be expected in the energy efficiency, durability, and adaptability of self-propelled farm equipment. These improvements are likely to come from R&D in a number of areas: 1) adiabatic and turbocompound engines, 2) electronic engine controls, and 3) onboard monitoring and control devices (Battelle, 1985).

Expenditures by farms on liquid fuels were $10 billion in 1982. Even modest improvements in energy efficiency in farm production will have a significant impact on the total cost of production in agriculture. However, these technologies will not be adopted unless they also deliver significant increases in productivity to individual farms. Farms are continuing to improve their energy efficiency by converting from gasoline-powered equipment to diesel-powered equipment at a rapid rate. Diesel fuel has more energy per dollar, and diesel engines extract more useful work from each calorie of fuel than do gasoline engines.

All conventional internal combustion engines, including diesel engines, are thermally inefficient because they must dispose of large amounts of heat by means of cooling systems. If engines can be constructed of special ceramics to withstand high operating temperatures, they would not need cooling systems and would be much more efficient. Engines of this type are called adiabatic engines.

Turbochargers are being widely used to increase the performance of gasoline and diesel engines by putting some of the exhaust gas energy to use. Even more work can be extracted from the exhaust gases by means of a device called a turbocompound unit. This device captures exhaust gas energy and applies it directly to the drivetrain of the vehicle instead of using the energy solely to compress intake air, as in the conventional turbocharger. Turbocompound units will be especially useful when installed on adiabatic diesels, owing to the high energy content of the exhaust gases from these engines.

Electronic engine controls are being introduced by some manufacturers in an effort to improve the efficiency of the fuel injection system on diesel engines. As with similar systems developed for automotive applications, this technology automatically works to optimize fuel delivery under changing conditions, based on information from engine sensors, implement sensors, and operator inputs. Minimization of tractor wheel slip by means of onboard monitoring and control technology also improves fuel efficiency. Other applications of this technology to onboard control of field operations is described in the section on monitoring and control technologies.

Considerable research has been conducted on the use of alternate fuels for agricultural applications. Much of this research was motivated by the oil embargo crisis and rapidly rising liquid fuel prices of the 1970s. None of the alternate fuels hold much promise to increase the fuel efficiency of conventional engines. This research has revolved around the use of onfarm production of ethanol for use primarily in gasoline-powered equipment and the onfarm pressing and refining of sunflower oil for use in diesel-powered equipment. Neither fuel is economically competitive with purchased liquid fuels in the absence of substantial subsidies. Moreover, both fuels are more difficult to use than fossil fuels. Ethanol-based fuels tend to absorb moisture and to separate in storage. Vegetable oil-based diesel fuels require special processing, which changes their chemical and physical characteristics, before they can be used reliably in unmodified diesel engines.

Crop Separation, Cleaning, and Processing Technology

New technologies being developed to separate, clean, and process crops offer many benefits in increased yield, quality, and value of crops. There are two major lines of research in this area: 1) improvements in separating and cleaning grain, and 2) in-field or onfarm processing of forages and oilseeds (Battelle, 1985).

The mechanization of grain harvesting and separation has been one of the most important factors in reducing the labor cost of grain production. Even small improvements in labor or capital efficiency have significant impacts on grain production because of the large total cost of producing the U.S. cash grain crop.

The basic methods of grain separation used in all combines are mechanical beating, aeration, and screening. While these methods have been continually refined, the same basic techniques have remained unchanged since antiquity.

Grain harvesting productivity has been improved over the past three decades by increasing the size and power of combines. Combines separate grain by beating and rubbing the grain stalk between a stationary surface and a cylinder rotating at high speed. The chaff and other debris are cleaned from the grain by blowing a large amount of air through the grain/chaff mix. The difference in the ability of the two materials to float on the airstream effects their separation.

Constraints on the total size and weight of combine equipment that can be transported over public roads limit the increases in general harvest productivity. Within this constraint, however, continued improvements in microprocessor-based monitoring and control technologies incorporated into grain combines will permit significant increases in capital and labor productivity. New electronic sensors will detect grain loss more accurately, allowing the operator to make adjustments quickly. Moreover, if enough of the internal monitoring and control of the grain separation process can be automated, and if grain losses are minimized, combine operators will be able to devote all of their attention to guiding the combine and can proceed at higher speeds. At present, the rate of travel must be held to 5 to 7 acres per hour so that the combine operator can monitor several functions of the combine.

Improvements in cleaning grain will result in a higher quality of grain and a reduction of dockage at the point of distribution or sale. New technologies will detect contaminants and remove them on the combine or as the grain is transferred into farm storage. Further improvements in grain cleaning will necessitate the use of automated aeration and screening processes.

Another way to increase the value of a crop is to do some of the processing in the field or on the farm. A good example of in-field processing is the extraction of leaf protein juice from alfalfa for use as high-value feed for pigs and poultry and as a food additive for humans. The residue of the process can be used as roughage for livestock.

Onfarm extraction of oil from oilseed crops such as soybeans and sunflowers has technical merit as a way for farms to produce a diesel fuel substitute or extender for tractors, combines, and other equipment. Onfarm production of vegetable oil fuel is more efficient than the conversion of grain to ethanol fuels. Moreover, the oilseed meal and glycerol byproducts from oilseed processing have substantial value as animal feed and chemical feedstocks. However, the principal technology employed uses highly volatile solvents and has a large requirement for capital, prohibiting its practical use on the farm. Moreover, present vegetable oil prices are approximately double the price of diesel fuel.

The adoption of onfarm processing of forages and oilseed is contingent on many domestic and international economic, political, and institutional factors that currently override technical considerations. On the other hand, most technologies to improve combine performance should be achieved by the end of the decade, at costs that will not significantly add to the total costs of today's combines.

CHAPTER 2 REFERENCES

Alexander, Martin, "Biological Nitrogen Fixation," paper prepared for the Office of Technology Assessment, Washington, DC, 1985.

Allen, Eugene C., "Regulation of Growth and Development," paper prepared for the Office of Technology Assessment, Washington, DC, 1985.

Bachrach, Howard L., "Genetic Engineering in Animals," paper prepared for the Office of Technology Assessment, Washington, DC, 1985.

Battelle Columbus Laboratories, "Communication and Information Management," paper prepared for the Office of Technology Assessment, Washington, DC, 1985.

Battelle Columbus Laboratories, "Crop Separation, Cleaning, and Processing," paper prepared for the Office of Technology Assessment, Washington, DC, 1985.

Battelle Columbus Laboratories, "Engine and Fuels," paper prepared for the Office of Technology Assessment, Washington, DC, 1985.

Battelle Columbus Laboratories, "Labor Saving Technology," paper prepared for the Office of Technology Assessment, Washington, DC, 1985.

Battelle Columbus Laboratories, "Land Management," paper prepared for the Office of Technology Assessment, Washington, DC, 1985.

Battelle Columbus Laboratories, "Monitoring and Control—Animal Agriculture," paper prepared for the Office of Technology Assessment, Washington, DC, 1985.

Battelle Columbus Laboratories, "Monitoring and Control—Plant Agriculture," paper prepared for the Office of Technology Assessment, Washington, DC, 1985.

Battelle Columbus Laboratories, "Telecommunication," paper prepared for the Office of Technology Assessment, Washington, DC, 1985.

Berry, Joseph A., "Enhancement of Photosynthetic Efficiency," paper prepared for the Office of Technology Assessment, Washington, DC, 1985.

Boersma, Larry, "Water and Soil-Water-Plant Relations," paper prepared for the Office of Technology Assessment, Washington, DC, 1985.

Broglie, R., et al., "Light-Regulated Expression of a Pea Ribulose-1,5-Bisphosphate Carboxylase Small Subunit Gene in Transformed Plant Cells," *Science* 224:838-843, 1984.

Brotman, H., "Engineering the Birth of Cattle," *The New York Times Magazine,* May 15, 1983, p. 106.

Browning, J.A., "Plant Disease and Nematode Control," paper prepared for the Office of Technology Assesment, Washington, DC, 1985.

Campbell, J.B., "Livestock Pest Control," paper prepared for the Office of Technology Assessment, Washington, DC, 1985.

Chemical and Engineering News, "News Focus," Dec. 24, 1984.

Croft, B.A., and Knight, A.L., "Evaluation of the PETE Phenology Monitoring System for Integrated Pest Management of Deciduous Tree Fruit Species," *Bull. Entomol. Soc. Amer.* 29:37-42, 1983.

Curtis, Stanley E., "Enviromment and Animal Behavior," paper prepared for the Office of Technology Assessment, Washington, DC, 1985.

Davis, Charles J., "Chemical Fertilizers," paper prepared for the Office of Technology Assessment, Washington, DC, 1985.

Feldberg, Ross, "Framing the Issue: The Social Implications of Biotechnology," *Science for the People* 17:3, May/June 1985, pp. 4-9 and 53.

Fischer, James R., "Crop Residue and Animal Waste Use," paper prepared for the Office of Technology Assessment, Washington, DC, 1985.

Foster, George R., "Soil Erosion, Productivity, and Tillage," paper prepared for the Office of Technology Assessment, Washington, DC, 1985.

Fraley, R., et al., "Expression of Bacterial Genes in Plant Cells," *Proc. Natl. Acad. Sci. USA* 80:4803-4807, 1983.

Fraley, Robert T., "Genetic Engineering in Plants," paper prepared for the Office of Technology Assessment, Washington, DC, 1985.

Francis, Charles A., "Multiple Cropping," paper prepared for the Office of Technology Assessment, Washington, DC, 1985.

Gage, S. H., et al., "Pest Event Scheduling System for Biological Monitoring and Pest Management," *Environ. Entomol.* 11:1127-1133, 1982.

Genetic Engineering News 3(6), 1983.

Hammer, Robert E., "Physiology of Transgenic Animals," paper presented at the Conference on Genetic Engineering of Animals: An Agricultural Perspective, University of California, Davis, Sept. 9-12, 1985.

Hansel, William, "Animal Reproduction," paper prepared for the Office of Technology Assessment, Washington, DC, 1985.

Herrera-Estrella, L., et al., "Chimeric Genes as Dominant Selectable Markers in Plant Cells," *The EMBO Journal* 2:987-995, 1983.

House Committee on Science and Technology, *Biotechnology and Agriculture*: Hearings before the Subcommittee on Investigations and Oversight, 99th Cong., 1st sess., Apr. 16-17, 1985.

Kalter, Robert J., Department of Agricultural Economics, Cornell University, personal communication, 1985.

Kalter, Robert J., Milligan, Robert, Lesser, William, Bauman, Dale, and Magrath, William, "Biotechnology and the Dairy Industry: Production Costs and Commercial Potential of the Bovine Growth Hormone," Cornell University, Department of Agricultural Economics, A.E. Research 84-22, 1984.

Kennedy, George G., "Management of Insects and Mites," paper prepared for the Office of Technology Assessment, Washington, DC, 1985.

Liebhardt, William, and Harwood, Richard, "Organic Farming," paper prepared for the Office of Technology Assessment, Washington, DC, 1985.

Lu, Yao-chi (ed.), "Forecasting Emerging Technologies in Agricultural Production," *Emerging Technologies in Agricultural Production,* U.S. Department of Agriculture, Cooperative State Research Service, 1983.

McWhorter, C.G., and Shaw, W.C., "Present Status and Future Needs in Weed Control," paper prepared for the Office of Technology Assessment, Washington, DC, 1985.

Murai, N., et al., "Phaseolin Gene From Bean Is Expressed After Transfer to Sunflower via Tumor-Inducing Plasmid Vectors," *Science* 222:475-482, 1983.

Murashige, T., "Plant Propagation Through Tissue Cultures," *Ann. Rev. Plant Physiol.* 25:135-166, 1974.

Nickell, Louis G., "Plant Growth Regulators," paper prepared for the Office of Technology Assessment, Washington, DC, 1985.

Osburn, Bennie I., "Animal Disease Control," paper prepared for the Office of Technology Assessment, Washington, DC, 1985.

Palmiter, R.D., et al., "Metallathionein-Human GH Fusion Genes Stimulate Growth of Mice, *Science* 222:809-814, 1983.

Pond, Wilson G., "Animal Nutrition," paper prepared for the Office of Technology Assessment, Washington, DC, 1985.

Suttan, A.L., et al., "Utilization of Animal Waste as Fertilizer," Purdue University, Cooperative Extension Service, Miner, ID, 101, 1975.

Vasil, I., et al., "Plant Tissue Cultures in Genetics and Plant Breeding," *Advance in Genetics,* vol. 20 (New York: Academic Press, 1979), pp. 127-215.

Wagner, Thomas E., "Gene Introduction and Regulation in Transgenic Animals," paper presented at the Conference on Genetic Engineering of Animals: An Agricultural Perspective, University of California, Davis, Sept. 9-12, 1985.

Welch, S.M., et al., "PETE: An Extension Phenology Modeling System for Management of Multi-Species Pest Complex," *Environ. Entomol.* 7:487-494, 1978.

Chapter 3

Impacts of Emerging Technologies on Agricultural Production

Contents

Tables

Figure

Chapter 3

Impacts of Emerging Technologies on Agricultural Production

Introducing to the marketplace the 150 emerging technologies forecasted in this study raises questions about the effects these technologies will have on crop yield, livestock feed efficiency, and future food production. Many people are concerned that the trends of major crop yields are leveling off and that the world may not be able to continue to produce enough food to meet the demands of its growing population. However, OTA's analysis indicates that the United States can continue to meet foreign and domestic demand for agricultural products *if* agricultural research is adequately supported and *if* economic and political environments are favorable. What this conclusion means in practice is the subject of this chapter.

OTA study participants arrived at this conclusion by first projecting where and under what economic and political conditions the various emerging technologies would be adopted and what the primary impacts of those technologies would be on net increases in production. Based on this information OTA projected the impacts of technology adoption on agricultural production on a per-unit basis (e.g., bushels of corn per acre) and then on an aggregate basis (e.g., million bushels of corn produced in the entire country).

TECHNOLOGY ADOPTION AND PRIMARY IMPACTS

OTA commissioned leading scientists, specialists in the 28 technological areas, to prepare state-of-the-art papers. Each paper: 1) defined and delineated the scope of a technology area, 2) identified four or five major lines of research where significant technologies were likely to emerge by 2000, 3) discussed the current state of technology development, 4) identified major breakthroughs in other science and technology areas that would be necessary for successful development of the technology in question, 5) discussed the institutional arrangements necessary for the research of the technology to be conducted or supported, 6) estimated the time in which a particular line of research would likely be completed and the resulting technology introduced commercially, and 7) estimated the potential primary impacts of each technology on crop and livestock production. These papers provided the basis for discussion in two technology workshops conducted by OTA.

The workshops—one for animal technology and the other for plant, soil, and water technology—were conducted to assess the impacts of emerging technologies on agricultural production. Workshop participants, carefully selected to include those with expertise in different stages of technological innovation, comprised physical and biological scientists, engineers, economists, extension specialists, commodity specialists, agribusiness representatives, and experienced farmers.

The participants provided data on: 1) the timing of commercial introduction of each technology area; 2) primary impacts, or net yield increases (by commodity), expected from each package of technologies; and 3) the number of years needed to reach various adoption percentages (by commodity).

The Delphi technique was used to obtain collective judgments from the workshop participants on the development and adoption of the emerging technologies.[1] To facilitate the proc-

[1]The Delphi technique is a systematic procedure for eliciting and collating informed judgments from a panel of experts. It has

ess of obtaining consensus, an electronic Consensor was used to help tabulate the ratings assigned by each expert. A detailed discussion of the methodology and workshop procedures is presented in appendix A.

The Timing of Commercial Introduction

Since the impact of a new technology on agriculture at a given time depends in part on when the technology is available for commercial introduction, workshop participants were asked to estimate the probable year of commercial introduction of each technology under three alternative environments:

1. Most likely environment—assumes to year 2000: a) a real rate of growth in research and extension expenditures of 2 percent per year, and b) the continuation of all other forces that have shaped past development adoption of technology.

2. More-new-technology environment (relative to the most likely environment)—assumes to year 2000: a) a real rate of growth in research and extension expenditures of 4 percent, and b) all other factors more favorable than those of the most likely environment.

3. Less-new-technology environment (relative to the most likely environment)—assumes to year 2000: a) no real rate of growth in research and extension expenditures, and b) all other factors less favorable than those of the most likely environment.

4. No-new-technology environment—assumes to year 2000: a) none of the emerging technologies identified in the study will be available for commercial introduction, and b) all the other factors are the same as those under the less-new-technology environment.

two distinct characteristics: feedback and anonymity. During the Delphi process, responses are collated and then referred to the experts for review. Each expert reevaluates his or her original answers after examining the summary of the group's responses. The iterative process of evaluation, feedback, and reevaluation continues until a consensus is reached. Since this is not a random sampling, the results obtained through the Delphi process depend heavily on the experts selected.

Table A-1 in appendix A shows in more detail the sets of assumptions made under the alternative technology environments.

The year of commercial introduction ranged from now—for genetically engineered pharmaceutical products; control of infectious disease in animals; superovulation, embryo transfer, and embryo manipulation of cows; and controlling plant growth and development—to 2000 and beyond, for genetic engineering techniques for farm animals and cereal crops. Of the 57 potentially available animal technologies, it was estimated that 27 would be available for commercial introduction before 1990, and the other 30 between 1990 and 2000, under the most likely environment. In plant agriculture, 50 out of 90 technologies examined were projected to be available for commercial introduction by 1990, and the other 40 technologies between 1990 and 2000. The major categories of animal and plant technologies are listed in appendix A, tables A-2 and A-3.

Primary Impacts

When a given package of technologies is adopted by a farmer and put into agricultural production, its immediate impact on plant agriculture is increased yields and/or increased percentage of planted acreage harvested.[2] To determine immediate impacts on animal agriculture, OTA considered feed efficiency for all animals and reproductive efficiency for beef cattle and swine, milk production per cow for

[2]It is often stated that U.S. agriculture needs cost-saving technology, not yield-increasing technology. Technologies can be classified into two general types according to their impact: 1) those that reduce the cost of production directly, and 2) those that increase productivity through yield increases. The first type of technology, such as nitrogen fixation and new crop varieties resistant to pest, disease, and environmental stress, saves costs of purchasing agricultural chemicals, at little additional expense. The second type of technology, such as pesticides, herbicides, plant-growth regulators, irrigation, and fertilizer, typically increase yields, but at additional expense. Regardless of the type of technology, all technologies reduce average costs if they are worth adopting. For example, a new variety of corn increases yields from 100 to 140 bushels per acre. Assuming no additional increase in the cost of purchasing the new variety of seeds, the total cost of production using the new variety will be shared by 140 bushels rather than 100 bushels. Thus, the new variety reduces the average cost 29 percent.

dairy cows, and the number of eggs per layer (producing hen) for poultry.

To estimate the net impact of emerging technologies on agricultural production, workshop participants were first asked to project the performance measures of crop and livestock production, such as crop yields and livestock feed efficiency, to 1990 and 2000 under the no-new-technology environment. Historical trend lines of the performance measures of crop and livestock production were provided to the participants as a basis for their projections. Through the Delphi process, participants collectively projected the performance measures for each of nine commodities for 1990 and 2000 (app. A, table A-5). The nine commodities included corn, cotton, rice, soybeans, wheat, beef cattle, dairy cattle, poultry, and swine.

Based on those estimates and on the information obtained from the presentations and from discussions with the authors of the commissioned papers, participants then jointly projected the net increases in crop yields, animal feed efficiencies, and other performance measures that could be expected if specific packages of technologies were commercially available and fully adopted by farmers. Generally, the 28 areas of technologies were grouped in "packages" according to their probable impacts on a commodity. Each package was further categorized as a 1990 version of the package or a 2000 version of the package, thus delineating those technologies that are expected to be introduced by 1990 and 2000, respectively. The packages of technologies are described further in appendix A.

Through the Delphi process, OTA obtained estimates for each package of technologies on each of the nine commodities under the three alternative environments. The results are shown in tables 3-1 and 3-2. In soybean production, for example, if technology package 1990A—which includes genetic engineering, enhancement of photosynthetic efficiency, plant growth regulators, plant disease and nematode control, and multiple cropping—is adopted by soybean producers, yields are predicted to increase 2.2 percent under the most likely environment, 15.2 percent under the more-new-technology environment, and only 1.2 percent under the less-new-technology environment. If package 2000A is adopted, soybean yields are predicted to increase 22.1 percent under the most likely environment, 23.9 percent under the more-new-technology environment, and 14.9 percent under the less-new-technology environment. Package 2000A increases soybean yields substantially more than package 1990A because it includes such major technologies as genetically engineered soybean plants, photosynthetic molecular biology and genetics, and genetically engineered pest-resistant plants, all of which would

Table 3-1.—Estimated Percentage Change in Crop Yield

Crop	Technology package	Technology environments: Less-new-technology 2000	Most likely 2000	More-new-technology 2000
Corn	Package A	15.6%	21.5%	28.5%
	B	8.8	14.4	20.8
	C	−31.2	−28.8	−28.0
Cotton	Package A	5.4	9.0	12.0
	B	2.3	2.8	3.1
	C	0	0	0
Rice	Package A	8.4	12.4	15.6
	B	8.8	14.4	18.6
Soybean	Package A	14.9	22.1	23.9
	B	4.9	7.2	7.5
	C	3.7	4.6	5.5
Wheat	Package A	24.0	24.0	24.0
	B	1.5	1.5	1.5
	C	5.0	5.0	5.0

SOURCE: Office of Technology Assessment.

Table 3-2.—Estimated Percentage Change in Animal Feed and Reproductive Efficiency

Animal	Technology package	Efficiency measure	Technology environments: Less-new-technology 2000	Most likely 2000	More-new-technology 2000
Beef	Package A	Pounds meat per lb feed	0	22.4%	30.4%
		Calves per cow	0	0	28.4
	B	Pounds meat per lb feed	5.8%	10.4	12.4
		Calves per cow	1.2	5.2	6.4
	C	Pounds meat per lb feed	1.8	4.5	5.8
		Calves per cow	1.2	2.0	3.2
	D	Pounds meat per lb feed	0.1	1.2	1.7
		Calves per cow	0	0.3	0.9
	E	Pounds meat per lb feed	1.4	2.8	3.3
		Calves per cow	2.3	5.3	6.6
	F	Pounds meat per lb feed	0	1.1	1.5
		Calves per cow	0	0	0
Dairy...............	Package A	Pounds milk per lb feed	5.8	13.2	15.2
		Pounds milk per cow	6.8	12.2	15.2
	B	Pounds milk per lb feed	7.6	11.0	13.0
		Pounds milk per cow	9.4	12.2	14.6
	C	Pounds milk per lb feed	7.8	12.4	15.2
		Pounds milk per cow	15.0	21.3	24.3
	D	Pounds milk per lb feed		25.6	25.6
		Pounds milk per cow		25.6	25.6
Poultry.............	Package A	Pounds meat per lb feed	7.3	9.2	11.3
		Eggs per layer per year	4.6	5.8	7.1
	B	Pounds meat per lb feed	2.5	3.1	3.9
		Eggs per layer per year	4.0	5.0	6.2
	C	Pounds meat per lb feed	1.3	1.6	2.0
		Eggs per layer per year	1.6	2.0	2.5
Swine..............	Package A	Pounds meat per lb feed	4.8	12.6	15.0
		Pigs per sow per year	14.4	27.6	50.0
	B	Pounds meat per lb feed	2.8	4.0	
		Pigs per sow per year	14.4	20.8	
	C	Pounds meat per lb feed		2.1	2.1
		Pigs per sow per year		0.8	2.4

SOURCE: Office of Technology Assessment.

not be ready for commercial adoption until after 1990.

Note that technology package C for corn production, which consists of only organic farming, received very low marks from the Delphi panel. If fully adopted, this technology will result in yield reductions ranging from 23 to 28 percent. Some organic farming specialists feel that the panel overestimated the negative impact. Harwood (1985) indicates that the best estimate from the published reports is about a 10-percent reduction. Since the cost of organic farming is lower, the economic efficiency for organic farming may be higher than that for conventional farming.

Adoption Profiles

The primary impacts estimated above assume that the technologies will be fully adopted by farmers and put into agricultural production. But when a new technology is introduced for commercial adoption, only a small number of farms, mostly the large and innovative ones, will adopt the technology initially because the possible payoff of the new technology is uncertain and because the potential adopters need time to learn how to use the new technology and to evaluate its worth. As early adopters benefit from using a new technology, more and more farmers will be attracted to it, increasing the speed of adoption exponentially. Eventually, as most potential adopters adopt a new technology, the percentage of adoption will level off and approach a maximum; thus, the adoption profile follows an S-shaped curve (Lu, 1983).

To derive an adoption profile of each package of technologies for each commodity under different economic environments, participants were divided into commodity groups according to their expertise in a particular commodity. There were four groups in the animal technology workshop (beef, swine, dairy, and poultry) and five in the plant, soil, and water technology workshop (wheat, corn, cotton, soybean, and rice). The participants were then asked the question, "If a specific package of technologies is introduced in the market today, how long will it take for farmers to have it adopted?" Based on their collective experience, the participants estimated the following for each package of technologies:

1. The maximum percentage of adoption.
2. The number of years it would take to reach 20-percent adoption.
3. The number of years it would take to reach 50-percent adoption.

Based on information from the commodity groups, a logistic curve was fitted for each package of technologies applied to each of the nine commodities under different scenarios. Figure 3-1 shows the estimated adoption curves for package A corn technologies, which consist of plant genetic engineering, plant disease and nematode control, management of insects and mites, water and soil-water-plant relations, communication and information management, monitoring and control, and telecommunications. The participants estimated that it would take 8 years to reach 20-percent adoption under the most likely environment, while it would take only 6 years to reach it under the more-new-technology environment, where the economic environment is more favorable for technology adoption. To reach 50-percent adoption, it would take 11 years under the most likely environment and 10 years under the more-new-technology environment. The maximum adoption rate projected is 80 percent under both environments.

Figure 3-1.—Logistic Adoption Curves for Corn, Package A

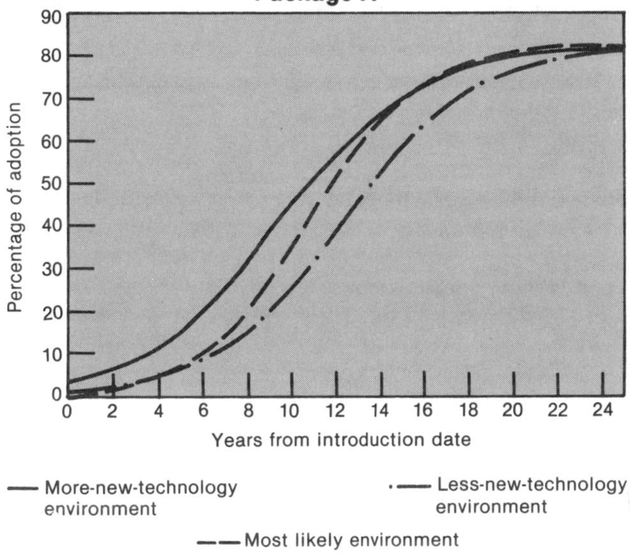

Source: Office of Technology Assessment

PROJECTION OF PER-UNIT CROP YIELDS AND LIVESTOCK FEED EFFICIENCIES

Based on the information obtained from the workshops on: 1) the years of commercial introduction, 2) the primary impacts, and 3) the adoption profiles, OTA computed the efficiency measurements for all animals and the average yield and percentage of planted acreage for all crops in 1990 and 2000 under alternative environments. The results are presented in tables 3-3 and 3-4.[3]

Under the most likely environment, feed efficiency in animal agriculture will increase at a

[3]For ease of presentation, the less-new-technology environment is not presented. Its estimates fall between the no-new-technology and most likely environments.

Table 3-3.—Estimates of Crop Yield and Animal Production Efficiency

	Actual 1982	No-new-technology environment 2000	Most likely environment 2000	More-new-technology environment 2000
Corn—bu per acre	113	124	139	150
Cotton—lb per acre	481[a]	511	554	571
Rice—bu per acre	105	109	124	134
Soybeans—bu per acre	30[a]	35	37	37
Wheat—bu per acre	36	41	45	46
Beef:				
Pounds meat per lb feed	0.070	0.066	0.072	0.073
Calves per cow	0.88	0.96	1.0	1.04
Dairy:				
Pounds milk per lb feed	0.99	0.95	1.03	1.11
Milk per cow per year[b] (1,000 lb)	12.3	15.7	24.7	26.1
Poultry:				
Pounds meat per lb feed	0.40	0.53	0.57	0.58
Eggs per layer per year	243	260	275	281
Swine:				
Pounds meat per lb feed	0.157	0.17	0.176	0.18
Pigs per sow per year	14.4	15.7	17.4	17.8

[a]Not actual—based on estimate from trend line.
[b]These estimates differ from those in table 2-2 of the first report from this study because of changes made at a later date by workshop participants in the adoption rate of some of the dairy technology packages.

SOURCE: Office of Technology Assessment.

Table 3-4.—Historical and Projected Rates of Annual Growth in Crop Yield

		1982-2000		
	1960-82	No-new-technology environment	Most likely environment	More-new-technology environment
Corn	2.6%	0.5%	1.2%	1.6%
Cotton	0.1	0.3	0.7	1.0
Rice	1.2	0.2	0.9	1.4
Soybean	1.2	0.8	1.2	1.2
Wheat	1.6	0.7	1.2	1.4

SOURCE: Office of Technology Assessment.

rate of from 0.2 percent per year for beef to 1.4 percent for poultry. In addition, reproduction efficiency will also increase, at an annual rate ranging from 0.6 percent, for beef cattle, to 1.1 percent, for swine. Milk production per cow per year will increase at 3.9 percent per year, from 12,300 pounds to 24,730 pounds per cow, in the period 1982-2000.

Major crop yields are estimated to increase from 1982 until 2000 at a rate ranging from 0.7 percent per year, for cotton, to 1.2 percent per year, for wheat and soybeans. Wheat yield, for example, is projected to increase at the rate of 0.7 percent per year, from 36 bushels per acre in 1982 to 41 bushels per acre in 2000, assuming no new technologies will become available before 2000. Under the most likely environment, wheat yields will increase at the rate of 1.2 percent per year to 45 bushels per acre. The difference in wheat yield between the two environments, 4 bushels per acre, represents the impact of new technologies under the most likely environment.

How do these rates of increase compare with historical trends? Will emerging technologies significantly change the trends? By far the most drastic increases in productivity will be in milk production, primarily because the products of genetic engineering will soon be available for commercial adoption by the dairy industry. One of the proteinaceous pharmaceuticals, bovine growth hormone, is alone expected to increase milk yields between 20 to 40 percent almost overnight via daily injections of the hormone into cattle.

From 1960 to 1982 milk production increased 2.6 percent per year, from 7,029 pounds per cow per year to 12,316 pounds. If no new technology is available from now until 2000, this rate of increase would not be maintained. Under such an environment milk production per cow per year is expected to increase at only 1.4 percent per year, from 12,316 pounds in 1982 to 15,700 pounds in 2000. However, if new technologies are adopted, the rate of increase in milk production would far surpass the historical rate, under the remaining technology environments. Under the more-new-technology environment, milk production is expected to reach 26,080 pounds in 2000, at an annual rate of 4.2 percent.

Application and adoption of new technologies will also increase the feed efficiency of other animals. Poultry feed efficiency has been increasing at 1.2 percent per year for the last 15 years. Under the most likely environment, feed efficiency will increase at 1.4 percent per year through 2000.

The feed efficiencies for beef and swine have not increased for the last 15 years. Beef feed efficiency declined from 0.093 pounds of beef per pound of feed in 1965 to 0.065 pounds in 1973 and then maintained at about 0.070 pounds in recent years. The introduction of new technologies will increase feed efficiencies. Under the most likely environment, the feed efficiency is projected to increase at an annual rate of 0.2 percent, reaching 0.072 pounds of beef per pound of feed in 2000. Swine feed efficiency has declined steadily from 0.19 pounds of pork per pound of feed in 1974 to 0.15 pounds in 1980. Under the most likely environment, feed efficiency will increase to 0.18 pounds of pork per pound of feed in 2000, at the rate of 0.4 percent per year.

Efficiencies in crop production will be less dramatic than those in animal production, primarily because development of biotechnology for plants is far behind that for animals. Most of the major plant biotechnologies will not be commercially available before 2000. Therefore, it will be difficult to maintain historical trends without infusion of new technologies. As shown in table 3-4, all major crops included in this study, except for cotton, have experienced phenomenal growth during the past 20 years. The average annual rates of growth range from 1.2 percent, for rice and soybeans (and 1.6 percent for wheat), to 2.6 percent for corn. Without new technologies, these trends cannot continue. Under the no-new-technology environment, the yields of major crops are expected to grow only at 0.2 percent per year for rice, to 0.8 percent, for soybeans. Even under the most likely environment, corn and wheat yields still could not keep up with past growth. Under the more-new-technology environment, the annual rates of growth of all major crops, except for corn and wheat, are expected to equal or exceed historical rates of growth. The growth rate of corn yields under the most favorable environment is expected to be 1.6 percent, which is far short of the historical rate of 2.6 percent per year.

New technologies could have a significant impact on cotton and rice yields. Cotton yields have not increased much during the last two decades. Instead, they have been fluctuating around the trend line, which has increased at the rate of only 0.1 percent per year from 1960 to 1982. Adoption of new technologies could shift the trend upward. Under the most likely environment, cotton yields are projected to increase at 0.7 percent per year, and under the more-new-technology environment, 1.0 percent per year.

Although rice yields have increased at an average of 1.2 percent per year since 1960, the yield curve has been flattened since 1967. During the 1960-67 period, rice yields increased at 4.1 percent per year, but the rate of growth has declined to only 0.2 percent per year since 1967. Introduction of new technologies into rice production could turn the yield curve upward. Under the most likely environment, rice yields are expected to increase 0.9 percent per year, and under the more-new-technology environment, 1.4 percent. This is the highest rate of growth estimated among all major crops.

PROJECTIONS OF AGGREGATE CROP AND LIVESTOCK PRODUCTION

OTA used the projected crop yields and percent of planted acres harvested for major crops, and the projected feed and reproductive efficiencies of livestock, to assess the collected impacts of the 28 areas of emerging technologies on the total production of various crop and livestock products. The primary tool used in the analysis was an econometric model which is an annual, partial equilibrium model consisting of a crop sector, a livestock sector, and a financial sector.[4] The model is a partial equilibrium model in that a general equilibrium solution is solved within the agricultural sector while a specified set of conditions are assumed to exist within the rest of the economy, such as population growth, income growth, export demand, and interest rates. The model was used in a 20-year simulation projecting the effects of technological change on the various crop and livestock commodities previously discussed. The results appear below.

[4]The model used was the Iowa State University econometric model developed by Earl Heady.

Crop Production

Applications of new technologies will increase aggregate crop production throughout the projection period—from 1981 to 2000. Table 3-5 shows projections to year 2000 of increased production for five major crops. Total U.S. crop production was determined by average crop yields and acres of crops harvested. Crop yields were projected to 2000 under the three technology environments from the results of the technology workshop. The projections took into account the timing, adoption profiles, and primary impacts of emerging technologies. Acres of crops harvested were determined by the model, based on expected returns from crop production, diversion payments, and other crop-specific considerations.

Although there will be a drop in the number of acres of corn planted, projected yield increases and increases in the proportion of planted acres actually harvested will cause corn production to increase over time under each environment. The increase will be greatest under the more-new-technology environment, a

Table 3-5.—Projections of Crop Production

Crop	Unit	1984	2000: No-new-technology environment	2000: Most likely environment	2000: More-new-technology environment
Corn:[a]					
Production	Billion bu	7.7	8.6	9.3	9.7
Growth rate	Percent		0.7	1.2	1.5
Cotton:					
Production	Billion lb	6.2	6.4	6.9	7.2
Growth rate	Percent		0.1	0.7	0.9
Rice:					
Production	Million cwt	137.0	153.6	163.4	169.2
Growth rate	Percent		0.7	1.1	1.3
Soybean:[a]					
Production	Billion bu	1.9	3.0	3.2	3.3
Growth rate	Percent		3.1	3.4	3.6
Wheat:[a]					
Production	Billion bu	2.6	3.3	3.5	3.5
Growth rate	Percent		1.5	1.9	2.0

[a]Projections shown for this commodity differ from those in table 2-3 of the first report from this study because the previous figures were preliminary.

SOURCE: Office of Technology Assessment.

situation that is also true for the other crops analyzed.

Unlike planted acres of corn, planted acres of soybeans will increase during the projection period. Increases in yields and increases in harvested acres will cause total U.S. soybean production to increase significantly over the 1982 through 2000 projection period. Because yields, planted acres, and proportion of planted acres harvested vary little across different environments, production increases do not vary much across environments. The rate of increase ranges from 3.1 to 3.6 percent per year for the no-new-technology and more-new-technology environments, respectively.

Planted acres of wheat are projected to increase under the no-new-technology environment but to decrease under the most likely and more-new-technology environments. Increases in average wheat yields will cause wheat production to increase over the projection period.

As shown in table 3-4, cotton yields are projected to increase relatively less than corn, soybean, and wheat yields. Planted acres of cotton are projected to increase under each of the technology environments, with only slight differences across environments. Increases in both yields and harvested acres will cause total U.S. cotton production to increase.

Planted acres of rice are also projected to increase under each technology environment. As shown in table 3-4, rice yields are projected to increase over time for each environment. Increasing yields and increasing harvested acres will cause total rice production to increase over time.

Livestock and Milk Production

Technology impacts are felt in the livestock sector through calving rate changes for beef and through feed input price differentials for beef and other livestock. Higher feed efficiencies and crop production levels under the more-new-technology compared with the no-new-technology environments result in lower corn, soybean meal, and wheat prices. The lower prices of these feed inputs cause livestock production to increase generally. The higher calving rates under the more-new-technology environment also tend to increase beef production. Increased production tends to depress livestock and meat prices if demand for livestock and meat does not increase proportionately.

The production of prime beef is determined by the number of feeder cattle slaughtered, the average fed cattle weight at slaughter, and the conversion ratio of live weight to carcass weight (dressing percentages).

As shown in table 3-6, prime beef production decreases over time for all technology environments. Due to higher calving rates and lower feed costs, beef production is highest under the more-new-technology environment. Under the most likely environment, beef production is projected to decline from 1984 to 2000 based on a weakness in consumer demand caused by changes in income levels, shifts in taste, and concern over potential health problems associated with the consumption of red meat, among other factors.

The impacts of technology on pork production are reflected only through differences in feed input prices. Differences in farrowing rates are not accounted for across environments. As shown in table 3-6, pork production is projected downward for all technology environments. The downward trend is attributed to higher feed input prices and higher retail pork prices resulting from lower production. Pork production under the most likely environment is projected to drop 15 percent from 1984 to 2000.

Chicken production is projected to increase over time for all technology environments, and the differences across the various environments are minimal.

Total milk production is determined by multiplying milk yield times milk cow numbers. Milk yield, as indicated earlier, is projected to increase through 2000, owing in large part to the anticipated emergence and adoption of biotechnologies in the dairy industry. Cow numbers are determined in the model as a positive function of the ratio of the blend price of Grade A and Grade B milk over the average ration cost

Table 3-6.—Projections of Animal Production

Livestock	Unit	1984	2000 No-new-technology environment	Most likely environment	More-new-technology environment
Prime beef:					
Production	Billion lb	16.0	12.5	14.1	15.7
Growth rate	Percent		−1.5	−0.8	−0.2
Poultry:					
Production	Billion lb	13.5	16.8	16.7	16.7
Growth rate	Percent		1.4	1.3	1.3
Pork:					
Production	Billion lb	13.8	10.7	11.7	13.0
Growth rate	Percent		−1.6	−1.0	−0.4
Milk:					
Production	Billion lb	135.4	126.1	192.1	201.8
Growth rate	Percent		−0.4	2.2	2.5

SOURCE: Office of Technology Assessment.

and a negative function of the cull price of dairy cows. The blend price falls slightly for each environment over the projection period. The average ration cost and cull cow price are exogenously projected to increase over the 1983-2000 period. As a result, cow numbers are projected to decline by at least 30 percent over the period, with only small differences across the environments.

Given the increases in milk productivity and the decreases in cow numbers, what will happen to total milk production over time? As shown in table 3-6, under the no-new-technology environment, milk production will fall at 0.4 percent per year from 1982 through 2000 because reductions in cow numbers more than offset increases in milk yield. Under the other two environments, milk production will increase despite the reductions in numbers of cows. The largest increases are projected to occur before 1990.

In the world agricultural marketplace, available information points to a periodic series of surpluses and deficits over the next two decades (Mellor, 1983; Resources for the Future, 1983). A Resources for the Future (RFF) study indicates that the global balance between cereal production and population will remain quite close until year 2000, indicating vulnerability to annual shortfalls resulting from weather, wars, or mistakes in policy. Over the next 20 years the world will become even more dependent on trade. There will be increasing competition for U.S. farmers in international markets. Much of this increased competition will come from developing countries selling farm commodities as a source of exchange to pay for imports such as oil. Despite this increased competition, exports of grain from North America are projected nearly to double by year 2000.

On the other hand, there is another school of thought that believes current studies such as that by RFF have not properly assessed the magnitude and impact of emerging technologies on farm production. Technologies such as genetic engineering and electronic information technology that are available now in various forms could mean rapid increases in yields and productivity. While such changes may improve the competitive position of American agriculture, they might create surpluses and major structural change—favoring, for example, larger, more industrialized farms.

Any conclusion regarding the balance of global supply and demand requires many assumptions about the quantity and quality of resources available to agriculture in the future. Land, water, and technology will be the limiting factors to agriculture's future productivity.

Agricultural land that does not require irrigation is becoming an increasingly limited resource. In the next 20 years, out of a predicted 1.8-percent annual increase in production to meet world demand, only 0.3 percent will come from an increase in quantity of land used in production (RFF, 1983). The other 1.5 percent will have to come from increases in yields—mainly from new technology. Thus, to a very large extent, research that produces new technologies will determine the future world supply/demand balance and the amount of pressure placed on the world's limited resources.

The OTA results indicate that with continuous inflow of new technologies into the agricultural production system, U.S. agriculture will be able not only to meet domestic demand but also to contribute significantly to meeting world demand in the next 20 years. This does not necessarily mean that the United States will be competitive or have the economic incentive to produce. It means only that the United States will have the technology and resources available to provide the production increases needed to export for the rest of this century.

SUMMARY AND CONCLUSIONS

OTA finds that emerging agricultural technologies, if fully adopted, will produce significant impacts on the performance of plant and animal agriculture. The most dramatic impacts will first be felt in the dairy industry, where new genetically engineered phamaceuticals (such as bovine growth hormones and feed additives) and information management systems will soon be introduced commercially. New technologies adopted by the dairy industry will increase milk production far beyond the 2.6-percent annual rate of growth of the past 20 years. Under the most likely environment, milk production per cow is expected to increase at an annual rate of 3.9 percent. Applications of new technologies will also increase the feed efficiency and reproductive efficiency of other agricultural animals.

Because development of biotechnology for plant agriculture is lagging behind that for animal agriculture, significant impacts from such technology will not be felt in plant agriculture before the turn of the century. The development and adoption of the new technologies under the most favorable environment will, in the short run, increase the rates of growth of major crop yields, except for corn, at about the level of the historical rates of growth. However, the impacts of these technologies will be substantially greater for plant agriculture after 2000.

The OTA study indicates that, with a continued flow of new technologies into the agricultural production system, major crop yields will continue to grow and U.S. agriculture will continue to provide enough food to meet domestic and foreign demand as long as agricultural research is adequately supported and economic and political environments are favorable.

CHAPTER 3 REFERENCES

Harwood, Richard, Program Officer, International Agricultural Development Service, personal communication, July 30, 1985.

Lu, Yao-chi, "Forecasting Emerging Technologies in Agricultural Production," in Yao-chi Lu (ed.), *Emerging Technologies in Agricultural Production,* Cooperative State Research Service, U.S. Department of Agriculture, 1983.

Mellor, John W., "Food Prospects for the Developing Countries," *American Economic Review,* May 1983, pp. 239-243.

Resources for the Future, "Meeting Future Needs for United States Food, Fiber and Forest Products," report prepared for the Joint Council on Food and Agricultural Sciences (Washington, DC: December 1983).

Part II

The Changing Structure of American Agriculture

Chapter 4

Dynamic Structure of Agriculture

Contents

Tables

Figures

Chapter 4

Dynamic Structure of Agriculture

Who will use a technology is as important a consideration as which technology will be adopted, for the distribution of technology affects both agricultural production and the socioeconomic structure of the entire agricultural sector.

The trend toward concentration of agricultural resources in fewer but larger farms will continue, although the degree of concentration will vary by region and by commodity. Indeed, in the future, 75 percent of the food and fiber in this country will probably be produced by only 50,000 of the 1 million farms in existence. Further concentration of resources will be most likely in those industries already highly concentrated, for example, the broiler, fruit and vegetable, and dairy industries.

Several factors contribute to the changing character of the agricultural sector: policies, institutions, economies of size, and new technologies themselves. This chapter provides a perspective for analyzing technology's distributional impacts on agricultural structure by surveying the characteristics of that structure and the factors that affect it.

PRESENT STRUCTURE OF AGRICULTURE

The heart of agriculture—the farm—is officially defined as a place that produces and sells, or normally would have sold, at least $1,000 worth of agricultural products per year. So defined, there were about 2.2 million farms in 1982. Farms in that year had an average net income from farming of $9,976 and an average off-farm income of $17,601, for a total of $27,577.

Perhaps the best known characteristic of U.S. agriculture is the trend toward larger but fewer farms. Currently, about 1 billion acres of land are in farms, resulting in an average farm size of about 400 acres. However, this average size has little meaning, since fewer than 25 percent of all farms fall within the range of 180 to 500 acres. Almost 30 percent of all U.S. farms have less than 50 acres, whereas 7 percent have more than 1,000 acres.

The number of farms reached a peak of about 6.8 million farms in 1935 and is now approximately 2.2 million. The rate of decline has slowed since the late 1960s, with a loss of about 100,000 farms since 1974.

Employment in farming began a pronounced decline after World War II, when a major technological revolution occurred in agriculture. The replacement of draft animals by the tractor began in the 1930s and was virtually complete by 1960, releasing about 20 percent of the cropland, which had been used to grow feed for draft animals.

The increased mechanization of farming permitted the amount of land cultivated per farm worker to increase fivefold from 1930 to 1980. The amount of capital used per worker increased more than 15 times in this period. Total productivity (production per unit of total inputs) more than doubled because of the adoption of new technologies such as hybrid seeds and improved livestock feeding and disease prevention. The use of both agricultural chemicals and fuel also grew very rapidly in the postwar period. Agricultural production began to rely heavily on the nonfarm sector for machinery, fuel, fertilizer, and other chemicals. These, not more land or labor, produced the growth in farm production. The resultant changes have greatly increased the capital investment necessary to enter farming and have generated new requirements for operating credit during the growing cycle.

One of the best ways to look at changes in the economic structure of U.S. agriculture is in terms of value of production as measured by gross sales per year. Farms can be usefully classified into the five categories of gross sales shown in table 4-1.

Small farms generally do not provide a significant source of income to their operators. This class of farms is operated by people living in poverty and by people who use the farm as a source of recreation.

Part-time farms may produce significant net income but in general are operated by people who depend on off-farm employment for their primary source of income.

Table 4-1.—Sales Classes of Farms

Class	Amount of gross sales per year
Small	< $20,000
Part-time	$20,000 to $99,999
Moderate	$100,000 to $199,000
Large	$200,000 to $499,999
Very large	≥ $500,000

SOURCE: Office of Technology Assessment.

Moderate-size commercial farms cover the lower end of the range in which the farm is large enough to be the primary source of income for an individual or family. Most families with farms in this range also rely on off-farm income. In general, farms in this range require labor and management from at least one operator on more than a part-time basis.

Large and *very large* commercial farms include a range of diverse farms. The great majority of these are family owned and operated. Most farms in these classes require one or more full-time operators, and many depend on hired labor on a full-time basis. Five percent of these farms are owned by nonfamily-owned corporations, a much higher percentage than in the other three classes. In general, the degree of contracting and vertical integration is much higher in these classes.

CHANGES IN THE STRUCTURE OF U.S. AGRICULTURE

In tables 4-2 to 4-5 changes in the structure of U.S. agriculture between 1969 and 1982 are presented in terms of four basic attributes: numbers of farms, gross income of farms, net farm income, and off-farm income. The information in each table has been adjusted to account for the impact of inflation and is presented in terms of constant 1982 dollars. Inflation in commodity prices over the 13 years between 1969 and 1982 has tended to move many farms from lower sales classes into higher sales classes. Farm numbers, sales, and income values have accordingly been redistributed to correct for this.[1]

Changes in Farm Size and Number

Major changes in the structure of U.S. agriculture can be seen in the changes in the number of farms shown in table 4-2. Even after the number of farms was redistributed toward the larger sales classes in the years prior to 1982, the real number of small farms declined by about 39 percent—a dramatic decline. Recent reports that the number of small farms has actually increased since 1978 refer primarily to farms that have less than 50 acres, not to farms with less than $20,000 per year in sales. The number of part-time farms has increased by about 57 percent. The number of moderate-size farms has increased greatly, by 111 percent. The numbers of large and very large farms have also increased very dramatically, by about 130 and 101 percent, respectively. The substantial increase in the real number of moderate-size farms appears to contradict many claims that the moderate-size farm is disappearing from the structure of American agriculture. However, as will be shown in the next two sections and in later chapters, changes in the number of farms is not, by itself, a good indicator of economic health or the ability of different classes of farms to survive financially.

[1]The redistribution to correct for inflation in terms of 1982 dollars has the effect of moving farm numbers, sales, and income from lower sales classes into higher sales classes in the years prior to 1982.

Table 4-2.—Number of Farms and Percent of Farms by Sales Class, 1969-82 (1982 dollars)

Sales class	Value of farm products sold	Number of farms 1969	1974	1978	1982	Percent of farms 1969	1974	1978	1982
Small	< $20,000	2,216,851	1,926,875	1,617,385	1,355,344	81.3%	70.9%	66.0%	60.5%
Part-time	$20,000-$99,999	371,180	559,076	573,976	581,576	13.6	20.6	23.4	26.0
Moderate	$100,000-$199,999	85,589	146,089	160,289	180,689	3.1	5.4	6.5	8.1
Large	$200,000-$499,999	40,691	67,091	75,891	93,891	1.5	2.5	3.1	4.2
Very large	≥ $500,000	13,800	19,200	21,500	27,800	0.5	0.7	0.9	1.2
All farms		2,728,111	2,718,331	2,449,041	2,239,300	100.0%	100.0%	100.0%	100.0%

SOURCE: Office of Technology Assessment. Compiled from data in *Economic Indicators of the Farm Sector: Income and Balance Sheet Statistics, 1983*. USDA, Economic Research Service, 1984. Data adjustment for inflation based on redistribution of farm numbers in the *Census of Agriculture*, 1969, 1974, 1978, 1982, Bureau of the Census, U.S. Department of Commerce. Price indices in *Agricultural Statistics*, 1983, USDA.

Changes in the Distribution of Sales and Income

Changes in the number of farms do not give the whole picture. Changes in the distribution of sales and income are more important to the economic structure of U.S. agriculture and more clearly show the direction in which U.S. agriculture is heading.

Changes in the distribution of gross farm income between 1969 and 1982 are shown in table 4-3.[2] As can be seen, the real value of total gross farm income increased significantly in the period 1969-78, then declined somewhat by 1982. The gross farm income of small farms decreased significantly between 1969 and 1978, then decreased greatly between 1978 and 1982, resulting in an overall reduction in the share of gross income, from 17 percent in 1969 to 6 percent in 1982. Gross income of part-time farms remained roughly the same over the period. Gross farm income of moderate-size farms increased from 15 to 19 percent. In the same period, the percent of sales from large and very large farms combined increased from 45 to 54 percent. Overall, the majority of market share shifted from the combined shares of the small, part-time, and moderate-size farms in 1969 to the combined shares of the large and very large farms in 1982.

The most telling changes of all have occurred in the distribution of net farm income, as shown in table 4-4. The large and very large farms not only have captured the majority of gross farm income, but also have controlled or substantially reduced their costs of production. As a result, their combined share of net income has increased from 51 percent in 1969 to 84 percent in 1982, after adjustment for inflation. Very large farms have been responsible for the majority of this growth in net income. This class of farms, which currently accounts for only 1.2 percent of U.S. farms, increased its share of net farm income from 36 percent in 1969 to 64 percent in 1982.

Examination of the amounts of net farm income in real terms shows that the total amount of net farm income for all farms increased greatly from 1969 to 1974, and then declined.

[2]Gross farm income includes cash receipts; net Commodity Credit Corporation loans; income from recreational, machine hire, and custom work; the value of home consumption of products produced onfarm, and gross rental value of farm dwellings.

Table 4-3.—Gross Farm Income and Percent of Gross Farm Income by Sales Class, 1969-82 (1982 dollars)

Sales class	Value of farm products sold	Gross farm income 1969	1974	1978	1982	Percent of gross farm income 1969	1974	1978	1982
Small	< $20,000	$ 21,791,756	$ 16,160,371	$ 17,694,223	$ 7,260,143	17.2%	12.7%	12.1%	5.5%
Part-time	$20,000-$99,999	28,012,247	30,844,011	35,623,571	28,783,908	22.1	24.3	24.3	21.9
Moderate	$100,000-$199,999	19,477,342	22,930,645	26,794,096	25,100,815	15.4	18.1	18.3	19.1
Large	$200,000-$499,999	19,566,095	22,233,997	26,180,305	27,680,560	15.4	17.5	17.9	21.0
Very large	≥ $500,000	37,835,967	34,704,598	40,311,553	42,764,189	29.9	27.4	27.5	32.5
All farms		$126,683,408	$126,873,622	$146,603,748	$131,589,615	100.0%	100.0%	100.0%	100.0%

SOURCE: Office of Technology Assessment. Compiled from data in *Economic Indicators of the Farm Sector: Income and Balance Sheet Statistics, 1983*. USDA, Economic Research Service, 1984. Data adjustment for inflation based on redistribution of farm numbers in the *Census of Agriculture*, 1969, 1974, 1978, 1982, Bureau of the Census, U.S. Department of Commerce. Price indices in *Agricultural Statistics*, 1983, USDA.

Table 4-4.—Net Farm Income and Percent of Net Farm Income by Sales Class, 1969-82 (1982 dollars)

Sales class	Value of farm products sold	Net farm income 1969	1974	1978	1982	Percent of net farm income 1969	1974	1978	1982
Small	< $20,000	$ 3,791,609	$ 1,802,327	$ (675,036)	$ (847,409)	10.3%	3.2%	−1.7%	−3.8%
Part-time	$20,000-$99,999	9,026,790	13,033,232	8,010,487	1,186,510	24.5	23.2	20.2	5.4
Moderate	$100,000-$199,999	5,400,579	11,384,523	7,720,282	3,218,012	14.6	20.3	19.4	14.6
Large	$200,000-$499,999	5,474,381	11,887,994	8,149,347	4,515,675	14.8	21.2	20.5	20.4
Very large	≥ $500,000	13,210,919	18,091,384	16,511,511	14,034,343	35.8	32.2	41.6	63.5
All farms		$36,904,279	$56,199,461	$39,716,592	$22,107,132	100.0%	100.0%	100.0%	100.0%

SOURCE: Office of Technology Assessment. Compiled from data in *Economic Indicators of the Farm Sector: Income and Balance Sheet Statistics, 1983.* USDA, Economic Research Service, 1984. Data adjustment for inflation based on redistribution of farm numbers in the *Census of Agriculture*, 1969, 1974, 1978, 1982, Bureau of the Census, U.S. Department of Commerce. Price indices in *Agricultural Statistics*, 1983, USDA.

Table 4-5.—Total Farm Income and Percent of Total Farm Income by Sales Class, 1969-82 (1982 dollars)

Sales class	Value of farm products sold	Off-farm income 1969	1974	1978	1982	Percent of off-farm income 1969	1974	1978	1982
Small	< $20,000	$37,936,097	$46,906,672	$33,712,998	$24,266,444	87.8%	85.5%	76.7%	71.8%
Part-time	$20,000-$99,999	2,898,500	4,852,067	6,697,664	5,593,893	6.7	8.8	15.2	16.5
Moderate	$100,000-$199,999	1,268,407	1,842,151	1,872,481	1,996,753	2.9	3.4	4.3	5.9
Large	$200,000-$499,999	802,790	981,677	1,103,743	1,256,672	1.9	1.8	2.5	3.7
Very large	≥ $500,000	285,377	282,039	575,800	687,778	0.7	0.5	1.3	2.0
All farms		$43,191,171	$54,864,605	$43,962,685	$33,801,541	100.0%	100.0%	100.0%	100.0%

SOURCE: Office of Technology Assessment. Compiled from data in *Economic Indicators of the Farm Sector: Income and Balance Sheet Statistics, 1983.* USDA, Economic Research Service, 1984. Data adjustment for inflation based on redistribution of farm numbers in the *Census of Agriculture*, 1969, 1974, 1978, 1982, Bureau of the Census, U.S. Department of Commerce. Price indices in *Agricultural Statistics*, 1983, USDA.

Changes in net farm income by sales class generally reflect this rise and fall in total net income. However, real net farm income has declined the least for very large farms, while all other classes of farms have had substantial declines in real net income. Moderate-size farms had an increase in percent of net farm income between 1969 and 1974. Since then their share of net income has declined. Farms in the large sales class increased their percentage of net income, from 16 to 20 percent in 1974, and basically held this share in 1978 and 1982. Moderate-size farms clearly have not been as successful as large and very large farms in controlling or reducing their costs of production.

Table 4-6 shows the average gross farm income, net farm income, off-farm income, and total income by sales class. As can be seen, the average net farm income of all classes of farms has declined substantially in real terms since the highly profitable years in the 1970s. But the comparison between 1969 and 1982 is even more telling. The average net farm income has declined. The average real net farm income of part-time and moderate-size farms has declined by a factor of 12 and 3.5, respectively. The net income of large farms has declined by a factor of 3, while the net farm income of very large farms has declined by a factor of 2.[3] In 1969 the average farm in the part-time sales class produced enough income to support a family. A farm that in 1982 is classed as moderate clearly had a substantial income in 1969. By 1982, the average part-time farm was extremely dependent on off-farm income, while even moderate-size farms required off-farm income to make ends meet.

Changes in the Sources of Income

Employment and the sources of income of U.S. farmers changed greatly in the 20th century, continuing at a rapid rate in the 1970s. The largest single source of change has been the tremendous increase in labor productivity made

[3]Table 4-6 must be interpreted in terms of 1982 dollars. Consequently, the values of earlier years are adjusted upward so that they are equivalent to the values in 1982. The sales class intervals are not adjusted. Therefore, the sales class names—small, part-time, moderate, large, and very large should be understood in terms of income-generating potential in 1982. For example, a farm in the part-time sales class in 1969 is roughly equivalent to a farm in the "moderate" sales class in 1982 in terms of income.

Table 4-6.—Average Gross Farm Income, Net Farm Income, Off-Farm Income, and Total Income of Farms, 1969-82 (1982 dollars)

Sales class	Value of farm products sold	Average gross farm income 1969	1974	1978	1982	Average off-farm income 1969	1974	1978	1982
Small	< $20,000	$ 9,830	$ 8,387	$ 10,940	$ 5,357	$17,113	$24,343	$20,844	$20,499
Part-time	$20,000-$99,999	75,468	55,170	62,065	49,493	7,809	8,679	11,669	13,216
Moderate	$100,000-$199,999	227,568	156,964	167,161	138,917	14,820	12,610	11,682	11,428
Large	$200,000-$499,999	480,846	331,401	344,972	294,816	19,729	14,632	14,544	12,834
Very large	≥ $500,000	2,741,737	1,807,531	1,874,956	1,538,280	20,679	14,690	26,781	24,317
All farms		$ 46,436	$ 46,673	$ 59,862	$ 58,764	$15,832	$20,183	$17,951	$17,601
		Average net farm income				**Average total income of farms**			
Small	< $20,000	$ 1,710	$ 935	($417)	($625)	$ 18,823	$ 25,279	$ 20,427	$ 19,874
Part-time	$20,000-$99,999	24,319	23,312	13,956	2,040	32,128	31,991	25,625	15,256
Moderate	$100,000-$199,999	63,099	77,929	48,165	17,810	77,919	90,538	59,847	29,238
Large	$200,000-$499,999	134,535	177,192	107,382	48,095	154,264	191,824	121,926	60,929
Very large	≥ $500,000	957,313	942,260	767,977	504,832	977,992	956,949	794,759	529,149
All farms		$ 13,527	$ 20,674	$ 16,217	$ 9,872	$ 29,359	$ 40,857	$ 34,168	$ 27,474

SOURCE: Office of Technology Assessment. Compiled from data in *Economic Indicators of the Farm Sector: Income and Balance Sheet Statistics, 1983*. USDA, Economic Research Service, 1984. Data adjustment for inflation based on redistribution of farm numbers in the *Census of Agriculture*, 1969, 1974, 1978, 1982, Bureau of the Census, U.S. Department of Commerce. Price indices in *Agricultural Statistics*, 1983, USDA.

possible by technological changes, resulting in a sharp drop in the demand for agricultural labor. During the 1930s the disposable farm income per capita was less than 40 percent of that in the rest of the economy. This income differential resulted in the large migration of the farm labor force out of agriculture and rural areas. This outmigration accelerated after the Great Depression of the 1930s as employment and per capita income opportunities increased greatly outside of agriculture. In general, the marginal productivity of labor was higher outside the agricultural sector from the 1930s to the early 1970s. Therefore, the migration of labor from farming to the nonfarm sector has contributed to national economic growth.

In the 1970s, the average income differential between farm and nonfarm households narrowed to about 88 percent, owing both to rapid increases in farm prices and a substantial increase in the number of farm jobs available from growth in rural industries. These two factors resulted in a slowing of the rate of outmigration.

Changes in off-farm income by sales class are shown in table 4-6. In 1982 the average income of farm and nonfarm households was quite close, at $27,578 and $28,638, respectively. However, two-thirds of the income of farm households comes from off-farm sources. The majority of farm operators today have some off-farm employment.

The average income statistics mask economic problems that exist for part-time and moderate-size farms. Farms in the part-time class are in serious trouble. There were about 580,000 farms in this class in 1982, at an average total income of about $15,000. The average net income from such farming is only $2,040. These farms are not large enough to generate much net farm income, and at the same time these farms have lower-than-average off-farm incomes. Moreover, the amount of off-farm income earned by part-time farmers decreased substantially between 1978 and 1982. Thus, part-time farms have a smaller share of total off-farm income now than in 1969. In contrast, households that operate farms with sales of less than $20,000 have substantial off-farm incomes and low or negative net farm income. Small farms have the largest share of off-farm income, and their share has increased the most since 1969. However, it should be noted that the socioeconomic structure of the small farm subsector is nonhomogeneous. This subsector contains a large number of subsistence-type farms whose operators live at or below the poverty level as well as a large number of affluent families to whom the farm is more a form of recreation than a source of income. So, while the average off-farm income of these households enables them to maintain this way of life, there are probably many small farms that may leave agriculture for economic reasons.

Moderate-size farms have sufficient off-farm income to maintain a household. However, this group may be under the most stress. To provide an adequate total income, moderate-size farms must earn almost as much off-farm as onfarm income. The total amount of off-farm income earned by moderate-size farms has declined in real terms since 1969. Since the number of these farms has increased in the same period, the average off-farm income of moderate-size farms has declined from $14,800 in 1969 (1982 dollars) to $11,400 in 1982. Farms with sales in excess of $200,000 have moderate off-farm incomes and moderate-to-very large net farm incomes. As a group, the households that own and operate these farms are well-off.

PROJECTIONS OF STRUCTURAL CHANGE IN U.S. AGRICULTURE TO YEAR 2000

The dramatic changes in the structure of agriculture that have occurred between 1969 and 1982 raise a new set of questions: if these trends continue what will the structure of agriculture be in 1990 and the year 2000? It is risky to extrapolate very far into the future on the basis of changes in the past, especially in a sector as dynamic as that of agriculture. However, the structural changes in agriculture are generally strong and consistent and warrant some extrapolation.

The most likely projection of farm numbers, based on a Markov chain projection using a 1969 through 1982 base, suggest that farm numbers are likely to decline from 2.2 million in 1982 to 1.8 million in 1990 and 1.2 million in 2000. The projections indicate that farm numbers will follow a bimodal or bipolar distribution—a large proportion of small and part-time farms, an increasing proportion of large farms, and a declining number of moderate farms (table 4-7). Small farms are projected to account for approximately 51 percent of all farms—down from 61 percent in 1982. In contrast, large and very large farms are projected to account for about 15 percent of all farms, three times their proportion in 1982. The number and proportion of moderate-size farms is likely to begin declining by the end of the century.

The projected decline in the number of small farms is dramatic but plausible, given the strong trend in this direction and the persistently negative farm income in this class. However, a substantial number of farms in the small size class are horse farms, small orchards, and vineyards that are primarily recreation or "hobby" type

Table 4-7.—Most Likely Projection of Total Number of U.S. Farms in 1990 and 2000, by Sales Class[a]

	Sales class	1982 (actual)	1990	2000
Small	<$20,000	1,355,344	991,609	637,597
Part-time	$20,000-$99,000	581,576	486,790	362,555
Moderate	$100,000-$199,000	180,689	126,205	75,011
Large	$200,000-$499,000	93,891	144,234	125,019
Very large	≥$500,000	27,800	54,087	50,008
Total		2,239,300	1,802,925	1,250,190
Percent of total farm numbers:				
Small	<$20,000	61	55	51
Part-time	$20,000-$99,000	26	27	29
Moderate	$100,000-$199,000	8	7	6
Large	$200,000-$499,000	4	8	10
Very large	≥$500,000	1	3	4
Total		100	100	100

[a]Based on a Markov chain projection using a 1969-82 base.

SOURCE: Office of Technology Assessment.

farms. The proportion of recreational farms is not known, but such farms may help stabilize the precipitous decline in the number of small farms.

The projections for the number and proportion of moderate-size farms show a decline, indicating that a farm of that size may not be economically viable. In the past there has been a steady increase in the number of these farms in real dollar terms, however, the outlook for financial survival of many of the moderate-size farms is not very good. In 1982, the average net farm income of $17,810 for moderate-size farms was less as compared with $63,099 in 1969 and $77,929 in 1974 (measured in 1982 dollars). During this period a large proportion of the growth of moderate-size farms was due to expansion of production by small farms and part-time farms into moderate-size farms. Survival of moderate-size farms will depend on the operator's ability to increase farm income or to provide sufficient off-farm income to compensate for low farm income.

An important implication of the projections is the further concentration of agricultural production in terms of total net farm income and total farm cash receipts. The share of total farm income by large farms has grown steadily from 51 percent in 1969 to 84 percent in 1982. If this trend continues, over 90 percent of net farm income will be earned by farms with sales over $200,000 by year 2000.

About 35 percent of total farm cash receipts were received by farms with sales over $100,000 in 1969. About 30 percent of the total farm production was produced by the largest 50,000 farms (2 percent of the total farms) and 50 percent by the largest 200,000 farms. This pattern will likely continue to the year 2000 when approximately 95 percent of total production is projected to come from farms with sales over $100,000. The 50,000 largest farms (sales over $500,000) will probably produce 75 percent of all farm products.

In general, if these trends continue, small farms are likely to disappear to the extent that the operators of these farms depend on them for income. The number of small recreational, or hobby, farms may increase or hold steady. Part-time farms could increase in number if the families that live on these farms are willing and able to earn the bulk of their income from off-farm sources. The number of moderate-size farms are likely to decrease and such farms will have a small share of total gross farm income and a declining share of net farm income. Large and very large farms will dominate agriculture.

Moderate-size farms comprise most of the farms whose owners depend on agriculture for the majority of their income. Traditionally, the moderate-size farm has been viewed as the backbone of American agriculture. As the numbers and economic importance of small and part-time farms decline, moderate, large, and very large farms all have an opportunity to increase their shares of farm income. However, large and very large farms are maintaining or increasing their shares of farm income, whereas the net income of moderate-size farms is decreasing both in absolute terms and in terms of their share of total farm income.

STRUCTURE OF U.S. AGRICULTURE BY MAJOR COMMODITY GROUPS

The preceding sections have provided a picture of the overall structure of agriculture for all commodities. This section provides a set of pictures of the structure of U.S. agriculture in terms of six major agricultural commodity groups: cash grains (primarily corn, wheat, and soybeans), cotton, dairy, poultry and eggs, cattle and calves, and pork. In particular, changes in the pattern of concentration of production, as measured by sales, will be described.

Figures 4-1 through 4-6 show the percent of commodity group sales in 1969 dollars by sales class for 1969 to 1982.[4] The general pattern is: the percent of sales by the lower sales classes declines, while the percent of sales of the upper sales classes increases.

Cash Grain Subsector

Figure 4-1 shows the percent of cash grain sales in real terms by sales class for 1969 to 1982. This figure clearly shows the dramatic decline in cash grain sales by farms with sales less than $100,000 and the great increase in sales by farms with sales over $100,000. The increase in market share from farms with sales in the $200,000 to $499,999 class is also especially striking. These farms increased their percent of total sales from 2 percent in 1969 to 34 percent in 1982. In the same period, farms with $100,000 to $199,000 in sales increased their share of gross sales from 6 to 18 percent. The combined sales of the top two sales classes of cash grain farms had increased to 50 percent of the total sales in 1982. Concentration of sales from farms with more than $500,000 in sales was lower than for most of the other commodity groups. However, the rate of growth of the market share of the top sales class was relatively high.

There is evidence that the structure of cash grain farms is bimodal in terms of sales by sales class. In both 1978 and 1982, farms in the $20,000 to $99,999 and $200,000 to $499,999 classes had more sales than farms in the middle class ($100,000 to $199,999 in sales).

Cotton Subsector

The cotton subsector includes all sales of cotton and cottonseed. Figure 4-2 shows the percent of cotton and cottonseed sales in 1969 dollars by sales class for 1969 to 1982. The growth in sales by cotton farms with more than $500,000 in sales has been very dramatic. The market share of these farms has increased from less than

[4]The discussion of national aggregate farm structure in the preceding section was presented in terms of constant 1982 dollars. In this section, the percents of sales and sales classes are presented in terms of 1969 dollars. This means that the sales class intervals used in the tables, figures, and text represent different real values. For example, farms with sales in the $20,000 to $99,000 interval in 1969 dollars as presented in this section would have sales in approximately the $45,000 to $225,000 range in 1982 dollars. Therefore, results for a given sales class in this section cannot be directly compared with results from a sales class in the previous section on national aggregate statistics. Since the sales class names used in the previous section—small, part-time, moderate, large, and very large—are defined in terms of the average income of farms in these classes in 1982 dollars, these names would be misleading if used in this section. Consequently, the sales classes in 1969 dollars are referred to in terms of the sales class interval alone.

Figure 4-1.—Cash Grain Sales by Sales Class, 1969-82 (1969 dollars)

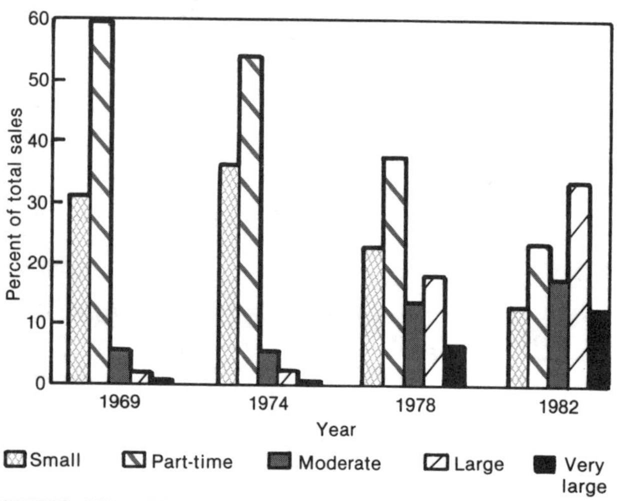

SOURCE: Office of Technology Assessment.

Figure 4-2.—Cotton Sales by Sales Class, 1969-82 (1969 dollars)

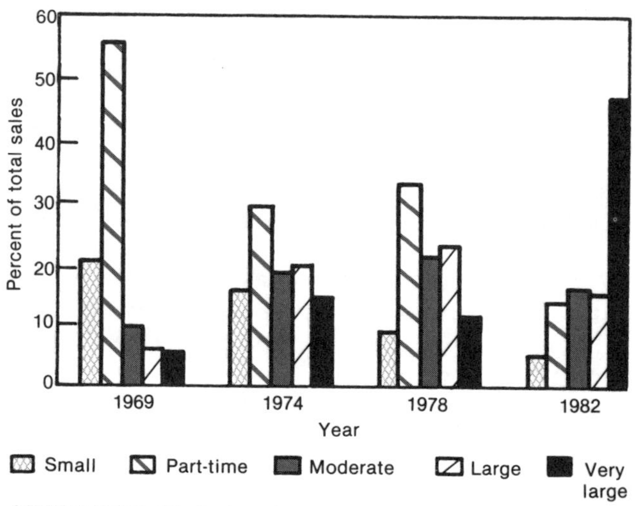

SOURCE: Office of Technology Assessment.

7 percent in 1969 to 48 percent in 1982. In the same period, sales from farms in the $20,000 to $99,999 class declined from 56 to 14 percent of the total. It is interesting to note that between 1974 and 1978 there was an upswing in the percent of sales from farms in the middle of the range ($20,000 to $499,999) and then a subsequent decline from 1978 to 1982. If the trend of the period 1978-82 continues, sales of cotton and cottonseed are likely to become even more heavily concentrated in the top sales class.

Dairy Subsector

Figure 4-3 shows the percent of dairy sales in real terms by sales class for 1969 to 1982. Farms in the $20,000 to $99,999 sales class completely dominated the production of dairy products in 1969 with about 66 percent of sales. By 1982 their share had declined to 41 percent. During the same period, dairy farms with sales in excess of $100,000 increased their share of production substantially. The most dramatic single change occurred in the period 1978-82, when dairy farms in the $200,000 to $499,999 class increased their market share threefold, from less than 5 to 14 percent. As with the other commodity groups, the trend in structural change is unambiguously in the direction of greater concentration of sales in the top sales classes. It is also clear that the subsector is likely to pass through a transition period in which there will be a bimodal distribution of sales among the five classes shown on figure 4-3. That is, dairy farms with sales less than $100,000 and more than $200,000 in 1969 dollars will both have greater shares of the market than farms in the $100,000 to $199,999 class.

Poultry Subsector

Figure 4-4 shows the percent of poultry and poultry products sales in real terms by sales class for 1969 to 1982. As with dairy farms, poultry farms in the $20,000 to $99,999 class dominated the structure of the subsector in 1969 with 61 percent of sales. Since 1969, the percent of sales from poultry farms with sales greater than $500,000 has increased at a very rapid rate, while the percent of sales from the $20,000 to $99,999 class has declined greatly. In 1982, poultry farms with sales in excess of $200,000 accounted for 77 percent of sales, compared with less than 25 percent of sales in 1969.

Cattle and Calf Subsector

Figure 4-5 shows the percent of cattle and calf sales in 1969 dollars by sales class for the years 1969, 1974, 1978, and 1982. Sales from cattle

Figure 4-3.—Dairy Sales by Sales Class, 1969-82 (1969 dollars)

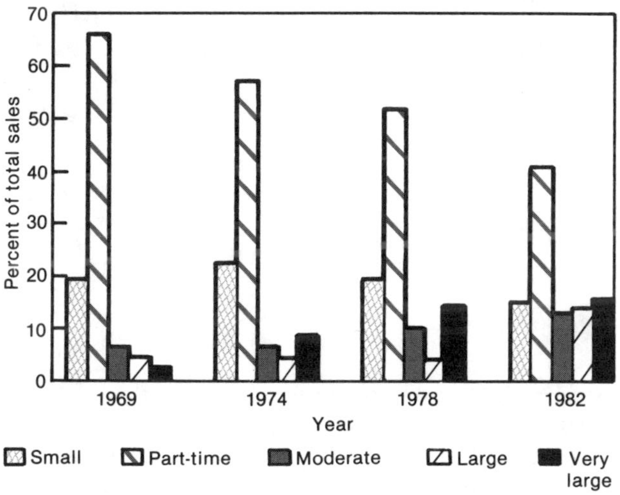

SOURCE: Office of Technology Assessment.

Figure 4-4.—Poultry Sales by Sales Class, 1969-82 (1969 dollars)

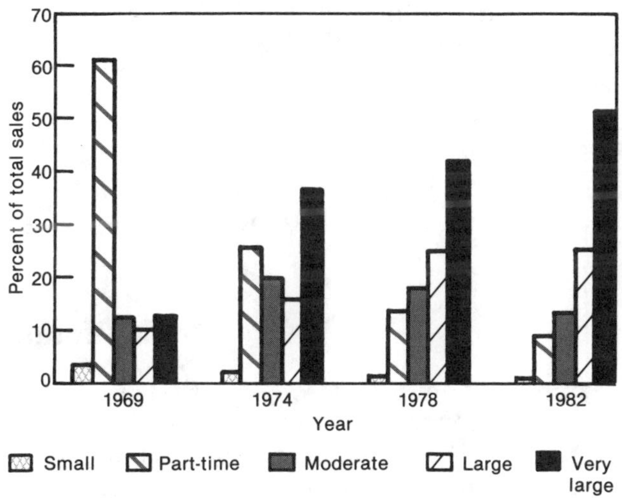

SOURCE: Office of Technology Assessment.

Figure 4-5.—Cattle Sales by Sales Class, 1969-82 (1969 dollars)

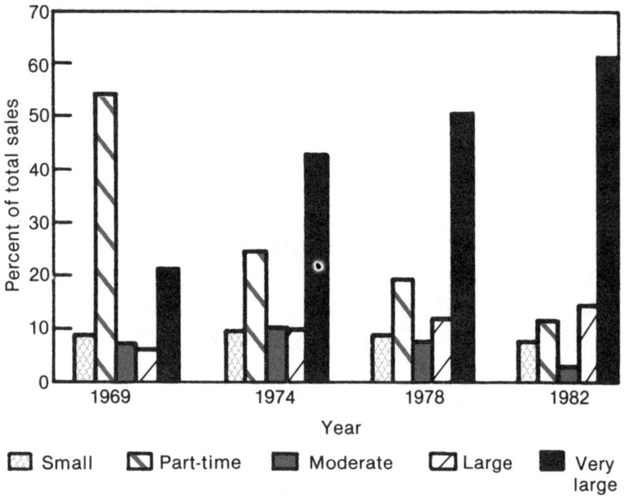

SOURCE: Office of Technology Assessment.

feedlots were excluded. The inversion of structure that has taken place over this time period is striking. In 1969, cattle operations in the $20,000 to $99,999 class had 56 percent of sales, and operations with sales in excess of $500,000 had 22 percent of total sales. The ranking of these two classes reversed in the 4 years between 1969 and 1974. By 1982, cattle operations with sales in excess of $500,000 had about 62 percent of sales, whereas the operations in the $20,000 to $99,999 range had fallen to 12 percent of total sales. Nationwide, cattle operations with sales greater than $200,000 per year had 77 percent of sales. This is remarkable in light of the broad distribution of cattle farms and the large numbers of cattle farms nationwide.

This subsector also clearly has a bimodal structure. While sales are skewed towards the largest cattle farms, both of the lower sales classes have a larger percentage of sales than the middle range ($100,000 to $199,999).

Pork Subsector

Figure 4-6 shows the percent of hog and pig sales in 1969 dollars by sales class for 1969 to 1982. As of 1982 this subsector did not yet have the same degree of concentration of sales in the upper sales classes that was apparent in the other commodity groups. As of 1982 there was a relatively high degree of equality of market share among the different sales classes. However, there have been tremendous structural changes in this subsector, and the direction of change is clear. Sales from farms in the $20,000 to $100,000 class have declined from 61 percent in 1969 to 28 percent in 1982. As a group, hog and pig farms with sales in excess of $100,000 in 1969 dollars had a majority of sales in 1982. Farms with sales in excess of $200,000 are gaining market share at the fastest rate. It is likely that these largest hog farms will soon have a majority of sales, if this has not already occurred.

Figure 4-6.—Hog and Pig Sales by Sales Class, 1969-82 (1969 dollars)

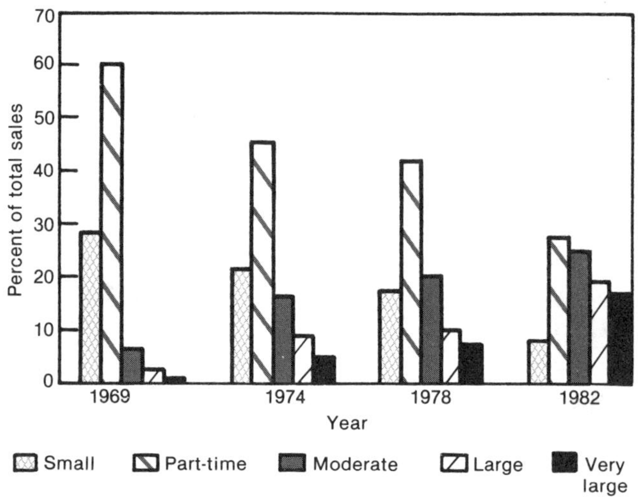

SOURCE: Office of Technology Assessment.

REGIONAL STRUCTURE

There is a common perception that U.S agriculture has become increasingly homogeneous from one part of the country to another. This is true in terms of many aspects of agricultural technology: machinery, crop varieties, livestock breeds, chemicals, and cultural practices have become standardized in many ways. However, there are still major differences in the structure of agriculture in the United States. These differences are seen in the predominance of certain commodities in different parts of the country, the extent to which the production of some commodities is concentrated in different regions, and the pattern of concentration of sales within regions. The basic intent of this section is to show how the structure of agriculture differs between the four major agricultural regions of the United States. The data on which this section is based came from the 1982 Census of Agriculture.[5] The basic units of analysis are the four regions of the United States shown in figure 4-7: the Northeast, South, North Central, and West. Alaska and Hawaii are included in the Western region.

Attention is concentrated on eight different commodity groups. These groups include the six commodity groups whose structure was considered in the national context in the previous section: cash grains (corn, wheat, soybeans, and other specialty grains); cotton and cottonseed; cattle and calves (except sales from feedlots); hogs and pigs; poultry and poultry products; and dairy products. Two additional commodities are included in this section on regional structure: 1) fruit and tree nuts, and 2) vegetables (including potatoes and melons).

[5]This data is the most current that is available on the regional structure of agriculture. Since the general trend has been toward increasing concentration of production in the large and very large sales classes of farms, it is likely that the distributions of sales by sales class described in this section underestimate the true structure of agriculture in 1985.

Figure 4-7.—Regions and Divisions of the United States

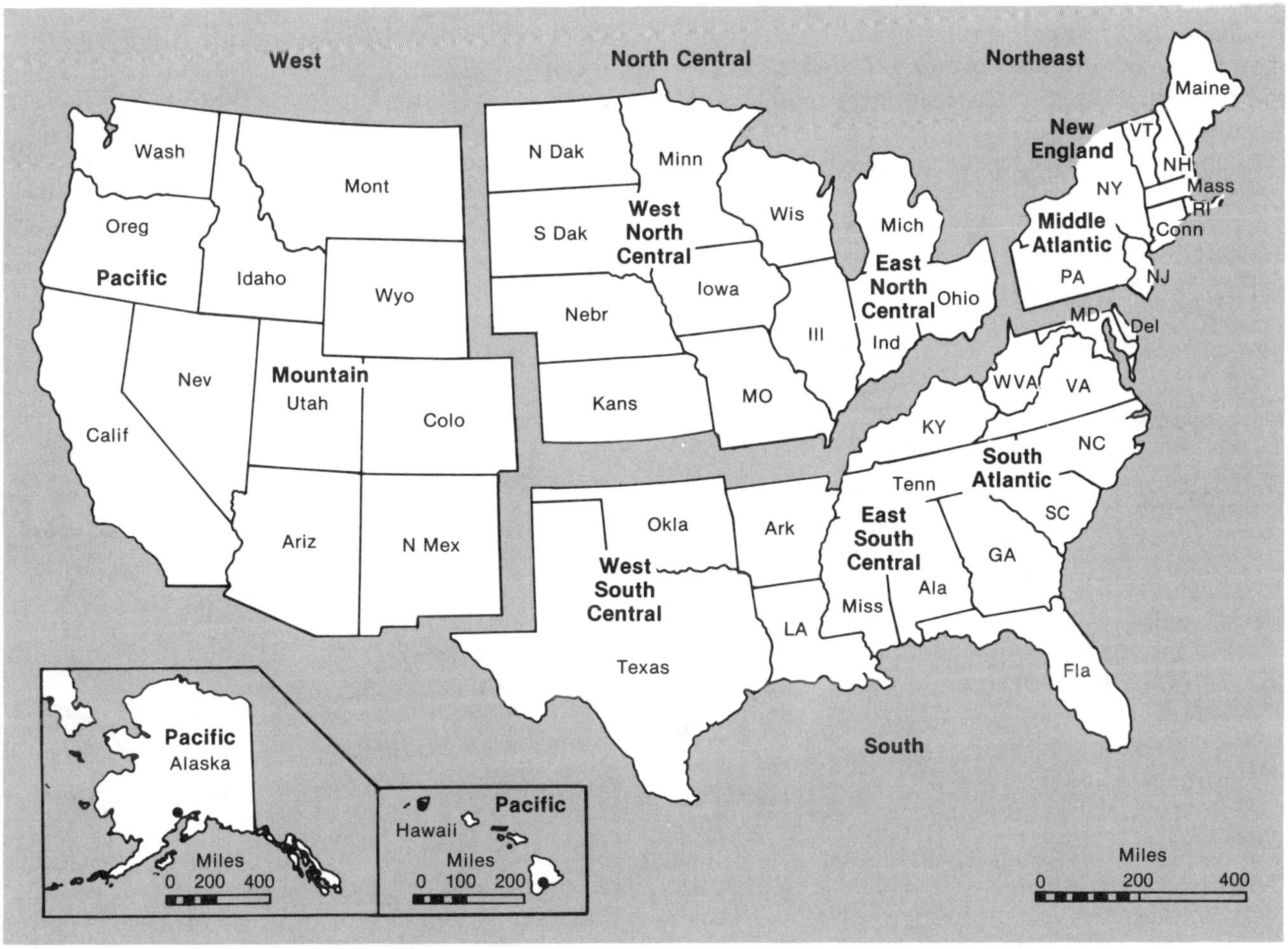

SOURCE: U.S. Department of Commerce, Bureau of the Census.

The first part of this section presents summary statistics on sales by region and commodity group. The second provides a more detailed look at the differences in the structure within and among the four regions in sales classes of farms. The use of sales classes as the unit of structural analysis is more useful than size of farm, since some commodities require much more land per dollar of sales than other commodities and since land values vary greatly from one part of the country to another. Information in this subsection is organized both in terms of sales by sales class and farm numbers by sales class.

Comparison Between Regions and Commodities

Table 4-8 shows the percent of combined total U.S. sales of the eight major commodity groups by group and region in 1982. Ranked in order of total sales, cash grains come first, and cotton is the least valuable commodity. The North Central region accounted for the largest share of sales of the combined commodity groups, at 47 percent. The Northeast region had only 5.2 percent of total sales in the United States in 1982.

Table 4-9 shows the distribution of total U.S. sales of each commodity among the four regions in 1982. The North Central region stands out as the predominant agricultural region of the United States. This region had the most sales in four of the eight commodity groups. It also had 80 percent of hog sales, the highest proportion of any region in any commodity. The West dominated the fruit and tree nut sales and vegetable and melon sales, with 65 and 58 percent, respectively.

A measure of the dependence on particular commodity groups by the agricultural sectors of the different regions can be seen on table 4-10, which shows the percent of each region's total sales of the eight commodities by commodity in 1982. The New England region had more sales from a single commodity group than

Table 4-8.—Percent of Total U.S. Sales of All Commodities by Commodity Group and Region, 1982

Commodity groups	Northeast region	Southern region	North Central region	Western region	Total United States
Cash grains	0.4%	6.5%	20.2%	3.4%	30.4%
Cattle and calves	0.3	7.4	12.0	6.5	26.3
Dairy	2.9	2.7	6.0	3.0	14.5
Poultry and eggs	0.8	5.7	1.5	1.1	9.1
Hogs and pigs	0.2	1.2	6.4	0.2	8.0
Fruit and tree nuts	0.4	1.2	0.3	3.3	5.1
Vegetables, melon, and potatoes	0.3	1.0	0.3	2.1	3.7
Cotton	0.0	1.5	.0	1.2	2.8
Total	5	27	47	21	100

NOTE: Totals may not add due to rounding.

SOURCE: Office of Technology Assessment. Compiled from regional data provided by the U.S. Department of Commerce, Bureau of the Census, Agriculture Division, *1982 Census of Agriculture.*

Table 4-9.—Percent of Total U.S. Sales of Each Commodity by Region, 1982

Commodity groups	Northeast region	Southern region	North Central region	Western region	Total United States
Cash grains	1.3%	21.4%	66.2%	11.0%	100.0%
Cattle and calves	1.3	28.2	45.6	24.9	100.0
Dairy	19.9	18.4	41.2	20.5	100.0
Poultry and eggs	8.6	62.2	17.0	12.2	100.0
Hogs and pigs	2.0	15.3	80.1	2.5	100.0
Fruit and tree nuts	6.9	23.4	5.2	64.5	100.0
Vegetables, melon, and potatoes	7.6	25.8	9.0	57.6	100.0
Cotton	0.0	54.9	1.4	43.7	100.0

NOTE: Totals may not add due to rounding.

SOURCE: Office of Technology Assessment. Compiled from regional data provided by the U.S. Department of Commerce, Bureau of the Census, Agriculture Division, *1982 Census of Agriculture.*

Table 4-10.—Percent of Total Regional Sales by Commodity, 1982

Commodity groups	Northeast region	Southern region	North Central region	Western region	Total United States
Cash grains	7.7%	24.0%	43.1%	16.1%	30 %
Cattle and calves	6.6	27.2	25.6	31.3	26
Dairy	55.4	9.8	12.8	14.3	15
Poultry and eggs	14.9	20.8	3.3	5.3	9
Hogs and pigs	3.1	4.5	13.7	1.0	8
Fruit and tree nuts	6.8	4.4	0.6	15.8	5
Vegetables, melon, and potatoes	5.4	3.5	0.7	10.3	4
Cotton	0.0	5.7	0.1	5.9	3
Total	100.0	100.0	100.0	100.0	100.0

NOTE: Totals may not add due to rounding.

SOURCE: Office of Technology Assessment. Compiled from regional data provided by the U.S. Department of Commerce, Bureau of the Census, Agriculture Division, *1982 Census of Agriculture.*

did any other region, with 57 percent of sales coming from dairy products alone. The North Central region ranked second, with 43 percent of sales in the cash grain group alone. The Southern and Western regions had relatively diversified agricultural sectors. However, both regions were more dependent on cattle production than on any other commodity. It is interesting to note that the West accounted for only 1 percent of national hog sales. This seems to be anomalous in light of the relatively large production of the other seven commodity groups in the West.

Distribution of Sales Within Regions and Among Regions

The data for this section is contained in appendix B, which shows the amount of sales of each commodity by sales class and region for 1982. Sales are expressed as a percent of the total regional sales of each commodity and as a percent of the national sales total for each commodity. Examination of these tables provides useful information on the distribution of production within regions and among regions. The extent to which agricultural production is concentrated in the large and very large sales classes is of particular interest because this information can contribute to an assessment of the rate of technology adoption and the impacts from technology adoption. However, in many cases the degree of concentration should also be considered in the context of the proportion of total national sales. Production of some commodities is highly concentrated in certain regions, but this production amounts to only a small percentage of the national sales of these commodities.

Cash Grains

Cash grain production was the least concentrated of the eight commodity groups within each region in 1982. With the exception of the West, sales of cash grains were concentrated in the part-time, moderate, and large sales classes. The West differed from the other regions in that its cash grain production was relatively skewed toward the larger farms. In the other regions, the moderate-size farm had the largest share of sales. Moderate-size farms in the North Central region had a relatively large share of national cash grain sales, 25 percent of the total. With the exception of the Western region, large farms also had higher sales than very large farms. The North Central region had 69 percent of the total number of cash grain farms in the United States, with small and part-time farms accounting for 57 percent of the total.

Cattle and Calves

The South had 159,000 small farms that raised cattle in 1982. These small farms accounted for 91 percent of the number of cattle farms in the region and 54 percent of the national total. However, these farms accounted for only 3.1 percent of national cattle sales. The other regions also had a disproportionate number of cattle farms in the small farm class. In general, these farms were either subsistence farms, whose owners had low incomes, or they were hobby farms, whose owners had sufficient income from other sources to subsidize this type of production.

The pattern of cattle and calf sales in the Northeast stands out in comparison with the other regions. The other regions had a high concentration of sales in the large farm class and fairly even distributions in the other classes. The Northeast had more sales in the small farm class (less than $20,000) than in any other class, and the sales from the very large farms and large farms were lower than those from the small, part-time, and moderate-size farms. However, the sales of cattle and calves from Northeast farms accounted for only 1.3 percent of the national total, whereas 24 percent of the Nation's cattle and calf sales were made by the very large farms in the North Central region.

Dairy

Three different distributions of dairy production are evident among the four regions. The Northeast and North Central regions had high concentrations of dairy production in part-time and moderate farms and very little production in very large farms. It is striking that the largest single national share of dairy sales were made by part-time farms in the North Central region. There were about 60,000 part-time dairy farms in 1982, 36 percent of all dairy farms in the United States. This large group of farms is especially at risk from rapid changes in the technology and cost structure of the dairy industry.

In contrast, the West had a moderately high concentration of production, 64 percent, in very large farms and relatively little production from part-time and moderate-size farms. The very large dairy farms of the West accounted for 13 percent of the Nation's dairy sales in 1982. This share is expected to increase rapidly.

The South falls between these two patterns, with 46 percent of production in large and very large farms combined and 38 percent in moderate-size farms. None of the regions had more than 2 percent of dairy production in small farms.

Poultry and Eggs

The South had the largest number of poultry and egg farms in the United States, with 28,000 operations in 1982. Twenty-five percent of all poultry and egg farms in the United States were moderate-size farms in the South.

Poultry and egg production was the most concentrated of the eight commodity groups in 1982. In all four regions, very large farms had the highest percentage of sales. The West had the highest degree of concentration, with 85 percent of sales from very large farms. The South had the least amount of concentration, with 39 percent of sales from very large farms and 54 percent of sales from moderate and large farms combined. However, the very large poultry and egg operations in the South had the largest single share of national sales, at 24 percent in 1982.

Hogs and Pigs

The North Central region had the largest number of hog farms in the country in 1982. Thirty percent of the Nation's hog farms were in the moderate size class in this region.

Next to the cash grains, hogs and pigs showed the least amount of concentration. The West was the most highly concentrated region, with 37 percent of sales from very large farms. However, the West had only 3 percent of the national sales of hogs. The North Central region had a low degree of concentration in the very large class, with only 17 percent of sales from these farms. Thirty-eight percent of sales came from moderate-size farms. However, since the North Central region accounted for 80 percent of national hog sales, the moderate-size farms in this region had the largest single share of national sales, 30 percent. Concentration of hog sales in the South was close to that of the West, 21 percent from moderate-size farms and 33 percent from very large farms. However, the very large farms in the South had only 0.9 percent of the national sales in 1982.

Fruit and Tree Nuts

The South had the most concentrated sales of fruits and nut crops; 65 percent of sales were from very large farms. Part-time, moderate, and large farms in the South all had nearly equal shares of 10 to 11 percent. It is interesting to note that 15 percent of the U.S. fruit and nut tree sales come from farms in the part-time sales class in the West. There are 4,462 fruit and nut farms in the part-time class in the West as compared with 7,247 in all 5 classes in the Northeast. The North Central region had the lowest concentration of sales; only 23 percent of sales were from very large farms. Twenty-four percent of sales in the North Central region came from moderate-size farms; however, these moderate-size farms accounted for only 1.3 percent of national sales in 1982.

Vegetables and Melons

The West has a high concentration of national and regional sales of vegetables and melons in the very large class of farms. In 1982 these farms had 83 percent of regional sales and 48 percent of national sales. Vegetable production is popularly associated with small and part-time farmers. The Northeast came the closest to meeting this concept, with 21 percent of sales from the part-time class of farms. However, none of the regions has more than 3 percent of national sales from small and part-time farms combined.

Cotton

The South produced 55 percent of the national cotton sales in 1982, and the West had 44 percent of sales. The Northeast does not produce any cotton, and the North Central region accounted for only 1.4 percent of national production in 1982. Very large farms in the West had 33 percent of total national sales of cotton in 1982. Cotton sales are highly concentrated in the West, with 76 percent of regional sales from very large farms. In contrast, most of the sales in the South came from moderate and large farms (with combined sales of 62 percent), accounting for 34 percent of the national total.

SUMMARY

Overall the trend toward concentration of agricultural resources in fewer but larger farms will continue but will differ by commodity and region. Farm numbers will continue to decline from 2.2 million in 1982 to approximately 1.2 million in 2000. They will follow a distribution of a large proportion of small and part-time farms, an increasing proportion of large farms and a declining number of moderate-size farms. Small farms will account for 50 percent of all farms—a decline from 60 percent in 1982. In contrast, large and very large farms will account for 15 percent of all farms, three times their proportion in 1982. The number and proportion of moderate-size farms will begin to decline by the end of the century.

An important implication of these projections is the further concentration of agricultural production. Over 90 percent of total net farm income will be earned by operators of large and very large farms by year 2000. And the 50,000 largest farms will produce 75 percent of all farm products sold.

However, the increased concentration of resources will differ by commodity and region. The four major agricultural regions differ in their total contribution to U.S. agriculture as a whole and in their contribution to the production of specific commodities. Major differences in structure are apparent when each region is considered in terms of the distribution of sales by sales class for each of the eight major commodity groups. Some regions, such as the West, have a high concentration of sales for several commodities in the large and very large sales classes and a low concentration of sales for other commodities. The North Central region is characterized by very large shares of regional and national production concentrated in moderate-size farms, especially in hogs, dairy, and cash grain sales. The Northeast stands out as a region that has little concentration of sales in the large and very large farms in any commodity, including dairy products, its largest single commodity.

In general, the Nation's agriculture cannot be considered structurally homogeneous even when examined on the basis of large geographical units. Differences in agricultural structure become even more extreme when the United States is considered at the subdivision and State level. As a consequence, agricultural policies that may be appropriate for one part of the country run the risk of being inappropriate when applied to another.

Chapter 5

Factors Contributing to Structural Change in Agriculture

Contents

Chapter 5

Factors Contributing to Structural Change in Agriculture

Traditionally, American agriculture has been dominated by farms in which the operators and their families provided most of the labor, made the management decisions, owned part of the resources, accepted most of the production and price risks, bought and sold in the open market, and depended on the farm as their major source of family income. Such farms have been revered since the days when Thomas Jefferson argued for national policies of public land distribution that favored small, independent landholders. In recent years, the dispersed, independent-farm, open-market system has become less dominant in American agriculture. Major questions are whether this system can compete for world markets and whether society should take steps to halt present trends that are gradually diminishing this system's prominence. Answering these questions entails viewing the causes of structural change—that is, how farm resources are organized and controlled—through economic and noneconomic perspectives.

THE ECONOMIC PERSPECTIVE

An economic perspective encompasses concentration and vertical integration in agriculture.

Concentration

Concentration refers to the proportion of production controlled by the largest firms. It is an important aspect to consider because the more highly concentrated the market, the greater the potential impact of a firm or group of firms on price.

Concentration of total production in agriculture compared with that in many of the other economic sectors is generally low. As discussed in chapter 4, concentration has occurred to the point where in 1982 about 28,000 very large commercial farms—1.2 percent of all farms—produced a third of the total value of U.S. farm products and accounted for over 60 percent of U.S. farm net income.

However, concentration in land resources is also occurring in agriculture.[1] Trends in the distribution of harvested cropland according to sales class show that these productive acres are rapidly becoming concentrated in the farms in the large and very large sales classes. Table 5-1 shows the percentage of total cropland harvested by the top two sales classes of farms for the census years 1969 and 1982 and projects them linearly to 1990 and 2000. If present trends continue, almost half of all cropland will be harvested by farms in these sales classes by 2000.

The degree of concentration varies by commodity. For example, beef cattle operators with

Table 5-1.—Historical and Projected Percentages of Cropland Harvested by Farms With Sales in Excess of $200,000

Sales class	Year 1969	1982	1990	2000
$200,000-$499,000	12.0	25.3	27.0	32.0
>$500,000	6.0	11.2	12.0	14.0
Total	18.0	36.5	39.0	46.0

Projection assumptions:
1. Growth in total harvested acres is linear, resulting in an increase of 2.4 million acres per year.
2. Growth follows the linear trend for the two sales classes and results in an increase of 2.7 million acres per year for the farms in the $200,000-$499,000 class and of 1 million acres per year for the >$500,000 class.
3. The linear projections are based on the acres harvested by sales classes, adjusted for inflation. Inflation in commodity prices tends to move acres from lower to upper sales classes. Since inflation in commodity prices is likely to continue, nominal growth in acreage harvested by these sales classes may be greater than projected.

SOURCE: Office of Technology Assessment.

[1] Land resources in the agricultural sector can be viewed in the general category of "land in farms," as defined by the Bureau of the Census, or in the "harvested cropland" category. The acreage of cropland harvested is a more accurate measure of productive agricultural resources than is the general category of land in farms.

sales over $500,000 per year in 1982 represented only 0.5 percent of all beef cattle operations and accounted for 55 percent of the total value of cattle sales. The 69 largest of these feedlots produced 21 percent of the fed cattle in 1980 (USDA, 1981). The largest cattle feeders were also some of the largest feed manufacturers and grain companies.

Higher levels of concentration exist for broilers (chickens). In 1977 the 16 largest broiler producers and contractors controlled about 50 percent of the production (Brooke, 1980). In vegetable crops, such as lettuce and celery, concentration is comparably high (Brooke, 1980).

On the other hand, concentration is still very low for most crop agriculture. Relative to other American industries, where the market share of the four largest manufacturers frequently exceeds 50 percent, concentration in agriculture, even for cattle feeding, broilers, lettuce, and celery, is low. However, attention is drawn to agriculture because of the rapidity with which certain industries, such as broilers and feed cattle, have gone from a diffused to a concentrated and integrated agriculture (Knutson, et al., 1983).

Concern exists that if extended over a period of time, the increasing concentration of agricultural production could lead to higher food prices (Breimeyer and Barr, 1972). This would result from increased merchandising and marketing costs, potential unionization of agricultural workers, and the lack of effective competition (Rhodes and Kyle, 1973).

Vertical Integration

Firms are vertically integrated when they control two or more levels of the production-marketing system for a product. Such control may be exercised by contract or by ownership.

Contract integration exists when a firm establishes a legal commitment that binds a producer to certain production or marketing practices. At a minimum, contract integration requires that the producer sell the product to the buyer. Additional commitments may bind the farmer to specified production practices and sources of inputs. While all forms of contract integration have created concern, the greatest controversy exists with contracts that control both production and marketing decisions of farmers. In addition, from a legal perspective, the producer may not even own the product being grown (Knutson, et al., 1983).

The extent of contract integration is not well documented. Ronald Knutson estimates that all forms of contract integration represented 32 percent of farm sales in 1981 (Knutson, et al., 1983). He makes the following observations on the extent of contracting:

1. Contracting used to be limited to perishable products; now it has expanded to virtually all commodities.
2. Production contracting appears to be associated with commodities where breeding and control of genetic factors play an important role in either productivity determination or quality control.

Ownership integration is a single ownership interest extended to two or more levels of the production-marketing system. It may involve either cooperatives or proprietary agribusiness firms. Knutson estimates that proprietary ownership integration accounts for about 6 percent of farm sales. Some proprietary agribusiness firms—such as Cargill (beef); Superior Oil (fruits, vegetables, and nuts); Coca-Cola (oranges and grapefruit); Tysons (broilers and hogs); Tenneco (fruits, vegetables, and nuts); and Ralston Purina (mushrooms)—have made substantial investments in agricultural production. In products such as broilers, eggs, cotton, vegetables, and citrus fruits, ownership integration is over 10 percent of total U.S. production (Knutson, et al., 1983).

Cooperative ownership integration is much more prevalent than proprietary ownership integration, accounting overall for 34 percent of farm sales. However, in only 13 percent of cooperative integration is there a legal commitment by farmers to market their commodities or to purchase inputs from the cooperative.

The economic implications and concern for structural change of vertical integration are debated. A principal problem in agriculture has been the difficulty of coordinating production with market needs. Vertical integration can make a substantial contribution to satisfying this need. For example, in broilers and turkeys, vertical integration has contributed to the uniform size and quality of poultry sold. It has also contributed to increased efficiency and reduced costs (Schrader and Rogers, 1978).

On the other hand, there are potentially adverse consequences of vertical integration. Contract integration with corporations, and sometimes cooperatives, radically changes the role of the traditional, independent farmer. More often than not, the farmer loses control of, if not legal title to, the commodities grown under a production-integrated arrangement. Payment to the grower is largely on a per-unit or piece-wage basis, and not necessarily related to product value.

It has been argued that in the long run, market power in integrated agriculture will become sufficiently highly concentrated that the consumer will pay higher prices for food. However, no definitive conclusion can be made. The above argument fails to take into account efficiency gains from integration. The extent to which these gains could be realized without the development of a vertically integrated system is open to question.

THE SOCIOLOGICAL PERSPECTIVE

Many concerns relating to structural change are of a sociological nature. They revolve around the impact of concentration and integration on the institution of the family farm, on rural communities, and on rural institutions.

Concern has been expressed that continuously increasing the concentration and integration will lead to the demise of the family farm as an institution. The term *family farm* has been associated with the existence of an independent business and social entity that shares responsibilities of ownership, management, labor, and financing. The family farm system leads to dispersion of economic power and has been associated with the perpetuation of basic American values and of the family as an institution. Increased concentration and integration tend to destroy the family farm institution. Very large farms lose many of the characteristics of the traditional family farm because their business and hired labor aspects clearly predominate. Most of the management functions traditionally associated with the family farm institution are removed by integration. With integration the farmer takes on more of the characteristics of a businessman.

Another concern is that concentration and ownership integration reduce the number of farms and make the integrator less dependent on the local community. As a consequence, small rural towns and their social institutions decline or vanish. Recent research conducted in California provides some evidence to substantiate such a relationship. Dean MacCannell (1983) has found that rural communities where a few large and integrated farms dominate are associated with few services, lower quality education, and less community spirit. (This is discussed in greater detail in chapter 11.)

Concerns are also expressed about the impact of structural change on the nature of the U.S. political system. Thomas Jefferson visualized the merits of a decentralized political system where power was highly diffused and where every individual had the opportunity to participate in public decisions. His philosophy placed a high value on independent farmers and landowners as a means of maintaining a democratic system of government.

Already there has been a marked departure from the decentralized power structure ideal visualized by Jefferson. The question is whether agriculture is basically unique and different from other sectors of U.S. society, as has long been maintained. Are there unique social, cultural, and traditional values in having landownership widely dispersed, or should agriculture join the mainstream where the other economic sectors have long been? As U.S. agriculture continues along the trends laid out in this report, it will increasingly take on the characteristics of the nonfarm sector. Some people will interpret this trend as progress; others will interpret it as a step backward.

CAUSES OF STRUCTURAL CHANGE

A number of factors have been identified by researchers as causes of structural change. However, there has been no delineation of the relative importance of each factor. One of the objectives of this study is such a delineation. Before moving to that analysis in the following chapters, however, it is important to understand why each of these factors is considered important to structural change.

Most observers of structural change cite three main determinants:

1. technology and associated economies of size, specialization, and capital requirements;
2. institutional forces; and
3. economic and political forces (figure 5-1).

Technological Forces

Certain farmers have a strong incentive to adopt new technology rapidly. The early innovator achieves lower per-unit costs and increased profits, at least for a short time, before other farmers follow his lead. For example, in

Figure 5-1.—Factors Influencing the Structure of Agriculture

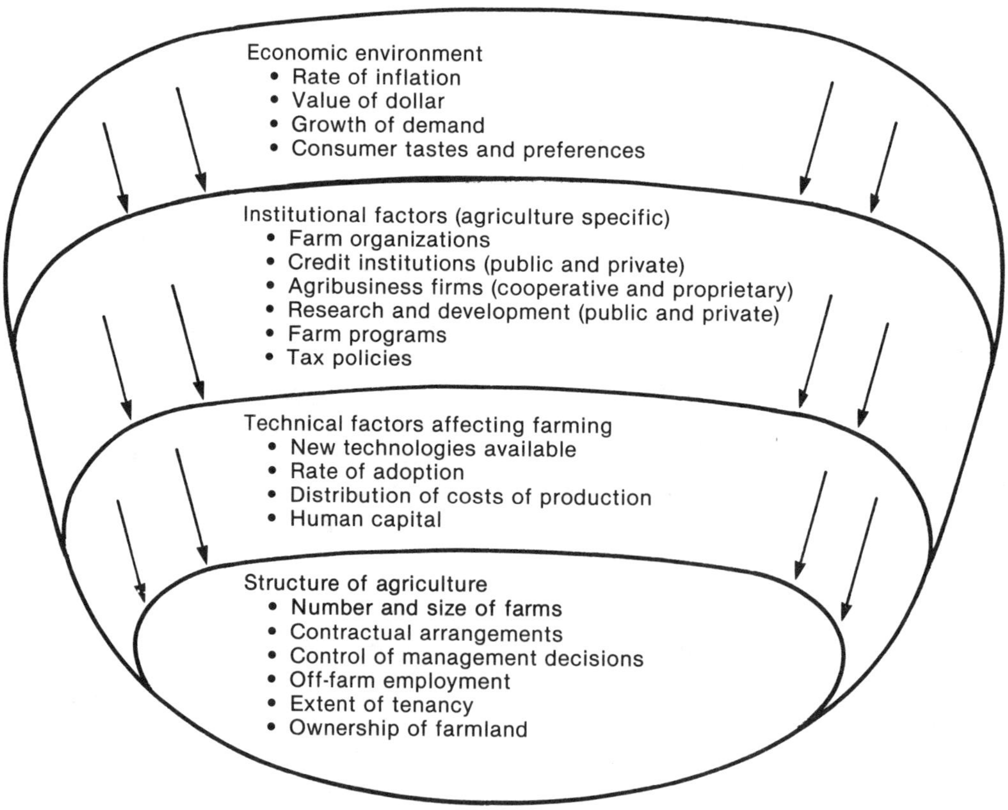

SOURCE: Office of Technology Assessment.

Washington State a winter wheat farmer with 2,500 acres can reduce average machinery costs by 9 percent per acre by replacing a conventional crawler tractor with a four-wheel-drive tractor. If he also expands the size of his farm to 3,900 acres, he can reduce costs by an additional 18 percent (Rodewald and Folwell, 1977). This nearly 60-percent increase in farm size can be made without additional labor. Once the innovative wheat farmer adopts the technology, other crop farmers generally have two options: purchase a four-wheel-drive tractor and expand the size of their farm, or accept a lower net income as market prices for their crops fall. In short, new technology can play an important role in determining acreage and capital requirements. Different farmers have different costs because they use different combinations of inputs, have different management skills, or have a different scale of operation.

Economies of Size

The relationship of scale of operation to cost is of particular significance to structure. If costs are relatively the same for all farm sizes, one would expect all farm sizes to have relatively little incentive to increase in size. In addition, with relatively even costs, consumers would clearly not benefit from increases in farm size. If, on the other hand, costs decline sharply as farm size increases, not only would there be strong incentives for farms to grow in size, but consumers would potentially realize lower prices for food. Of at least equal importance to policymakers, if costs decline sharply as farm size increases, efforts to prevent this change from occurring—e.g., to preserve the family farm—would not only be difficult, but could be counterproductive from a consumer perspective. Smaller farm operators could exist in a cost-declining environment only if they were willing to accept lower returns to contributed labor, capital, and management, and/or had an off-farm job.

Past studies of the relationship between average production costs and farm size support two major conclusions: First, most economies of size are apparently captured by moderate-size farms. Second, while the lowest average cost of production may be attainable on a moderate-size farm, average cost tends to remain relatively constant over a wide range of farm sizes. Thus, farmers have a strong incentive to expand the sizes of their farms in order to increase total profits. (This phenomena is explored in detail in chapters 8 and 9.)

Earlier studies on economies of size have several limitations. External economies gained from buying and selling in large volumes and from access to credit have usually been ignored. Common ownership of related farm and nonfarm activities has not been considered. There is some evidence that inclusion of such pecuniary economies would lower the average production costs for large farm units and would shift the conclusion about the size of the most competitive farm (Smith, et al., 1984).

Specialization

Technology has also influenced specialization and regional production patterns. Cotton production has moved westward, for example, into areas of broad, flat fields where larger machinery can be used to optimum advantage. Specialization in crop production is also due in part to technology. Farmers who once relied on crop rotation and diversification to conserve soil fertility, prevent soil erosion, and control pests have replaced these practices by chemical fertilizers, insecticides, and herbicides, with questionable long-run effects. Such farmers can thus grow one crop exclusively year after year, specializing in commodities that are the most profitable. Similarly, the development of new disease control techniques has given poultry and livestock farmers unprecedented opportunities to specialize. The vertically integrated broiler industry of today would have been impossible without scientific advances in breeding, feeding, housing, and medicine, which have reduced the real cost of broilers by as much as 50 percent over the past 30 years.

These scientific breakthroughs have generally enabled both small and large farmers to specialize more. However, improvements in farm machinery have perhaps been most important in fostering large-scale, specialized operations.

A decision to invest in a specialized piece of equipment means that an operator will emphasize production of the commodity for which the machine is intended, quite likely at the expense of some other commodity. And insofar as a machine is most economical on a particular size of operation, expansion to that size is encouraged. Thus specialization and farm growth occur simultaneously.

Capital Requirements

Agriculture is one of the most capital-intensive industries in the American economy. As a result, the requirements for credit to finance new capital investments, production, or storage are high. Technology has made barriers to entry more formidable. The cost of machinery raises capital requirements for beginning farmers. Technologies that allow individuals to farm increasingly larger acreages have added to the competition for land, resulting in high land prices, the single greatest expense in farming today. The average investment in 1980 in a farming operation with gross sales between $40,000 and $60,000 ranged from $350,000, for fruit and nut farms, to over $800,000, for livestock ranches.

Institutional Forces

Institutional factors have their primary influence on the costs of inputs used in production, the prices of products, and the generation of new technology for agriculture. These institutions may be either in the private or the public sector.

The costs of inputs are primarily a function of competition between private sector agribusiness firms. Input costs do not have to be the same for all farmers. Input suppliers may offer farmers discounts for larger volume purchases of fertilizer or chemicals. Likewise, larger scale farmers may receive higher prices for products marketed through the use of crop contracts or futures markets.

Research and Extension

New technologies are generated in both the public and private sector. Basic agricultural research is primarily a public sector function performed by the U.S. Department of Agriculture (USDA) and the land-grant universities. Applied research functions are shared between the public and private sector, with the private sector dominating development activities. Extension activities assist in evaluating and transferring technological innovations into practice. An integral part of the agricultural research and extension policies involve the generation of higher levels of training and expertise embodied in human capital. The result is more skilled farmers, agribusinessmen, scientists, and agricultural policymakers.

Research and extension have had different impacts on farms, farm workers, rural communities, and even entire regions, depending on the characteristics and type of technology developed. Some technological innovations, particularly mechanical innovations, have favored and hence fostered larger farms. Technological innovations that could be applied on farms of any size are often first adopted by larger farms (Paarlberg, 1981; Perrin and Winkelman, 1976). By being the first to adopt new technologies, larger farms receive greater benefits than those not adopting the technologies (typically, smaller farms).

A major effort of extension is to disseminate timely information through public meetings. The topics covered in publications and public meetings are heavily influenced by current research results. Any bias toward larger farms that is embodied in research results would most likely be carried over into meetings and publications.

Even though extension personnel make information available to all farmers, those farmers that make the most use of the research results and extension information can generally be characterized as the more innovative, more aggressive, and better managers, usually of larger farms (Paarlberg, 1981). Such farmers are also generally more vocal, providing feedback to research and extension personnel on the usefulness of the information received. Even though no overt effort is made to exclude particular groups, such as operators of small farms, the net result is that many research and extension programs become more oriented toward those select groups that generally avail themselves of the information (Paarlberg, 1981).

This lack of structural neutrality was recognized in 1979 by Secretary of Agriculture Bergland when he questioned the use of Federal funds for research projects having the objective of producing large-scale, labor-saving technology and set up a special task force to investigate the impact of research and extension on structure. At the same time, Congress earmarked research and extension funds for increased work with small farms and for projects involving direct marketing from farmers and consumers. However, no special programs were developed for moderate-size farms.

The Bergland initiative on research was deemphasized with the change in administration in 1981. It has, however, been rekindled by the announcement of joint initiatives in biotechnology research between private sector companies and universities. Questions have arisen as to whether the primary beneficiaries of the initiatives will be the private sector firms or the initial adopters of the resulting new technology.

Public Policy

Many public policies affect the structure of agriculture by influencing resource use, capital requirements, technology development and adoption, freedom of decisionmaking, exchange arrangements, risks, and costs and profits. Some policies are oriented specifically to the farm sector, such as price and income policy (commodity programs). Others affect agriculture directly but are more broadly oriented, such as tax policy. Still others are general—e.g., national macroeconomic policy—and affect agriculture indirectly.

Public policies offer viable ways to maintain or alter the structure of the agricultural sector. In this section, areas of public policy involvement that affect the structure of agriculture are briefly examined.

Commodity Programs.—Beginning with the Agricultural Adjustment Act of 1933, a series of commodity programs have evolved to deal with price and income problems in farming. These programs have covered such commodities as wheat, feed grains, cotton, wool, sugar, rice, peanuts, tobacco, and dairy products. To stabilize and increase farm prices and incomes, a variety of program tools has been used: price supports, direct payments, acreage allotments, set-asides, conservation reserves, surplus disposal, and stock accumulation.

There is widespread agreement that these programs, in the short run, held farm incomes above the long-run income effects. Price stability from these programs has enabled farmers to adopt new and improved technologies. And logic suggests that the higher the level at which prices are stabilized, the more rapid and widespread will be technological adoption in farming. But it does not follow that high and stable prices necessarily speed resource concentration in farming. The high and stable prices may help the weak and inefficient stay in business. Little is known about how different levels of price and income support affect the rate of resource concentration in farming (Cochrane, 1983). Thus, the question becomes whether policymakers who want to change the rate at which productive resources in farming are concentrated in fewer and fewer hands should support and stabilize product prices at low levels, at high levels, or somewhere in between—or whether they should do something different instead. (The effect of commodity programs on resource concentration is analyzed in chapters 8 and 9.)

Tax Policy.—Tax laws and provisions are widely recognized as being a determinant of agricultural structure. There is, however, no agreement about the relative importance of tax policy because of its interactions with other structural determinants. Some tax laws and provisions can be directly related to structure (i.e., estate and corporate tax law), while others (i.e., investment tax credits, depreciation provisions, capital gains, and cash accounting) are indirectly related and often interact with credit and commodity policies.

In animal agriculture, tax factors such as cash accounting, current deductibility of costs of raising livestock, and capital gains treatment for sales of breeding livestock, together with investment tax credits and accelerated depreciation, influence livestock investments and can affect structure. Tax policy issues in animal agriculture include tax shelter and nonfarm investments, tax provisions as a factor in economies of size, and the legal structure of agriculture. The cattle sector provides one example.

The income tax advantages of cattle feeding were packaged as limited partnership syndicates in the late 1960s and early 1970s and sold to nonfarm investors. The growth of nonfarm investment in cattle feeding was closely associated with the movement of cattle feeding out of the Midwest and with the growth of large-scale feedlots in the High Plains area. Other factors also played a role, but limited empirical evidence suggests that tax-induced investment in cattle feeding through limited partnerships was related to structural change (Carman, 1983).

For mechanical technology, current tax laws favor the substitution of capital for labor and may speed the adoption of mechanical systems. Two tax factors are at work: payroll taxes, which increase the cost of labor, and provisions for investment tax credit and accelerated depreciation, which decrease the cost of machinery (Carman, 1983).

It is conventional wisdom that tax provisions are an important consideration in the adoption of capital-intensive innovations, since investment tax credit and accelerated depreciation do have a significant impact on after-tax costs. Such innovations include large four-wheel-drive tractors, circle irrigation systems, minimum tillage systems, and large-scale and improved harvesters.

An important implication can be drawn about structural change from the above discussion. Small farms and very large farms have more off-farm interests against which to offset farm losses than do moderate-size farms. This could be a significant factor in accounting for the decline of the moderate farm. (The effect of tax policy on structural change in agriculture is examined in chapters 8 and 9.)

Agricultural Credit Policy.—Public policy directly influences the supply of capital to farmers through the Farmers Home Administration (FmHA) of USDA and the Farm Credit System, which includes the Federal Land Bank, the Production Credit Association, and the Bank for Cooperatives. The original capital for the Farm Credit System was supplied by the Federal Government, but the system is now wholly owned by its borrowers. However, the Farm Credit System is still accorded agency status whereby interest costs on its bonds and discount notes are lowered. The FmHA is a Government agency that has a mandate from Congress to make low interest loans to family farmers who cannot obtain credit elsewhere. The FmHA and the Farm Credit System together account for approximately 40 percent of the total farm debt outstanding (8 and 33 percent, respectively) (Barry, 1983).

The general intent of farm credit policies has been to ensure appropriate capital availability for agriculture. Policies established by these agencies and their attendant programs are thought to have influenced the structure of the farm sector, although the extent of their impact has not been studied thoroughly. (Chapter 7 explores the relationship between credit policy and structural change in agriculture.)

Economic and Political Forces

Agriculture operates in a broader overall economic and political environment. This environment determines the rate of interest, the rate of inflation, and the value of the dollar—all of which influence the costs and prices of farm products. The increased importance of these effects has made macroeconomic policies that influence the overall economic environment within which agriculture operates more important to farmers.

THE DYNAMICS OF STRUCTURAL CHANGE

A study of this type cannot possibly analyze all of the technical, economic, and institutional factors that influence the structure of agriculture. This study therefore concentrates on those factors that appear to be the most critical in affecting structure and that also relate to current farm policy decisions. These factors include:

- the technical factors influencing the costs of production as related to farm size;
- the major farm program elements; and
- the institutions that lead to the development and assimilation of new technology.

These factors interact in a dynamic fashion to influence the structure of farming. New technology continuously infused into agriculture is adopted by the most progressive farmers. While the initial adopters assume increased risk in applying a new technology, they generally also gain substantially higher returns. Farm programs that reduce price risk help assure higher returns.

As more farmers realize the advantages of new technology, the adoption process becomes more general. As this happens, supplies increase, with the tendency to force down market prices. If Government policies prevent market prices from falling, surpluses build up, as they have in the dairy industry or did before the payment-in-kind (PIK) program. If market prices fall, Government payments rise.

Wider adoption of technologies also changes the nature of costs as farm size increases. If larger farms are the first adopters, their costs are substantially lower. The laggers in adoption realize much higher costs. By not adopting, they become, in effect, left behind—eventually being either forced off the farm altogether or forced to take an off-farm job.

These consequences often lead to suggestions of turning off the technological wheels of progress. Such a strategy, however, would have a devastating impact on the competitive position of American farmers in world markets. Instead of just some people being left behind, the whole American farm system would be left behind.

SUMMARY

This chapter has viewed structural change from both an economic and sociological perspective and identified the major forces of this change. These include technological, institutional, economic, and political factors. Technology and associated economies of size, specialization, and capital requirements have had an important influence on structural change in agriculture. Likewise, private and public institutional factors, which include research and extension, credit institutions, farm programs, and tax policies, have played a significant role. And the economic and political environment including the value of the dollar, rate of inflation, growth in demand, and consumer tastes and preferences are becoming even more important factors.

There has been a marked departure from the decentralized power structure ideal visualized by Thomas Jefferson that causes many people to be concerned for a variety of reasons. The question is whether agriculture is basically unique and different from other sectors of U.S. society. Are there unique social, cultural, and traditional values in having landownership widely dispersed, or should agriculture join the mainstream where other U.S. economic sectors have long been? As American agriculture continues along the trends laid out in this report, it will increasingly take on the characteristics of the nonfarm sector. Some people will interpret this as progress; others will interpret it as a step backward.

To help policymakers better understand how and why agricultural structure is changing, this study focuses on the major technical, economic, and institutional factors which influence structural change. They are: 1) technology, 2) major farm program and tax elements, 3) financial institutions, and 4) institutions that lead to the development and assimilation of new technology. In the chapters to follow the impact of each of these factors, as well as the dynamic interactions between them, will be studied.

CHAPTER 5 REFERENCES

Barry, Peter J., "Credit Policy and Structural Issues in Agriculture," paper prepared for the Policy, Technology, and Structure of Agriculture Workshop, Office of Technology Assessment, Washington, DC, March 1983.

Breimeyer, Harold F., and Barr, Wallace, "Issues in Concentration Versus Dispersion," *Who Will Control U.S. Agriculture?* North Central Regional Extension Publication 32 (Urbana, IL: University of Illinois, August 1972), pp. 13-27.

Brooke, Donald L., "Changing Structure of Florida Vegetable Farms 1945-1974," *Farm Structure—A Historical Perspective on Changes in the Number and Size of Farms,* U.S. Senate, Committee on Agriculture, Nutrition, and Forestry, April 1980, pp. 363-379.

Carman, Hoy F., "Tax Policy, Technology, and the Structure of Agriculture," paper prepared for the Policy, Technology, and Structure of Agriculture Workshop, Office of Technology Assessment, Washington, DC, March 1983.

Cochrane, Willard W., "Farm Commodity Policy, Technological Advance and Structural Change," paper prepared for the Policy, Technology, and Structure of Agriculture Workshop, Office of Technology Assessment, Washington, DC, March 1983.

Knutson, Ronald D., et al., *Agricultural and Food Policy* (Englewood Cliffs, NJ: Prentice Hall, Inc., January 1983).

MacCannell, Dean, "Agribusiness and the Small Community," paper prepared for the Policy, Technology, and Structure of Agriculture Workshop, Office of Technology Assessment, Washington, DC, March 1983.

Paarlberg, Don, "The Land Grant Colleges and the Structure Issues," *American Journal of Agricultural Economics* 63:129-134, 1981.

Perrin, Richard, and Winkelman, Don, "Impediments to Technical Progress on Small vs. Large Farms," *American Journal of Agricultural Economics* 58:888-894, 1976.

Rhodes, James, and Kyle, Leonard R., "A Corporate Agriculture," *Who Will Control U.S. Agriculture?* North Central Regional Extension Publication 32-3 (Urbana, IL: University of Illinois, March 1973).

Rodewald, Gordon E., and Folwell, Raymond J., "Farm Size and Tractor Technology," *Agricultural Economics Research* 29:89, July 1977.

Schrader, Lee F., and Rogers, George B., "Vertical Organization and Coordination in the Broiler and Egg Subsectors," *Vertical Organization and Coordination in Selected Commodity Subsectors,* North Central Regional Project No. 117, W.P. 20 (Madison, WI: University of Wisconsin, August 1978).

Smith, Edward, et al., *Cost and Pecuniary Economies in Cotton Production and Marketing: A Study of Texas Southern High Plains Cotton Producers,* Texas A&M University, Texas Agricultural Experiment Station Bulletin B-1475, August 1984.

U.S. Department of Agriculture, *Cattle on Feed,* MTAN 2-16 (Washington, DC: Economic, Statistics and Cooperative Service, January 1981).

Part III

Analyses of Technology, Public Policy, and Agricultural Structure

Chapter 6

Emerging Technologies and Agricultural Structure

Contents

Tables

Chapter 6

Emerging Technologies and Agricultural Structure

New technologies have, historically, had significant impacts on structural change. New disease control techniques gave poultry and livestock farmers unprecedented opportunities to specialize and vertically integrate. Improvements in farm machinery fostered large-scale, specialized farm units.

Like their predecessors, the emerging technologies examined for this study will make a considerable impact on farm structure, especially by year 2000. Biotechnologies will have the greatest impact because they will enable agricultural production to become more centralized and vertically integrated. Although in the long run the use of new technologies will not increase the farmer's overall need for capital, there will be trade-offs: biotechnology will entail less capital; information technology will entail more. This chapter discusses how the emerging technologies are expected to affect these and other elements of agricultural structure.

The chapter evaluates the new technologies' impacts on agricultural structure. It covers: 1) the methodology and assumptions for evaluating the impact of technologies, 2) the analysis of technology's impact on structure, 3) relative adoption rates by size of farm, and 4) relative effectiveness of policy in achieving a structure.

METHODOLOGY

To assess the relationship between technology and structure, OTA conducted a 2-day workshop with a panel of 14 experts.[1] This workshop will be referred to as the Agriculture Structure Group (ASG). The panel members represented a broad range of backgrounds and regions within the United States. (The names, affiliations, and disciplinary specialty of each workshop participant are included in app. C.) A major portion of the first day was devoted to briefings by experts on the 28 previously defined technologies. The initial discussions of the ASG involved the potential impact of these technologies on capital and labor at the farm level and on the structural elements of vertical coordination, market access, and barriers to entry. The second day's discussions were concerned with the distributional impacts of the technologies and the broad categories of public policies that might achieve a predetermined structural change over time.

[1]This chapter is based on the results of the workshop as analyzed by Thomas Sporleder in the OTA paper "Agricultural Structure Impacts of Emerging Technologies in American Agriculture" and reviewed by the workshop participants.

The assessment of distributional impacts was by broad category of technologies. The panel members considered both rate of adoption and change in physical output by farm size, focusing on the flow of impacts from various technologies to structure.

As a final item, the ASG addressed the reverse flow—from structure to technology development and adoption. Thus the potential causality between various farm structures and the development and adoption of technology groups was also addressed by ASG.

Group discussion was unstructured except for materials prepared for the ASG that provided an agenda of discussion topics. The Delphi procedure was used to help the group reach consensus about the potential impacts of technology on selected economic variables.

The assumptions made by the work group that applied throughout the assessment of technological impacts to the year 2000 were:

1. Economic variables such as tax policy, in-

terest rates, inflation rates, and prices are held constant at 1984 levels.
2. Impacts of technology groups are assessed separately, assuming that only one technology group exists and is 100 percent adopted in 1984. Thus, potential interaction effects among the technology groups were not assessed.
3. Capital and labor were defined broadly so that they were the only factor inputs in the production process. Capital was defined as all nonlabor, nonmanagement resources, including land, while labor was defined to encompass both management and labor.
4. The focus was production agriculture—farms and ranches. An important element of structure was defined as the number and size of farm firms. Although the group recognized the importance of other levels in the commodity marketing and food distribution channel, the focus remained on the farm level for reasons of manageability.

TECHNOLOGY GROUPINGS

The ASG judged the 28 study technologies to be sufficiently similar in their impacts on market structure to permit assessment of them within groups rather than individually. Thus the technologies were grouped into three broad categories for animal agriculture and four broad categories for plant agriculture (table 6-1).

Table 6-1.—Groups of Technologies Analyzed, by Animal and Plant

Technology group/technologies included
Animal:
Biological.—Genetic engineering, Animal reproduction, Regulation of growth and development, Animal nutrition, Disease and pest control
Mechanical.—Environment of animals, Animal behavior, Crop residues, Animal waste use, Robotics
Information.—Monitoring and control, Communication and information, Telecommunications
Plant:
Biological.—Genetic engineering, Enhancement of photosynthetic efficiency, Plant growth regulators, Plant disease and nematode control, Management of insects and mites, Weed control, Biological nitrogen fixation
Mechanical.—Robotics, Engines and fuels, Crop separation—cleaning—processing
Information.—Monitoring and control, Communication and information, Telecommunications
Management techniques.—Water and soil-water-plant relations, Soil erosion, Soil productivity and tillage, Multiple cropping, Organic farming, Land management

SOURCE: Office of Technology Assessment.

The biological group essentially consists of technologies that use living organisms or their isolated components for manipulating plant or animal production. The mechanical group encompasses technology development in machinery used to produce and/or harvest the results of crop or animal production. The information group includes the technologies of sensors, controllers, and actuators, along with broad developments in computer technology applicable to the collection and analysis of information for producer-level management decisions.

Whereas the other categories apply to both crop and livestock production, the technologies within the management techniques group apply only to crop production. This group includes technologies that assist in a more optimal, long-term combination of inputs at the producer level, and each involves cultural or management practices.

STRUCTURAL ELEMENTS

The structural dimensions assessed for technological impacts were capital and labor, vertical coordination and control, market access, barriers to entry, and regional impacts. These elements are not necessarily the only relevant ones for judging the impact of various technologies. However, in the interest of manageability of scope and time, these were judged most important.

Capital is viewed broadly as all nonlabor and nonmanagement inputs, including land, while labor was viewed broadly as both management and labor. *Vertical coordination* is defined as coordination of quality, quantity, and timing across producer/first handler markets. *Control* is primarily the ability of producers to exercise authority over production and marketing decisions. *Market access* refers to whether producers have access to most or all buyers at a particular stage of the marketing channel. *Barriers to entry* are defined as the inability, for whatever reason, of new firms to enter a particular industry.

These structural elements are the common ones normally viewed as important in agricultural commodity markets. For any of the structural elements, it is difficult to judge or measure precisely the magnitude of impact from various technologies. Often, the direction of impact (positive or negative) is easier to judge. The procedure used by workshop participants was to discuss direction of impact, then use the Delphi process to judge magnitude of impact within some predetermined range (e.g., 0 to 10 percent, 10 to 20 percent).

IMPACTS OF TECHNOLOGY ON STRUCTURAL ELEMENTS

The structural elements provide some indication of the potential for a technology group to induce a change in farm size and number over time. The directional impact of the technology groups on each selected structural element is summarized in tables 6-2 and 6-3. The tables apply to both animal and plant agriculture across all technologies.

Capital and Labor

Technology can affect the capital and labor used in production of either animals or crops. The absolute and relative change induced by technology in capital and labor was addressed and is depicted in table 6-3.

Biological Group

Several primary effects of technologies in the biological group on capital were identified. It was assumed that reproduction and genetic engineering technology would be adopted by farmers via contracting for a specialized service. The adoption would slightly reduce (<5 percent) both capital and labor necessary for a given size herd. An example is the expected dramatic savings in time that will result from hiring a specialized service to check a dairy herd for estrus.

Growth hormone technology is also expected to decrease both capital and labor needs slightly in animal production. Animal nutrition and disease control technologies are expected to decrease slightly or have no impact on capital and labor in the long run.

Because both capital and labor are expected to decrease slightly as a result of these biological technologies, no significant change (i.e., <5 percent) is expected in the capital/labor ratio. Capital is viewed as decreasing slightly less than labor, on the average, but this difference is not viewed as significant.

The biological group of technologies is viewed as having a long-run neutral impact on capital input at the farm level, primarily because the majority of these technologies have become available as new plants or seeds. Thus the technology is imbedded into the factor input without a separate purchase of it. Potential price increases in plants or seeds were viewed as being offset by productivity gains.

A slight decrease in labor input is expected from the technologies in this group for plant agriculture. This was expected mostly from the potential for weed, insect, and mite control technologies reducing some labor input to the pro-

Table 6-2.—Potential Directional Impact of Technology Groups on Structural Elements at the Producer Level, by Animal and Plant

Area and technology group	Potential additional direction of impact induced by technology group by year 2000: Vertical coordination and control	Market access	Barriers to entry
Animal:			
Biological group	Closer coordination encouraged	Slight reduction	No significant change
Mechanical group	No significant change	No significant change	No significant change
Information group	No significant change	Slight increase	Slight-to-definite reduction
Plant:			
Biological group	Slight encouragement of closer coordination	No significant change	No significant change
Mechanical group	No significant change	No significant change	Slight increase
Information group	No significant change	Increase	No significant change
Management techniques group	No significant change	No significant change	Slight-to-moderate

SOURCE: Office of Technology Assessment.

Table 6-3.—Potential Impact of Technology Groups on Capital and Labor at the Producer Level, Assuming Adoption, by Animal and Plant

Area and technology group	Potential additional change induced by technology group by year 2000: Capital	Labor	Capital/labor ratio
Animal:			
Biological group	Slight decrease (<5%)	Slight decrease (<5%)	No significant change
Mechanical group	Moderate increase (5-10%)	Slight decrease (<5%)	Moderate increase (5-10%)
Information group	Moderate increase (5-10%)	Slight increase (<5%)	Moderate increase (5-10%)
Plant:			
Biological group	No significant change	Slight increase (<5%)	No significant change
Mechanical group	Moderate increase (5-10%)	Slight increase (<5%)	Moderate increase (5-10%)
Information group	Moderate increase (5-10%)	Slight increase (<5%)	Moderate increase (5-10%)
Management techniques group	Slight increase (<5%)	Moderate increase (5-10%)	No significant change

SOURCE: Office of Technology Assessment.

duction process. This is also the potential impact of the plant disease and nematode control technologies.

The capital/labor ratio is not expected to change significantly as a result of the technologies in the biological group for crops. The decrease in labor anticipated from some technologies within the biological group is not expected to be important enough to change the capital/labor ratio significantly in the long run.

Mechanical Group

The mechanical group for animals is viewed primarily as housing and lighting control for animals that might influence breeding or growth. The other technologies within this group are viewed primarily as improvements in mechanical methods for crop residue or animal waste processing.

Almost by definition, this technology group is composed of technologies that require capital equipment expenditures if adopted, resulting in a moderate increase (5 to 10 percent) in the amount of capital. Another expected consequence of these mechanical technologies is a slight decrease in labor, primarily because of the potential of these technologies to reduce stress on livestock, which in turn may reduce management input. The decrease in labor for livestock operations, however, is anticipated to be less than 5 percent in the long run. It is further expected that the moderate increase in capital and the slight decrease in labor will increase the capital/labor ratio moderately (5 to 10 percent) for livestock producers.

The adoption of the mechanical technologies group for plant agriculture, as with this technology group for animal production, is expected to result in a moderate increase in capital input. The cost of engines is expected to rise, and engines are expected to have a higher horsepower for the same size. The major technologies in this group are expected to be capital-intensive.

The potential for labor reduction from the technologies for crops is expected to be similar to that for livestock. The expectation is for a slight decrease (<5 percent) in labor input attributable to this technology group. The capital/labor ratio is expected to increase moderately (5 to 10 percent), owing primarily to increased engine costs and small labor reductions. The expectation for the capital/labor ratio for crop production is similar to that for livestock production.

Information Group

The technologies in the information group are viewed as capital-intensive and are expected to have impacts that are similar for both crop and livestock producers. Actuators, sensors, and controllers would require additional capital expenditures for production units such as feedlots, dairy barns, or crop fields. The consensus is that these technologies would require a moderate increase (5 to 10 percent) in capital for either crop or livestock operations.

Because the total amount of data and information available to managers would increase as a result of this technology, a slight increase (<5 percent) in managerial time is expected. The managerial input results from an increase in information that would be generated by the technology. Such information would require more analysis time from managers. This situation is similar for either livestock or crop operations.

The capital/labor ratio is viewed as increasing moderately (5 to 10 percent) as the net result of these technologies. No significant differences are anticipated between livestock and crop production as a result of information technologies. The capital/labor ratio is expected to increase primarily from an increase in capital equipment items used in production of either crops or livestock.

Management Techniques Group

The management techniques group represents various management regimes useful in crop production.

The organic farming technology within this group is viewed essentially as a substitution of mechanical factor inputs (more cultivation) for chemical factor inputs (such as fertilizer or insecticides). This substitution would lead to some capital expenditure decreases and some labor expenditure increases in crop production in the long run.

Other technologies in this group are viewed as relatively more capital-intensive—such as soil erosion, tillage, and general land management. As a result, the net impact of this technology category on capital used in crop production is a slight increase (<5 percent).

Many of the technologies within this group are expected to require additional management input. Almost by definition, the adoption of technologies in this group is expected to demand greater management in the production process. As a result, labor (encompassing management) is expected to increase moderately (5 to 10 percent).

The capital/labor ratio attributable to this technology group is not expected to change significantly. The increase in capital and the increase in labor are expected to be sufficiently similar that no significant change is induced in the capital/labor ratio for crop production in the long term.

Vertical Coordination and Control

The consideration of vertical coordination and control is that if technology induces tighter vertical coordination by an integrator, it may simultaneously induce a shift in control over production from the farmer to the integrator. This, in turn, could affect the number and size of farms in the long run. The poultry industry is an example of a commodity marketing channel that exhibits relatively tight vertical coordination, loss of producer control, and a consequent shift to fewer but larger production units.

In general, the emerging technologies are expected to allow more control over end-product characteristics. Examples include less fat per unit of lean meat in animals or a specific color characteristic in corn, with the implication that more homogeneity within a type of product may result but that more end products will have engineered characteristics. This situation entails some shift away from sorting or grading as a way to achieve greater homogeneity and a shift toward more control over the production process.

An anticipated economic consequence of this increased control over production practices will be in the area of contracting. Contracting allows husbandry practices or cultural practices to be monitored and controlled closely during the production process, resulting in products that adhere to uniform specifications. Controlled diversity would result from this arrangement. That is, greater process control would lead to uniform product differentiation.

Biological Group

The biological group of technologies is expected to encourage closer coordination in livestock production compared with the situation in 1984. The technologies in this group would encourage greater process control, which would be manifested in more contracted livestock production. One example is swine producers who contract with meat packers to produce pork of uniform specifications and are paid for their labor and facilities at a predetermined fixed rate.

Another example is the potential from these technologies for modifying milk at the cow rather than at the processing plant. This technology group holds promise for producing more highly unsaturated fats in milk. If such technology is adopted, it would require close coordination at producer/first handler markets and additional process control at the production level.

The expectation for this technology group is that it will encourage closer coordination in crop production, as well. However, even though the direction of impact was viewed similarly between livestock and crop production, the magnitude is expected to be relatively less for crops compared with livestock. Part of the reason is that relatively more process control in livestock than in crops is expected from adopting the technologies.

Mechanical Group

No significant change in vertical coordination is expected from the technologies in the mechanical group for either crop or livestock production. The technology for this group is essentially embodied in capital equipment items such as tractors and other machinery. As a consequence, the adoption and use of the technology is a decision made by individual production units, with no real implication for vertical coordination or control.

Information Group

The technologies within the information group are not expected to have any significant impact on vertical coordination or control for either crop or livestock production. The situation for this technology group is similar to that for the mechanical group. Adoption and use of these technologies are decisions of individual farm and ranch managers.

One possible impact from the information technologies on vertical coordination is the potential for the technology to encourage more open markets for commodities rather than contractual arrangements. The technology is viewed as having some potential for coordination across markets without integration. This impact would be attributable to better communication of buyers' needs to production-level managers. This potential is slightly more important for livestock production than for crops.

Management Techniques Group

The technologies in this group are expected to be neutral in impact on vertical coordination and control, again mostly due to individual managers' decisions on adoption and use. Also, this technology group primarily reflects production decisions that are more likely to change quantity of output rather than quality of output.

Market Access

Market access as a structural element of agricultural marketing channels refers to the ability of sellers or potential sellers to gain access to buyers or potential buyers. The extent to which producers have a number of alternative buyers or marketing arrangements for their commodities is the essence of market access. If alternatives are few, market access is low. Foreclosure to participation is not necessary for market access to be low, but it may be one reason for lack of market access.

Biological Group

The biological group is expected to reduce market access slightly for livestock producers in the long run. This impact is expected because the biological technologies will allow targeting of certain product characteristics to specialized end-use markets, narrowing the range of alternatives for producers adopting the technology. Thus foreclosure to other market segments is one impact expected from the technology.

The impact of the biological group on market access for crop production is expected to be neutral. Even though some potential for market segmentation by end-use characteristics is possible from this technology group, no significant change is expected.

With today's technology of production, there are a number of possibilities for producing specialized crops or for sorting commodities for particular end-use markets. One example is sorting and grading soybeans on the basis of oil yield. For storable commodities, especially, the sorting may well occur after production rather than through exercising process control during production. To the extent this happens, the technology group would have a negligible impact on market access over time.

Mechanical Group

Mechanical group technologies are not expected to have any significant long-term impact for either crop or livestock production. The technologies in this group are viewed as neutral on market access for reasons similar to those for vertical coordination and control.

Information Group

Some differences are expected between plant and animal agriculture from information group technologies. The direction of impact from this technology group is the same for both areas—to increase market access. The magnitude, however, is expected to be relatively more for crop producers.

This technology group encompasses increased information available to managers for both production decisions and marketing decisions. The marketing information component is expected to be important to all farm managers, but relatively more important to crop producers. As marketing information increases among buyers and sellers, improvement in market access is expected. If market information is asymmetrically held by buyers and sellers, the technology should result in more equality among buyers and sellers. The potential significance of this improvement is expected to be slightly greater for crops than for livestock.

Management Techniques Group

No significant change in market access is expected as a result of the technologies within the management techniques group for crops. Again, reasons are similar to those provided under vertical coordination and control.

Barriers to Entry

A variety of barriers restrict the ability of new firms to enter an industry. For example, use of specialized capital-intensive and managerially sophisticated technologies for production within an industry can represent a barrier to new entrants in the long run. Unequal access to information or significant amounts of proprietary information within an industry are also conventionally regarded as discouraging to new entrants.

Biological Group

No significant impact on barriers to entry is expected from the biological technology group for either crop or livestock production. This technology group may increase the level of sophistication and/or specialized knowledge necessary for production. However, the expertise necessary would be available to nearly all production units. This expertise is seen as being available through firms that specialize in providing the necessary expertise that may be required for successful adoption and use of the biological technologies. Thus, impact on barriers to entry was viewed as negligible.

Mechanical Group

The impact of mechanical technologies on barriers to entry in the long term is expected to be a slight increase for crop production but no significant increase for livestock production. This difference is attributable to the expected differences in the relative importance of mechanical technologies in the two areas.

Mechanical technologies for crops are viewed as relatively more important and capital-intensive per dollar of output than are mechanical technologies for livestock. The capital-intensive nature of this technology group for crops is expected to impose slightly increased barriers to entry in long-run production.

Information Group

Information group technologies have different impacts on barriers to entry for crop and animal production. This technology group is not expected to change barriers to entry significantly in crop production. However, slight-to-definite reductions in barriers to entry in livestock production are expected from information technologies.

This group holds the potential for significantly increasing the amount of information on markets available to livestock producers without entailing large increases in capital expenditures for adopting the technology. In addition, monitoring and control devices are expected to be relatively more cost-effective for livestock producers than for crop producers. This implies the potential for increased productivity with, perhaps, a lower quality of management. Both the production and marketing impacts from information technologies combine to encourage new entrants into livestock production—or reduce barriers to entry in the long run.

Management Techniques Group

Some portions of the management techniques group are expected to be capital-intensive. For example, land management strategies and multiple cropping are considered to be technologies that will require expanded capital expenditures for adoption. The potential to create an additional barrier to entry from this technology group stems from the capital-intensive nature of the technology. If multiple cropping or land management practices were the norm for crop producers, they would discourage new entrants into crop production in the long run.

RELATIVE ADOPTION RATES BY SIZE

The rate of adoption by size of firm for each technology group was addressed. Adoption rates were estimated for the year 2000, assuming that firms would adopt at least one of the technologies within the group in some significant way (table 6-4). Size categories used for farms were annual sales in 1984 dollars:

1. less than $20,000;
2. $20,000 to $99,999;
3. $100,000 to 499,999; and
4. $500,000 and over.

Relative adoption rates of biological technologies for both plant and animal agriculture are expected to be significantly higher by year 2000 than that for any of the other technologies. This is especially true in the smallest size category of farm. About 40 to 50 percent of the smallest crop production units are expected to adopt at least one technology, whereas 10 to 20 percent of the smallest animal production units are expected to adopt.

This perceived difference in adoption rates between crop and livestock producers is attributable to the form in which the technology will be available for adoption. Many biological technologies available to crop producers are expected to be embodied in seeds, fertilizers, or other in-

Table 6-4.—Percent Adoption Rate of at Least One Technology Within a Technology Group by Year 2000, by Size of Farm

Area and technology group	Adoption rate range (percent), by sales category (1984 constant dollars)			
	<$20,000	$20,000-$99,999	$100,000-$499,999	>$500,000
Animal:				
Biological	10-20	30-40	60-70	80-90
Mechanical	0-10	10-20	40-50	70-80
Information	0-10	10-20	55-65	80-90
Plant:				
Biological	40-50	60-70	85-90	90-100
Mechanical	0-10	10-20	40-50	70-80
Information	0-10	15-25	55-65	75-85
Management techniques	10-20	30-40	55-65	70-80

SOURCE: Office of Technology Assessment.

put items that normally would be purchased by crop producers. Many crop producers are expected to adopt some simple techniques and information technologies as a normal practice. This accounts for the perception that these will be relatively more widely adopted.

Relative adoption by size is expected to be greatest for larger farms. Generally, 70 percent or more of the largest farms are expected to adopt some technologies from each technology group. This contrasts with only 40 percent for the second largest and about 10 percent for the smallest two categories. The economic advantages from the technologies are expected to accrue to early adopters, and a larger proportion of large farms are anticipated to be the early adopters.

Among the largest production units, biological technologies would be adopted by a relatively higher proportion of producers than would mechanical technologies. Adoption of mechanical technology entails more capital expenditures, whereas the biological technologies were anticipated to be available from a service company on a fee basis. Thus large and small firms may have these biological technologies more readily available to them.

RELATIVE EFFECTIVENESS OF POLICY IN ACHIEVING STRUCTURE

Policies Considered

The potential for achieving a particular distribution of farm size through various broad types of policy was assessed. Eight types of policy were defined regarding their potential use in changing relative size distributions of farms in the future: commodity, tax, credit, research and extension, trade, monetary, fiscal, environmental, and regulatory. The discussion focused on which of these policies might be most useful in achieving some public policy-determined farm structure distribution (number and size of firms). As before, a Delphi procedure was used to rank the relative effectiveness of each of the major policy categories. The relative effectiveness of these policy categories was then assessed by the panel (table 6-5).

Relative Effectiveness

Commodity policy and tax policy are expected to be most effective in achieving a particular size distribution in the future. That is, specific policies under these two broad types could be designed that would either significantly decrease or increase the trend toward fewer but larger production units.

Table 6-5.—Relative Effectiveness of Types of Policy in Attaining a Desired Specific Structure

Policy type by rank order, most to least effective
1. Commodity
2. Tax
3. Credit
4. Macro policy—monetary and fiscal
5. Regulatory
6. Trade
7. Research and extension
8. Environmental

SOURCE: Office of Technology Assessment.

Credit and macroeconomic policy are also expected to be effective in changing structure. Environmental policy and policies involving research and extension are expected to be the least effective in changing the size distribution of production units.

Commodity policy is viewed as the single most effective policy category for achieving some predetermined level of size and number of farm firms. For example, if a commodity policy were put in place that had a $10,000 limit on payments, the effect would be to slow or stop the move toward larger farms. Similarly, adopting a commodity policy that has no payment limitation would encourage the trend toward larger size firms. Also, even though commodity policies primarily affect dairy and major crop producers, aspects of commodity policy could be designed to encourage participation. Such items include differential support rates by size of farm. Smaller firms would have higher support levels, thereby encouraging broader participation across all farm sizes.

Tax policy was considered the second most effective policy category for changing the number and size of farm firms to some desired level. The major items discussed were investment tax credits or other tax items that encourage capital to flow into or out of agriculture for tax shelter considerations. Treatment of capital gains, conversion of capital gains, and value-added taxes were considered to be critical items that could influence size distributions.

In the credit policies category, several specific items were considered to be a direct influence on the movement over time to larger size firms. They include interest rates, availability of credit for agricultural production, capital rationing, and subsidized credit.

STRUCTURAL IMPACT ON TECHNOLOGY

The majority of this analysis is directed toward technological impacts on structure. Another area considered by ASG is the potential causality of a particular farm structure on the direction and magnitude of development and adoption of technology. The notion is that structure and technology are simultaneously related through time. Not only will technology influence structure, but structure will influence technology. As a final item, the panel considered the potential relationship from structure to the various technology groups.

The assessment of structure on technology considered each previously defined technology group separately for both animal and plant agriculture. The question addressed by ASG is the direction and magnitude of impact that a bimodal farm structure would have on the development and adoption of the technology groups. The impact is relative to the structure that currently exists. The bimodal distribution assumed to exist for this portion of the assessment was a total of 1.1 million farms, 528,000 in an annual sales category of less than $20,000; 143,000 in a category between $20,000 and $99,999; 120,000 in a category between $100,000 and $499,999; and 309,000 in a category above $500,000. This bimodal distribution has 48 percent of all farms in the smallest category and 28 percent of all farms in the largest category. Thus these two categories account for 76 percent of all farms.

As before, the Delphi technique was employed to gain a consensus on the influence of structure. Overall, the influence of structure on the development and adoption of technology is substantial (table 6-6). A bimodal structure is viewed as having moderate or large increases on development and adoption for each of the technologies for both animals and plants. This is primarily because of the proportion of total farms

Table 6-6.—Potential Directional Impact of Farm Structure on Development and Adoption of Technology Groups, by Animal and Plant

Area and technology group	Potential direction of impact induced by bimodal structure by year 2000
Animal:	
Biological group	Moderate increase
Mechanical group	Moderate to large increase
Information group	Moderate to large increase
Plant:	
Biological group	Moderate increase
Mechanical group	Moderate increase
Information group	Moderate to large increase
Management techniques	Large increase

SOURCE: Office of Technology Assessment.

in the largest size category (28 percent in the bimodal distribution compared with less than 2 percent actual in 1982).

Large farmers have relatively greater adoption rates, and a significant proportion of large farms would also encourage development of various technologies beyond what would otherwise be developed. Differences among technology groups are slight. The greatest impact of structure on development and adoption is the management techniques group for crop producers. Larger crop farms would tend to adopt technologies such as multiple cropping, soil erosion, or tillage practices; or soil-water-plant management techniques considerably quicker than would smaller farms.

SUMMARY AND CONCLUSIONS

All technology groups are expected to have considerable economic impact on farm structure by 2000. Biological technologies will have a more important impact than that of the other technology groups. A historical perspective on mechanical technologies is that they have been vitally important in shaping the livestock and plant production structure that existed in 1984. However, they are not expected to have as important an impact on future structure.

Capital and Labor

For animal production, both the information and mechanical groups are expected to increase capital moderately, with a slight decrease in capital induced by the biological group. Both biological and mechanical groups are expected to generate slight decreases in labor, with a slight increase from information technologies. No significant change is expected in relative capital and labor from the biological group. Moderate increases in capital intensity (the increase in the capital/labor ratio) are expected from both the mechanical and information groups.

For plant production, moderate increases in absolute capital are expected from both the mechanical and information groups, a slight increase from the management techniques group, and no significant change from the biological group. In terms of absolute labor and management, all but the biological group are expected to increase labor slightly to moderately. The only labor-decreasing technology is expected to be in the biological group. No significant change in the relative amounts of capital and labor are expected from the biological or management techniques groups. The mechanical and information groups are expected to induce a moderate increase in capital relative to labor. None of the technologies is expected to induce a decrease in the capital/labor ratio.

Comparison of the technology groups for plant and animal production reveals that, in general, the technologies are expected to be similar in terms of impact on capital and labor. Neither the biological nor management techniques groups are expected to induce any significant change in the capital/labor ratio for plant production. The biological group is not expected to change the capital/labor ratio, whereas all other techniques are expected to increase this ratio moderately in the long term.

Other Structural Elements

The potential direction of the marginal impact induced by the technology groups on vertical coordination and control, market access, and barriers to entry was also assessed. The biological group is expected to encourage closer vertical coordination (i.e., more contracting), with a slight reduction in market access as a consequence. This would subsequently encourage the trend toward fewer but larger farms.

In the opposite direction, the information group is expected actually to reduce barriers to entry and to increase market access without any significant change on vertical coordination or control at the producer/first handler level. The mechanical group is expected to be neutral on all structural elements analyzed.

Expected impacts on these structural elements for plant production is similar to that expected for animal production. However, the impact of the biological group on crop agriculture is expected to be less than that on animal agriculture. Less impact on vertical coordination and market access is expected from biological technologies for crops. The management techniques group, a group unique to crops, is not expected to change vertical coordination or market access, but is expected to increase barriers to entry slightly to moderately in crop production in the long run.

The potential for regional shifts from technology was seen as most likely from the livestock biological technology group, particularly as it affects beef production. Biological technologies that increase the efficiency of beef cattle forage utilization may have an important regional dimension. The potential for increasing pasture conversion of beef cattle is likely to favor shifts in production away from the higher opportunity cost of agricultural lands in the Midwest to those in the South and West.

Relative Adoption Rates by Size

Relative adoption rates of biological technologies for both plant and animal agriculture are expected to be considerably higher by year 2000 than they are for any other technology group. This is especially true for small farms.

Relative adoption rates of all technology groups are expected to be greatest for larger farms. Generally, 70 percent or more of the largest farms are expected to adopt some technologies from each technology group. This contrasts with only 40 percent for the middle-size farm units and about 10 percent for the smallest farms. The economic advantages from the technologies are expected to accrue to early adopters; a large proportion of large farms are anticipated to be early adopters.

Policies Effective in Achieving a Structure

Commodity policy and tax policy are the two broad categories of policy that are expected to be most effective in achièving a particular size distribution of farms. Specific policies under these two policy categories could be designed either to enhance or to slow the historical trend toward fewer but larger production units.

Chapter 7

Impacts of Agricultural Finance and Credit

Contents

Tables

Figures

Chapter 7

Impacts of Agricultural Finance and Credit

The severe financial stress of a large proportion of farmers and the recent regulatory and competitive changes in financial markets have combined to change forever the financial framework of farming. The farm of the future will be treated financially like any other business—it will have to demonstrate profitability before a bank will finance its operation. Managing a farm efficiently and profitably, which will necessitate keeping technologically up-to-date, will be the key to access credit.

The cost of credit, however, will be higher and more volatile. Interest on loans may be variable rather than fixed. Moreover, given the concentration in the banking industry, decisions about extending credit will more likely be made at large, centralized banking headquarters far removed from a loan applicant's farm. Loan decisions will thus be less influenced by the considerations of neighborly goodwill that frequently shaded the decisions of the more local banks.

Congress will have to consider all of these factors because the availability of capital will continue to be an important factor in agricultural production in general and in the adoption of agricultural technologies in particular. Readily available capital at reasonable rates and terms, plus technologies that aid profitability, provides a favorable environment for technology adoption. For the most part, the emerging technologies will pass the test for economic feasibility.

This chapter considers the relationships between technology adoption, financing consequences, and the structure of agriculture. The major financing focus is on the credit component of financial capital, and on how the regulatory and competitive changes in U.S. financial markets during the 1980s will influence structural change as well as the cost, availability, and other terms of credit for agricultural producers. In the following sections, some background information on capital and credit markets and institutions is reviewed, and an analytical framework is established for understanding the relationships between credit, technological change, and agricultural structure. Then, various changes in the regulatory environment affecting farm lenders are reviewed, and implications are given for technology adoption and structural change.

PRESENT FINANCIAL SITUATION IN AGRICULTURE

Before considering the long-run impacts of technological change and of financing consequences, it is important to consider the present deteriorating financial situation in agriculture. Financial conditions of many farmers and farm lenders have deteriorated significantly over the past 4 years. Large supplies and weak export demand have squeezed farm income and reduced the net worth of farmers. Many farmers face insufficient cash flow, declining asset values, problems of access to credit, and forced liquidation, foreclosure, and bankruptcy.

A substantial proportion of the U.S. farm sector is under severe financial stress, which can be measured by use of the debt-to-asset ratio. Approximately 11 percent of all farms (243,000 farms) have debt-to-asset ratios of 40 to 70 percent. These farms are "highly leveraged," tend to have serious cash shortfalls, and together owe one-third of all farm debt. Another 143,000 farms have debt-to-asset ratios above 70 percent. These "very highly leveraged" farms make up about 7 percent of all farms, but they owe almost 25 percent of all farm debt (table 7-1).

Table 7-1.—Distribution of Farms by Debt-to-Asset Ratio and Sales Class, January 1984

	Highly leveraged (debt-to-asset ratios of 40 to 70%)			Very highly leveraged (debt-to-asset ratios over 70%)		
Sales class	Percent of class	Number of farms	Percent of debt	Percent of class	Number of farms	Percent of debt
>$500,000	17.4	5,200	4.8	15.3	4,500	4.9
$250,000-$499,999	19.0	17,600	5.1	12.6	11,000	4.2
$100,000-$249,999	18.1	52,800	10.5	9.2	26,400	5.9
$50,000-$99,999	14.7	44,000	6.2	8.7	26,400	3.9
<$50,000	8.3	123,200	5.8	5.0	74,800	4.8
All farms	11.1	242,800	32.5	6.6	143,100	23.7

SOURCE: U.S. Department of Agriculture, *The Current Financial Condition of Farmers and Farm Lenders*, Economic Research Service Bulletin No. 490, March 1985.

Many short-run programs are being considered to alleviate this current financial situation. However, as discussed later in this chapter, these programs will not allow for the adjustments needed to solve the problem adequately. In chapters 8 and 9, alternative short-term policies are analyzed along with other policy changes.

IMPACTS OF MONETARY AND FISCAL POLICY

The agricultural sector is closely linked to national and international economies. Thus the public sector policies and programs that influence these economies also influence technology adoption and structural change in agriculture. The potential influences of monetary and fiscal policies on agriculture are identified in the following sections.[1]

Monetary Policy

The amount of money and credit in the economy, and its rate of change, are the primary concerns of monetary policy. The Federal Reserve System (FRS) is the primary regulatory authority that determines the direction of monetary policy in the United States. The objectives of FRS are to promote domestic economic growth, avoid excessive inflationary or recessionary pressures, maintain a sound U.S. balance of payments, and promote full employment. The simultaneous achievement of these goals is extremely difficult, and FRS is often faced with selecting which policy objective has highest priority.

[1]This section and the next are based on a paper by David A. Lins, "Overview of Capital and Credit Markets Serving Agriculture: Their Impact on Technology Adoption and Structural Change," prepared for the Office of Technology Assessment, Washington, DC, March 1985.

FRS influences the amount of money and credit in the economy through a variety of instruments. Discussion of these instruments in detail is beyond the scope of this chapter. To determine its success in controlling the amount of money and credit in the economy, FRS uses indicators, the most commonly used being: 1) interest rates, and 2) the rate of growth in the money supply (this is also used as an instrument by FRS).

For many years FRS used the level of interest rates as a key indicator of the success of monetary policy. Nominal interest rates were controlled within a fairly narrow range. However, during the 1970s the inflation rate began to rise, while interest rates were controlled by FRS actions. The net effect was a fall in real interest rates. Figure 7-1 identifies the estimated real interest rate on 3-month Treasury bills from 1962 through 1984.

From 1962 through 1972 the real interest rate was generally positive, in the range of 1 to 2 percent. From 1972 through 1979, real interest rates were usually negative, suggesting that investors in Treasury bills lost money in real terms. FRS actions to control interest rates in the face of rising inflation were primarily responsible for this outcome.

Figure 7-1.—Real Interest Rate on 3-Month Treasury Bills

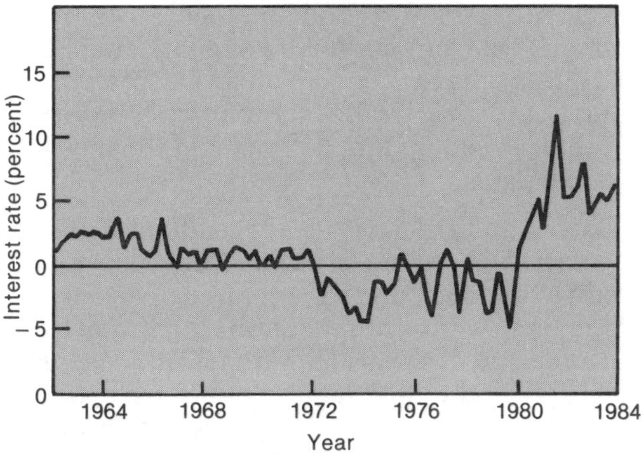

SOURCE: Office of Technology Assessment.

Recognizing that savings are strongly discouraged by negative real interest rates, FRS in 1979 shifted from a policy of controlling interest rates to a policy of controlling the rate of growth in the money supply. The result was a rapid increase in the real interest rate as well as increased variability in nominal interest rates.[2]

Since 1983, the real interest rate on 3-month Treasury bills has generally been in the range of a positive 5 to 7 percent. The level of real interest rates is likely to be a major determinant of investment in agricultural assets, particularly for nonfarm investors. With high real interest rates there is less incentive to borrow money to invest in new technologies. Consequently, actions taken in the pursuit of monetary policies have a major impact on the agricultural sector.

The strength of the U.S. dollar also has a major impact on the agricultural sector. The level of interest rates in the United States compared with those in other countries is a major determinant of the strength of the dollar. Since monetary policies have a direct influence on the level of interest rates, they also have a direct impact on the strength of the dollar. In 1984 the U.S. dollar reached a 12-year high against many major foreign currencies. A strong dollar decreases the level of agricultural exports, thereby reducing incomes of producers of export-dependent products. While the incomes of other producers may actually increase in such a situation, the overall level of income for the agricultural sector would probably decline. As incomes of agricultural producers decline, less capital is available for investment in new technologies, and credit may be used more to overcome shortfalls in income than to finance new investments or transfer resources.

Many of the emerging agricultural technologies appear to be those that will require expenditures on operating inputs such as genetically enhanced seeds, chemicals, embryo transplants, and other products normally financed with the farm operator's capital or short-term credit. Although the decisions on the purchase of these inputs is affected by the level and variability of interest rates (which in turn are influenced by monetary policy), it seems more likely that the decisions to adopt these new technologies will be more strongly influenced by the expected returns from adoption. Some technologies may be so profitable to adopt, at least in the short run, that the level of interest rates has little impact on the decision process.

Fiscal Policy

Fiscal policy involves the taxation and spending policies of the Federal Government. The objectives of fiscal policy are to carry on the functions of Government, promote economic growth and full employment, and maintain price stability. Fiscal policy instruments used to achieve these objectives include both automatic and discretionary taxation and spending alternatives.

Automatic taxation instruments include the progressive income tax structure, which raises taxes as incomes increase and decreases taxes when incomes fall, even if Congress has made no explicit changes in tax rates. One automatic spending instrument is unemployment compensation, which automatically changes Government expenditures when unemployment changes.

Discretionary items include those taxation and spending patterns that require specific congressional action to change. In the area of taxa-

[2]Interest rates not adjusted for inflation.

tion, for example, depreciation rates and investment credit change only as the result of legislative changes. Likewise, numerous spending programs require legislative action before the level of expenditure is changed. A major problem in meeting the objectives of fiscal policy is that much Government spending falls into the category of "entitlements," leaving little that legislators can do to change the total level of Government expenditures.

Federal budget deficits, the excess of Government spending over tax revenues, are frequently cited as a major determinant of interest rates. Some argue that large budget deficits create such a strong demand for credit that Government borrowing will "crowd out" the demands of the private sector for credit if the deficit is not funded by expanding the money supply. Others suggest that budget deficits occur primarily as a result of high unemployment and recessions, which reduce tax revenues and create more expenditures on income transfer programs. If true, this latter view suggests that Federal deficits have little impact on interest rates. In fact, statistical studies show a very low correlation between budget deficits and the level of interest rates.

To the extent that fiscal policies affect the level and variability of interest rates, they also affect credit availability and the adoption of new technologies in agriculture. Again, the impact on the adoption of new technologies may depend on whether the new technologies are capital-intensive. Fiscal policies that result in large budget deficits will have a more deleterious effect on capital-intensive technologies than on technologies that require little capital investment and that reduce costs of production.

CAPITAL SOURCES FOR AGRICULTURE

It is useful to separate capital used for agriculture into two broad categories—debt capital and equity capital. Debt capital is defined as funds that are borrowed and must be repaid with interest. In contrast, equity capital represents an ownership interest in the business. Net income and capital gains reflect the returns to equity capital.

As shown in table 7-2, approximately 20 percent of the total capital used in agriculture is in the form of debt capital. Debt capital as a percent of total capital in agriculture increased from about 10 percent in 1950 to 20.7 percent by 1985. To the extent that new technologies require the use of borrowed funds for adoption, lenders as well as farm operators must be convinced of the value of new, and perhaps untested, technologies. Educational efforts to acquaint lenders with new technologies will become increasingly important if such technologies are to be financed with debt capital.

Equity capital accounts for the majority of funds used in agriculture and may come from a variety of sources, including initial investment

Table 7-2.—Balance Sheet of the Farming Sector, Jan. 1, 1985

Item	(billions of dollars)
Assets:	
Physical assets:	
Real estate	$ 749.2
Non-real estate:	
Livestock and poultry	50.4
Machinery and motor vehicles	106.5
Crops stored onfarm and off-farm	38.2
Household equipment and furnishings	26.0
Financial assets:	
Deposits and currency	18.7
Savings bonds	3.7
Investments in co-ops	29.7
Total assets	$1,022.4
Claims:	
Liabilities:	
Real estate debt	$ 110.4
Non-real estate debt to:	
CCC	8.3
Others	93.0
Total liabilities	211.7
Proprietors' equity	810.7
Total claims	$1,022.4
Debt-to-asset ratio	20.7

[a]Preliminary.

SOURCE: U.S. Department of Agriculture, Economic Research Service, *Agricultural Finance: Situation and Outlook*, AFO-25, December 1984.

and retained earnings of farm owners and operators. Importantly, much of the equity capital in agriculture is the result of asset appreciation. Equity capital is also provided by investors who are not farmers but do have an ownership interest through shares of stock, partnership interests, or other forms of equity investment.

Much of the equity capital in agriculture is invested in farm real estate, although not necessarily by farm operators. For example, a significant portion of this equity capital, some 42 percent nationwide, is rented (but substantial regional variation exists). In addition, a growing amount of machinery and equipment is leased to take advantage of tax regulations. Overall, the significant amount of leasing of agricultural assets suggests that a considerable amount of equity capital in agriculture is controlled by individuals or institutions that may not be actively engaged in farming operations.

DEBT CAPITAL SOURCES FOR AGRICULTURE

American agricultural producers borrow from a wide variety of lending sources. The financial institutions that serve agriculture are in a constant state of change, in part because of changes in the regulatory environment under which they operate. Indeed, recent changes in the regulatory environment have altered the nature and operating characteristics of these financial institutions. Savings and loan associations, as well as Sears, American Express, and other nontraditional sources, are more likely now to provide financial services to farmers. The financial institutions that serve agriculture, and the changes within those institutions, are described below.

The Farm Credit System

The Farm Credit System (FCS) is a cooperative that is owned and controlled by member borrowers. The system began in 1916, when the Federal Land Banks were established to help farmers and ranchers gain access to long-term farm loans under more favorable rates and terms than were available from other sources. In 1923 the Federal Intermediate Credit Banks were formed to provide discounting services for short- and intermediate-term loans. Production Credit Associations (PCA) and the Banks for Cooperatives were started in 1933. The cooperative nature of this system is the central focus of its organization and operation.

Originally, the system was partially capitalized by the Federal Government; however, all Government capital has since been repaid. Although the system is directed by the borrowers and their elected representatives, it is supervised by the Farm Credit Administration, an independent agency in the executive branch of the Federal Government. Unlike most other private lenders serving the farm sector, FCS is restricted to making loans only to farmers, fishermen, agricultural cooperatives, and rural residents who meet eligibility standards set by law. Commercial banks and life insurance companies, in contrast, face no such restrictions.

FCS acquires funds to lend through the sale of bonds and discount notes in the national money market. The system has agency status in selling its bonds and discount notes. (Agency status has been shown to reduce the cost of issuing bonds.) In recent years, agency status for FCS has come under attack as an unfair competitive advantage, and will be discussed in a later section of this chapter.

Today, virtually all loans from FCS are on a variable interest rate. As a result, interest rate risks have been passed on to borrowers. Despite the charging of variable rates, the system has achieved a fairly stable pattern of rates. However, some farmers have experienced interest rate increases so high that anticipated profitability was not achieved. In recent years some parts of the system have begun to offer fixed-rate financing alternatives through financial leasing of machinery and 5-year, fixed-rate loans on real estate. Such fixed-rate alternatives may help risk-averse farmers finance the purchase of new technologies not previously available through variable-rate loans.

At present, there appears to be a move within the system to consolidate administrative units and to offer a broader range of financial services to farmers. These actions may have little direct impact on credit availability or on the adoption of new technologies. However, they may make the system more efficient and cost-effective, thereby reducing the cost of credit. Such a reduction in cost would likely provide some small impetus to technology adoption.

Commercial Banks

Commercial banks are a major source of both real estate and non-real estate loan funds for agriculture. Historically, they have been the largest institutional source of non-real estate farm loans. While not all commercial banks are actively involved in making long-term farm mortgage loans, most provide referral services that help farm operators obtain farm mortgage funds from other lenders.

In general, most of the loans to farmers and ranchers made by commercial banks come from small and intermediate-size banks serving a relatively small geographic area. While large banks also lend directly, they serve agriculture through correspondent services through smaller rural banks and via loans to agribusiness firms.

Each State regulates the extent to which both State-chartered and nationally chartered commercial banks can branch within a State. The alternatives include: 1) unit banking, 2) limited branching, and 3) statewide branching. Twenty-three States now allow statewide branching, 16 allow limited branching, and 11 are unit banking States. Some States also allow multibank holding companies, whereas others do not.

A fundamental change in the structure of commercial banks appears to be taking place. Many small and intermediate-size banks are being acquired by larger banks or bank holding companies. As a result, the number of banks is expected to decline and the average bank size is expected to increase. It remains to be seen how this situation will affect credit availability, technological adoption, or structural change in agriculture. However, one change appears to be the growing aggressiveness of banks in seeking farm real estate loans, suggesting that larger banking units may more aggressively seek new lending opportunities, including those for technology adoption. However, metropolitan banks that acquire smaller rural banks may be reluctant to finance new technologies if they are not familiar with agricultural lending. The movement of the decisionmaking process from the local scene to a more metropolitan center, and the need for specialists in agricultural lending, may make lender education on new technologies more important.

Insurance Companies

Several major life insurance companies have been actively engaged in farm mortgage lending for many years. Five companies (Equitable, John Hancock, Prudential, Travelers, and Metropolitan) have accounted for over 75 percent of the total farm mortgage lending by insurance companies. Insurance companies have tended to focus on real estate loans for owners of larger farms. Since these farmers may have been early adopters of new technology, insurance companies may have had a greater role in the adoption of land-intensive technologies than their market share of farm debt would indicate. However, the market share of insurance companies as a whole has diminished over time.

Insurance companies offer a variety of loan terms and financing plans. Fixed-rate loans with a relatively long amortization period, but with balloon payments after 10 or 20 years, used to be common. However, the inflationary environment of the late 1970s and early 1980s caused insurance companies to shorten substantially the period before which interest rates could be renegotiated.

Some insurance companies have also experimented with shared appreciation mortgages (SAMs). A SAM works in the following manner. In exchange for a fixed interest rate at below market rates, the lender shares in a designated portion of capital gains. At the end of a designated period, normally 5 or 10 years, the land is either sold or reappraised, with the lender's share of the gain due. The amount due the lender can be handled either as a lump sum payment or, more likely, as an increase in the loan balance.

The insurance industry has undergone substantial changes in its sources of funds and in the products and services that it offers to agriculture. Equity participations appear likely to flourish in the future, either in the form of direct investment or in the form of shared-appreciation mortgages. Thus insurance companies may become more actively engaged in equity financing than in debt financing.

Government Lending Agencies

The Federal Government provides loan funds to agriculture primarily through the Farmers Home Administration (FmHA) and the Commodity Credit Corporation (CCC). The Small Business Administration (SBA) no longer lends to farm firms.

The FmHA offers insured and guaranteed loans. Insured loans are made and serviced by FmHA personnel and represent about 80 to 90 percent of FmHA's total loan volume. Guaranteed loans are made and serviced by other lenders, but are guaranteed against default by FmHA. To be eligible for these loans, farm owners must demonstrate that they are unable to obtain adequate loan funds at reasonable terms from other lenders. As a result, FmHA is usually considered a lender of last resort.

Congress controls the extent of FmHA lending programs in two major ways: appropriations and lending authorization. Appropriations are used to cover losses and administrative expenses. By controlling the level of appropriations, Congress also controls the extent to which administrative expenses and loan losses can be incurred.

Lending authorizations specify the maximum amount that FmHA can lend out under various programs. Set annually, lending authorizations are designed to control the nature of the programs offered. For example, if Congress wishes to encourage guaranteed loan programs over insured programs, it can raise the lending authorizations for guaranteed loans while reducing the authorization for insured loans.

FmHA farm loan programs have focused on farm operators with limited resources and on those affected by disaster. The impact of these programs has probably been to slow the concentration of resource ownership and control in large farms by keeping smaller farms in agriculture. It is less clear what impact, if any, FmHA programs have had on technological adoption. However, to the extent that adoption of new technologies is based on the ability of farm operators to control larger units (e.g., by using four-wheel drive tractors), FmHA loan programs may have slowed the rate of adoption by preventing additional land from coming onto the market through foreclosure.

CCC is part of the U.S. Department of Agriculture (USDA). It provides financial assistance to farm operators through four channels: 1) deficiency payments, 2) disaster payments (although these have essentially been replaced by multiple-peril crop insurance), 3) crop loans, and 4) storage facility loans. The programs offered by CCC are part of the Government farm programs designed to improve and/or stabilize the incomes of agricultural producers.

CCC loan programs and the associated farm commodity programs have likely had a significant influence on the structure of U.S. agriculture. The stabilization and improvement of incomes generated by such programs have probably reduced risks and encouraged the adoption of new technologies.

Merchants and Dealers

The term "merchants and dealers" refers to farm suppliers of feed, seed, chemicals, fertilizer, petroleum, machinery, and equipment. These firms are an important source of non-real estate loan funds for agriculture. For operating inputs, such credit often takes the form of accounts payable. For capital inputs, credit may be extended for a period of 3 to 5 years.

Dealer credit is often viewed as a method of promoting sales. To some degree, merchant-dealer credit programs have helped foster the adoption of new technologies. This is particularly true for new technologies associated with

Figure 7-2.—Market Shares of Farm Real Estate Debt

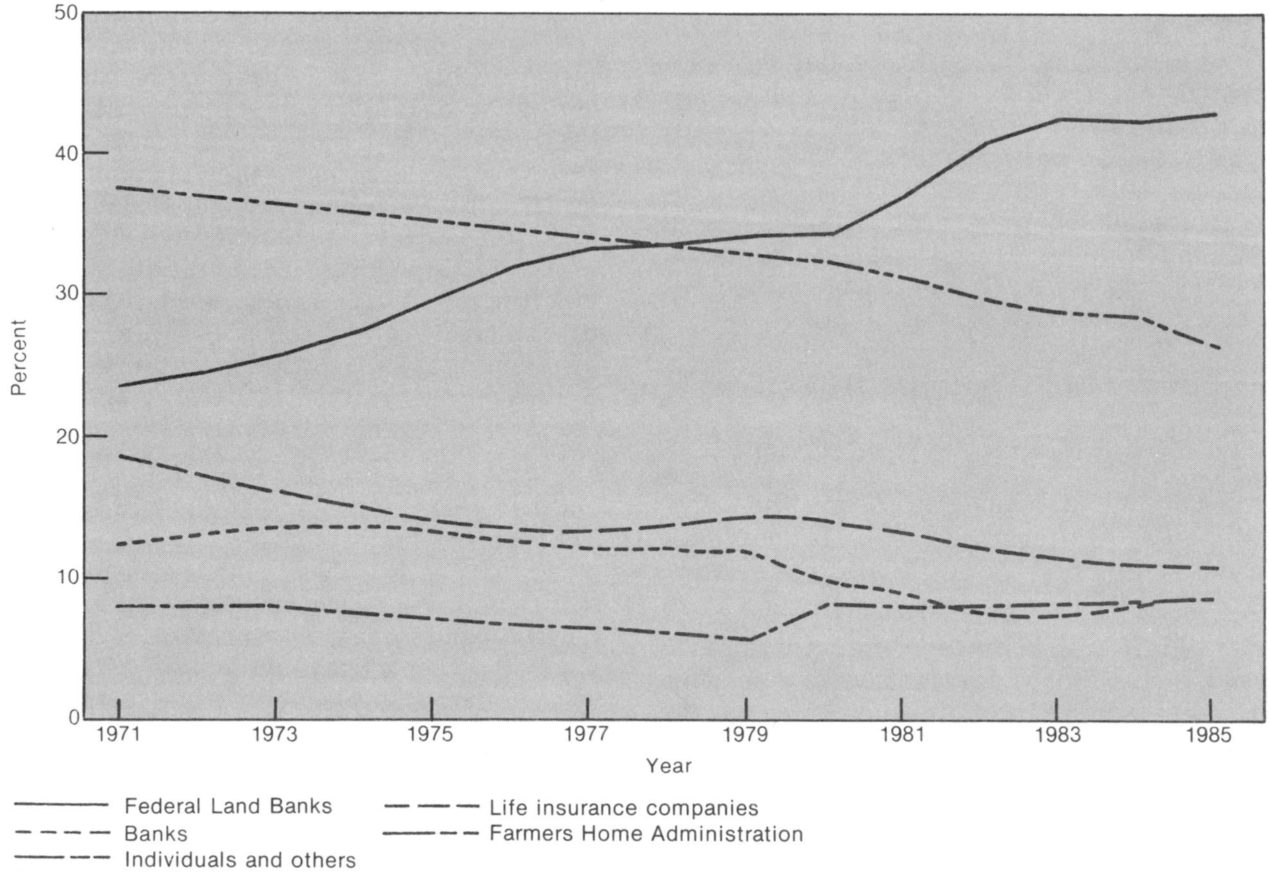

SOURCE: Office of Technology Assessment.

high-cost capital items. Dealer credit programs would appear to have less impact on the adoption of new technologies associated with operating inputs.

Market Shares of Farm Debt

The "market share of farm debt" refers to the percentage of the total volume of lending by a particular lender. It is useful to distinguish market shares of farm real estate debt from market shares of non-real estate farm debt.

Trends in the market share of farm real estate debts are shown in figure 7-2. By 1978, the Federal Land Banks had become the dominant source of farm real estate loans, surpassing those provided by individuals. In contrast, the market shares for individuals, life insurance companies, and commercial banks have decreased. The market share of farm real estate debt for FmHA has remained fairly constant.

The growing dominance of the Federal Land Bank System has implications for the future structure of agriculture. Policies adopted by the system will tend to dictate how transfers of land ownership will be financed. However, the changing market shares of farm real estate debt would appear to have little direct impact on the adoption of new technologies.

Market shares of non-real estate farm debt are illustrated in figure 7-3. The most notable feature of this graph is the rather sharp decline in market share for commercial banks and the increase in market share for CCC and FmHA. Market shares for PCAs and others (primarily merchants and dealers) have remained fairly constant.

Figure 7-3.—Market Shares of Non-Real Estate Farm Debt

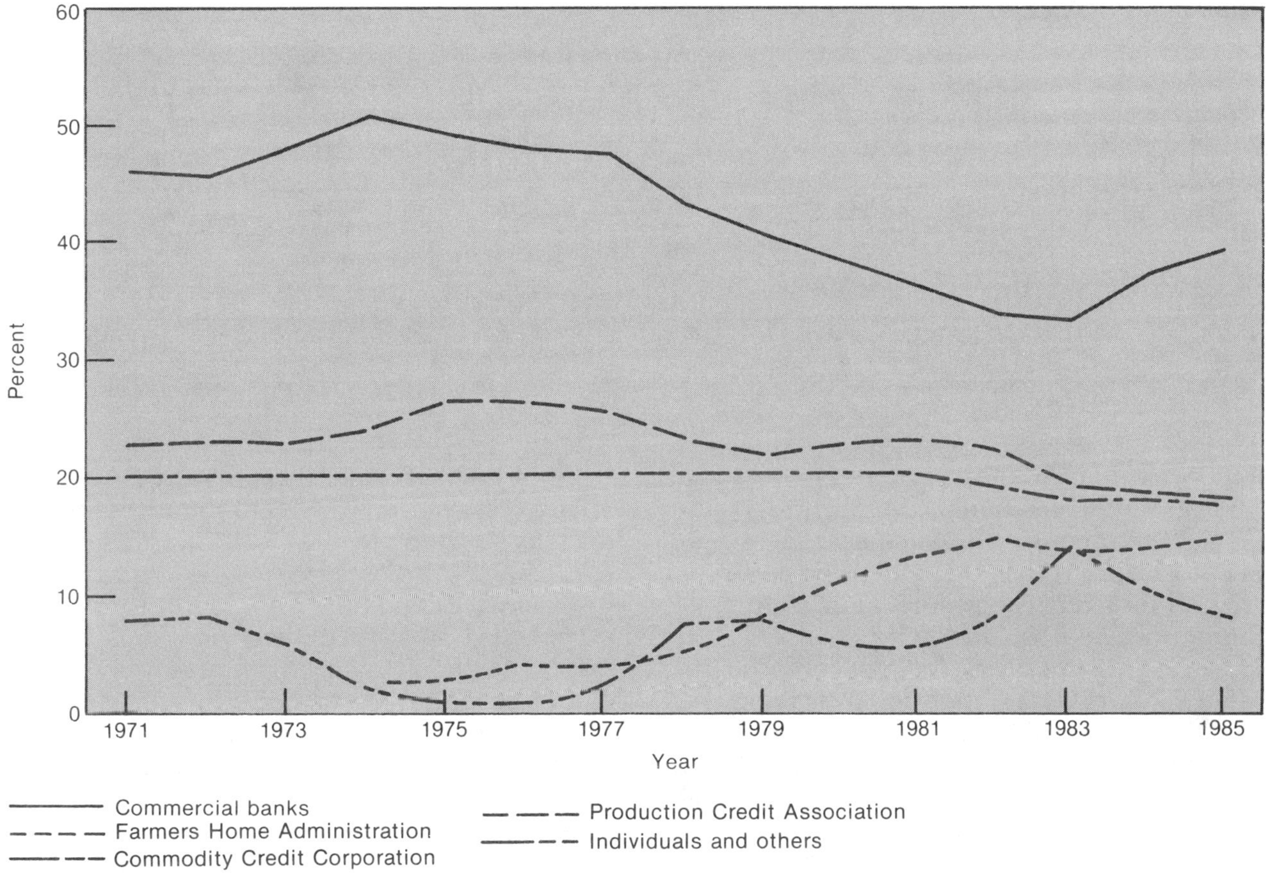

SOURCE: Office of Technology Assessment.

The increase in market shares for Government lending institutions (FmHA and CCC) reflects the expansion of Government programs to support agricultural prices and to deal with economic and natural disasters. The decline in market shares of commercial banks resulted from many factors, including the problems that some commercial banks have had in attracting deposits. Legislative constraints periodically prevented commercial banks from offering competitive rates to savers. These constraints are being phased out by the Monetary Control Act of 1980.

EQUITY CAPITAL SOURCES FOR AGRICULTURE

Equity capital, the dominant form of capital used by U.S. agriculture, arises from three primary sources: 1) net farm income and unrealized capital gains, 2) off-farm income, and 3) the infusion of new investment capital from persons or institutions not actively engaged in agriculture. Differences in the relative importance and access to these forms of equity capital have a direct impact on the availability of debt capital and on the structure of U.S. agriculture. Each of the three sources of equity capital is described in more detail below.

Net Farm Income and Capital Gains

Table 7-3 identifies the USDA estimate of net farm income and capital gains achieved by the farm sector since 1940. Net income can be bro-

Table 7-3.—Income and Capital Gain Returns for the Farming Sector

Year	Imputed return to operator's labor and management (percent)	Residual income to equity (billions of dollars)	Real capital gains[a] (billions of dollars)	Return as a percentage of equity value (percent): From residual income	From real capital gains	Total
1940-49 average	8.25%	$ 4.32	$ 1.67	6.90%	3.36%	10.26%
1950-59 average	9.12	4.24	2.87	3.57	2.38	5.95
1960-69 average	7.17	6.06	5.19	3.44	3.06	6.50
1970-79 average	9.00	21.55	30.05	4.46	7.86	12.32
1980	10.80	9.80	−4.50	1.30	−0.60	0.70
1981	12.50	17.00	−75.30	2.10	−9.20	−7.10
1982	10.70	9.90	−61.30	1.30	−7.00	−6.50
1983	9.40	4.00	−22.90	0.50	−3.10	−2.60

[a]The change in the real value of physical farm assets (after subtraction of real net investment) plus the changes in the real values of currency, demand deposits, and farm debts.

SOURCE: U.S. Department of Agriculture, Economic Research Service, *Economic Indicators of the Farm Sector: Income and Balance Sheet Statistics*, ECIF 2-2, September 1984.

ken down into two components—1) returns to operators' labor and management, and 2) residual income to equity in farm assets. The decades of the 1940s and 1950s marked a period in which returns to operators' labor and management nearly doubled the income return to equity. During the 1960s, the two components of net income were about equal. During the 1970s the income return to equity exceeded the return to operator labor and management. This pattern reflects the significant substitution of capital for labor that occurred over the last four decades.

Net income, whether in returns to operators' labor and management or in returns on equity capital, is in the form of cash. Estimates of the amount of net income retained in the farming sector are not readily available, but it does seem likely that the majority of these funds would be used for other purposes, especially for family living.

Real capital gains reflect the return to equity capital from an appreciation in asset values that is greater than the rate of inflation in the general economy. In the decades of the 1940s and 1950s, real capital gains were on average positive and were 40 to 70 percent as large as the residual income to equity. During the 1960s, real capital gains nearly equaled the residual income to equity; during the 1970s, real capital gains on average far exceeded the residual income to equity. Thus from 1940 to 1980, landowners came to expect significant real capital gains from the ownership of agricultural assets.

Since 1980, real capital gains have been negative every year. By 1984 real wealth of the sector was down by over $160 billion from what it had been in 1979. This massive reduction in the real wealth position of agriculture has had a dramatic impact on the economic and psychological attitudes toward new investment. Purchases of capital assets such as machinery have been postponed or delayed as long as possible by many operators. Land sales languish from an overabundance of parcels offered. The problems created by this massive reduction in real wealth have been most strongly felt by farm operators with heavy debt loads. While forced sales have not yet reached substantial proportions, most observers believe that a major restructuring of asset ownership could occur as a result of the economic conditions of the early and mid-1980s. In particular, land will likely be redistributed from the highly leveraged operators to those with a strong financial position and low leverages.

Income returns as a percentage of equity value were relatively high during the 1940s, but dropped to an average of under 5 percent for the next three decades. Since 1980, income returns as a percent of equity value have been extremely low. Total returns, measured as income and real capital gains, were relatively high in the 1970s, but have been negative in recent years.

Since unrealized capital gains can be monetized only through the sale of assets or by borrowing, the magnitude of such returns may have limited direct impact on the purchases of operating inputs. However, the impact on the ability to borrow and the psychological impacts of declining real asset values will likely adversely affect the ability and desire to adopt new technologies that are costly or uncertain.

Off-Farm Income

Off-farm income is a major source of income for farm firms, as indicated in chapter 4. For example, in 1983 off-farm income accounted for nearly 60 percent of the total income per farm firm. However, as farm size (measured by annual gross sales) decreases, the relative importance of off-farm income increases.

If technological adoption occurs first on the very large-scale farms (over $500,000 in gross sales), then the impact of off-farm income on technological adoption may be low, since such income is a relatively small component of total income for the largest farms. In contrast, if technological adoption occurs first on small or moderate farms, off-farm income may be an important source of income for financing technology adoption.

New Equity Investment Capital

Capital from the sales of stocks in corporations or shares of partnership interest has been a rather limited source of equity capital for agriculture. In the past, shares of partnership interest have generated significant amounts of new equity capital for large cattle feeding and poultry operations. These investments were motivated by favorable tax laws—laws that have since been changed.

At present, farmers are considerably interested in the possible infusion of new equity capital to assist financially distressed operators with large debt loads. Interest by investors, however, is rather limited. Nevertheless, significant amounts of new equity capital have been raised for investing in new technologies in agriculture. For example, much of the equity investments for embryo transplants in dairy cattle have come from nonfarm investors. Thus, while new equity capital may be a small component of the total equity capital in agriculture, it may be used to finance some of the new technologies whose risks and payoffs are expected to be high.

THE ROLE OF CREDIT

Credit Policy and Structural Issues in Agriculture

The impacts of credit and credit policies on structural change in agriculture can be viewed from two vantage points: 1) a broad view of the farm production sector as an aggregate unit structured to achieve desired social objectives, and 2) an intrasector view that considers changes in the sector's makeup over time.[3] Viewed from the broad vantage point, credit arrangements and policies of the past are believed to have contributed to maintaining a structure of the farm production sector that, compared with many other sectors, largely has a small-scale, pluralistic, noncorporate, competitive market organization of ownership, management, and control. These characteristics presumably have been consistent with social objectives for agriculture, including low-cost, abundant, and reliable supplies of food and fiber, although empirical verification of this situation needs further testing. Some examples of these past credit arrangements and policies include:

- creation and evolution of the Cooperative Farm Credit System;
- maintenance of a dual system of commercial banking (basically, large and small banks) with some special provisions for agricultural financing;
- creation of government credit programs for agriculture—FmHA and CCC at the Fed-

[3]This section and the next are based on a paper by Peter J. Barry, "Regulatory and Performance Issues for Financial Institutions: Their Effects on Technology Adoption and Structural Change in Agriculture," prepared for the Office of Technology Assessment, Washington, DC, December 1984.

eral level and various credit programs at the State level;
- actions and policies taken by Federal and State governments to discourage or impede the flow of outside equity capital into the agricultural production sector;
- laws to protect the interests of tenants and thus encourage the traditional leasing arrangements for farm land; and
- encouragement of seller financing of farm land that keeps the financing function within local communities.

Given this broad view of the farm production sector, credit arrangements and policies have also facilitated various structural changes within the sector. Included, among others, are the mechanization and modernization of farm units, greater capital intensity, growth in farm size (and reductions in farm numbers), greater leverage from debt and leasing, and greater market coordination. Credit also plays an important risk-bearing role through providing the liquidity to cope with risk and through the various alternatives in debt management for restructuring and rescheduling farmers' financial obligations. However, special credit programs and concessionary terms are also believed to have highly sensitive, adverse effects on resource allocation, asset values, and risk positions. That is, these credit programs may, on occasion, tend to overfacilitate changes or to hamper long-term resource adjustment.

In general, then, a reasonable consensus of past studies and observations on the relationships between credit and structural change in agriculture is twofold (U.S. Department of Agriculture, 1980; Farm Credit Administration, 1980; Lins and Barry, 1980; Lee, et al., 1980). First, the availability of credit is a necessary condition for undertaking the investments and other activities (including adoption of new technology) that result in structural change. However, credit availability is not a sufficient condition—basic economic incentives are needed as well. Second, credit and credit policies can be facilitating instruments for structural change in agriculture, although not very effective ones, since the unintended negative effects may outweigh the intended positive effects. That is, the special credit policies may sometimes result in too much use of credit, too much risk-taking, higher land values, and slower mobility of some resources.

Credit Determinants and Relationships to New Technologies

The availability and cost of credit to agricultural producers are based on a number of determinants that may change over time and that may differ among financial institutions. Some credit determinants originate in the financial markets. These include both macro and micro conditions. Macro conditions reflect monetary and fiscal policies, inflation, savings rates, and other forces, both domestic and international, affecting interest rates, money supplies, and credit use. Micro conditions reflect the responses of both financial institutions and borrowers to changes in the market and regulatory environment.

Other credit determinants originate in agriculture through the macro effects of supply-demand conditions for commodities and resources, and through factors affecting the creditworthiness of individual borrowers. Creditworthiness is based on those fundamental factors that lenders use to evaluate a borrower's ability to meet his financial obligations.

In this chapter, the primary focus is on the relationships between credit terms from the major farm lending institutions and changes over time in these institutions' regulatory and competitive environments. For the above credit determinants, this focus primarily involves the micro conditions of these lending institutions, although the interrelationships with various macro financial forces are important, too. Also important are the impacts of new technologies on the creditworthiness of farm units with different structural characteristics, and the implications for the cost and availability of credit.

From a creditworthiness standpoint, most of the new technologies projected by OTA for adoption involve refinements in production processes without requiring large capital outlays by agricultural producers. This is especially true for technologies involving genetic engineering, diseases and pests, and fertility and nutrition. Such technologies should largely be embodied in the operating inputs used by crop and livestock operations to carry out production. Credit for acquiring most of these technologies will probably come from short-term operating loans, with loan repayment occurring from the sale of products being produced. For crops, the sales may occur at harvest or over a marketing year as stored inventories are liquidated. For livestock the sale time is based on the market readiness of the animals and on the byproducts involved. Most of these new technologies should be financed by short-term, self-liquidating loans that are highly preferred by most lenders. Moreover, as normally occurs in production loans, lenders will take security interests in the products being produced (e.g., growing crops, marketing contracts, feeder livestock, milk products) in order to provide the necessary loan collateral. In many cases, security interests will also be taken in the borrower's capital assets (e.g., machinery, facilities, breeding livestock) in order to provide a broader collateral base, especially when the same lender finances both operating inputs and intermediate-term capital assets.

For those new technologies involving fixed capital, as with systems for environmental control, irrigation and water management, performance monitoring, and information and communication, the capital outlays will be greater and the economic payoff periods will be longer. Credit arrangements for these technologies will likely involve intermediate or longer term loans, with security interests in the capital assets serving as loan collateral. Important considerations are the length of payoff period for these technologies and the time pattern of returns. Some of the assets may be highly specialized, with low liquidity and high transactions costs in the event liquidation must occur. Others will be more easily transportable, with lower transactions costs and thus greater liquidity.

In general, then, according to evaluations of creditworthiness based on repayment expectations and collateral alone, the new technologies should not encounter financing limits or other loan terms that differ much from those for other types of agricultural assets. However, a more important lender response will likely involve the management skills and risks associated with using these technologies. Clearly, more complex technological systems will demand greater skills in both management and labor for their effective use. In some cases, considerable investments in human capital by agricultural producers may be needed to provide the necessary management skills. Complementary investments in computers and information processing technologies may also accompany the adoption and use of new technologies. Both of these factors may involve financial requirements and thus influence the borrower-lender relationship.

From the risk standpoint, considerable uncertainty may arise about the proper use and payoffs from these technologies, especially in the early stages of adoption and use. Moreover, the market values of some new technologies could drop rapidly, owing to obsolescence or to lower production costs as sales increase. Thus lenders will place greater emphasis in credit evaluations on the ability of agricultural producers to demonstrate rigorously that they have the necessary resources and skills in management and labor to use the new technologies effectively, and that the risks are not excessive. Moreover, lenders themselves must be able to understand the new technology and to communicate clearly with borrowers about its adoption, use, and financial consequences. These features will likely favor those lending institutions that have the size, expertise, funding capacity, and other characteristics to make a substantial commitment to agricultural finance.

REGULATORY ISSUES

Regulatory and Performance Issues Affecting Depository Institutions

During the 1980s virtually all of the major farm lenders have experienced significant changes in their competitive environment, owing to the combined effects of numerous factors. Among these factors are the following: 1) the high, volatile inflation rates of the 1970s and early 1980s and the related pressures on interest rates; 2) the strong growth in competition for funds and financial services from new entrants to the financial services industry (Sears, Merrill Lynch, J.C. Penney, money market mutual funds, and others); 3) the new technology in financial markets, involving electronic transfers of funds and cash management services; 4) the financial stresses affecting many borrowers; and 5) the regulatory changes affecting financial institutions, with heavy emphasis on deregulation. The regulatory changes are considered in the following sections.

At the beginning of the 1980s, the major areas of regulatory change affecting commercial banks and other depository institutions involved four areas: 1) the decontrol of interest rate ceilings on deposits and loans, 2) controls on ownership forms and geographic scope—the branching and holding company issues, 3) the range of products and services these institutions can offer, and 4) the adoption of uniform Federal Reserve requirements for all depository institutions. The major pieces of legislation enacted by 1984 included the Depository Institutions Deregulation and Monetary Control Act of 1980 and the Garn-St. Germain Depository Institutions Act of 1982. These acts focused primarily on the decontrol of interest rates, changes in reserve requirements, and aid for ailing thrift institutions. In addition, several bills under consideration by the House and Senate in late 1985 could affect the range of products and geographic liberalization.

Interest Rate Regulation

The deregulation of interest rate ceilings on bank deposits—called Regulation Q—was largely complete by 1984. It has made the pricing environment more homogeneous among depository institutions and has greatly reduced the historic insulation of rural banking markets from national and even international forces. The levels and volatilities of banks' costs of funds have increased, and virtually all of the funding sources for banks have become rate-sensitive. In response, banks of all types and sizes have adopted more market-oriented pricing policies for loans, funds acquisition, and services, and have moved toward improved methods of managing assests and liabilities. Greater emphasis has been placed on the use of such techniques as floating rates, risk assessment and pricing, spread and gap management, matching maturities, interest rate hedging, cost accounting, loan documentation, and market analysis. The traditional loan-deposit relationship at the customer level is changing too, with more emphasis on revenue generation from borrowers rather than reliance on deposit balances and related lending terms. Most of these new banking practices were initially undertaken by larger banks and holding company systems, although their use by smaller banks has increased as well.

In the early stages of interest rate deregulation, most small banks were able to maintain strong profit performance. Banking data indicate, for example, that the average annual after-tax rate of return on equity capital for about 4,300 "agricultural" banks (banks with ratios of farm loans to total loans of 0.25 or above) was 14 percent for the 1970s. This figure climbed to 16 percent in 1980 and then declined, falling to 11 percent in 1983. Most of the decline appears attributable to higher loan losses, including those on farm loans, rather than on narrower margins between loan rates and cost of funds.

But the full story is probably not yet available on banks' profitability responses to both financial stress in agriculture and financial deregulation. These two phenomena may be closely related, since banks have responded to higher, more volatile costs of funds by passing risks on to borrowers through floating loan rates and other loan repricing methods. This in turn has caused greater financial distress for many borrowers, which then reverts to the lender through higher loan risks, more delinquencies, and greater loan losses. Moreover, the bank's practice of responding to these credit problems by spreading the increased lending risk to other borrowers through higher risk premiums in loan rates has likely widened the incidence of credit problems in agriculture. This, of course, reflects the strong market power of most banks in local credit markets. However, it also means that the profit position and lending capacity of many agricultural banks could deteriorate further in the future as lagged responses occur to farmers' stress positions, and as the competitive pressures of financial deregulation become more intense.

An offsetting factor to these interest rate conditions for banks and borrowers is that interest rate deregulation has relieved the disintermediation pressures of the past and reduced the likelihood of periodic credit crunches in which the bank's availability of loan funds is dried up. Thus the past risk of swings in credit availability, and the attendant liquidity problems for banks and borrowers, has shifted strongly to swings in interest rates. This in turn gives clearer signals about changes in financial markets and improved the efficiency of financial markets.

Banks, Products, and Services

The second line of deregulation is the focus on possible changes in the authority of banks and other depository institutions to offer various products and services. Many banks are seeking greater authorities to offer insurance, real estate brokerage, securities underwriting, equity participations, and other nonlending activities. In addition, some banks are becoming more active in adopting, using, and merchandising information processing activities that meet their own needs for information (e.g., credit evaluations), while offering information services to customers (e.g., accounting systems). The products and services area will receive careful scrutiny and much debate in the policy arena. Nonetheless, additional liberalization of banking powers seems likely, given the thrust of competitive market forces. The effects on rural credit may not appear significant, although indirect effects may occur if new banking products have favorable profit prospects relative to lending.

Geographic Structure Issues

The third major line of regulatory change involves the geographic scope of banking. A longstanding U.S. philosophy has been to let individual States determine branching and holding company activities within their boundaries. Various laws have prohibited national branching, given State branching authority to each State, and prevented bank holding companies from crossing State lines unless agreed to by the States involved. The result has been a diverse set of State limitations on branching and holding companies.

Considerable attention has focused on liberalizing these geographic restrictions. But except for savings and loan associations and other thrift institutions, Congress began to address these issues only in 1984. The approach in the recent past mostly involved letting individual States initiate geographic liberalization using reciprocal authorities granted in existing legislation. In addition, greater discretionary relaxation by the various regulatory agencies has occurred. This approach essentially allows the drift of market forces to work, creating a climate in which many banks and banking systems have exhibited considerable aggressiveness. Examples of these movements have included the development of regional banking markets, especially among States in the Northeast and Southeast, the creation of nonbank banks (banks that do not simultaneously make commercial loans and take deposits), and the rapid expansion of multibank holding companies in States that have eased restrictions on these activities.

Moderate deregulation should affect smaller institutions more heavily than larger ones; thus, the number of banking entities in the United States should decline significantly—perhaps by one-third by the mid-1990s. However, public pressures will likely continue to provide various types of protection for smaller community banks that have been so prominent in States that have prohibited branch banking. Moreover, the financial stresses being faced in many unit banking States during the 1980s may accelerate the trend toward reciprocal banking agreements between States in order to broaden the market for failed and stressed banks.

The surviving banks will be higher performing community banks that are well managed, well capitalized, and strongly localized in their services. They will serve portions of the financial markets that are not well suited to the scale and technology of the larger banking systems. These banks will give considerable attention to the competitive pricing of products and services and to market segmentation, including specialization in activities like agricultural lending.

In general, geographic liberalization should bring greater competition in all phases of banking. This will put downward pressure on bank earnings, but will contribute positively to the availability, cost, and usefulness of financial services for customers. Banks may take on greater risks but have greater risk-carrying capacity through increased diversity in loan portfolios, larger resource bases, greater depth and breadth in management, and the discipline exerted from market factors rather than from regulations. For agricultural finance, geographic liberalization should enhance the availability of credit services, although more along the lines of commercial lending procedures for commercial-scale farmers and consumer lending procedures for small, part-time farmers.

A continued swing will occur toward greater financing from larger, more sophisticated banking systems, with these larger systems seeking the business of larger farm units and agribusinesses. Smaller, independent banks with strongly localized customer orientations will make substantial use of funding and service relationships with larger banking systems. This arrangement will be similar to the correspondent arrangements of the past, although the correspondent institutions themselves will be operating in larger markets. In the near term some banks may seek to develop funding and loan participation arrangements further with various units of the Farm Credit System, although over the long term, bankers prefer a reliable, cost-effective source of nonlocal funds within the banking industry. The funding mechanism provided by MASI, Inc. (a division of Mid-America Banking Service Co., MABSCO) is a step in this direction. This mechanism will allow participating banks in more than a dozen States to discount acceptable farm loans with a funding source in the national-international financial markets. This future funding should also include the ability to make long-term real estate loans in a fashion that will not jeopardize bank liquidity or increase interest rate risks.

In light of these developments, the location of credit control and loan decisions may continue to shift away from the local rural community; however, the availability of experienced, well-trained farm lenders in rural areas should maintain an emphasis on local servicing of farm loans while still fostering greater uniformity in loan documentation, risk assessment, and other lending practices. This standardization should benefit both the financial institutions and farm borrowers.

Regulatory and Performance Issues Affecting the Farm Credit System

The major legislative authority of FCS is the Farm Credit Act of 1971 (as amended). In general, the system's legislative authority defines its mission as one of providing appropriate credit and related services to eligible, creditworthy agricultural borrowers throughout the United States during all phases of the economic cycle in order to improve their income positions and overall well-being. FCS is specialized in financing agriculture. Thus local associations and individual districts are vulnerable to the problems affecting their agricultural borrowers. Moreover, because the system is a cooperative organization, much of its equity capital is owned by farmers who in turn financed this equity contribution with funds borrowed from the system. However, a number of factors at the systemwide level help counter the risks associated with this mandated specialization: 1) the system's national structure of full-service agricultural lending; 2) diversification of loans across borrowers, associations, districts, and farm types; 3) loss-sharing and participation agreements between the various banks and associations; 4) a strong financial position and excellent credit history; 5) efficient operations with low per-unit costs of funds management, loan administration, and the like; and 6) a systemwide emphasis on risk management. These characteristics, along with regulatory privileges in funding (see below), have enabled FCS to grow significantly and to become the largest farm lender in the 1980s, especially in long-term lending.

Like other lenders, FCS has been significantly affected by the financial stresses of agriculture in the early 1980s. Most indications during the early 1980s were that unless farm losses became extremely heavy and widespread, the FCS should come through the stress times in reasonably good shape. Loan volume had declined for some units, higher loss rates were occurring, some borrowers were discontinued, more associations were merging, and intra-system assistance packages were developed for some units. Moreover, the FCS had taken several actions to strengthen its liquidity and build its risk management. Some of these actions involved continued restructuring of the system's capital positions, operations, and management through greater centralization of these functions at the system, district, bank, and association levels. In general, the overall financial structure of FCS remained relatively strong through the mid-1980s and the system's capacity to sell securities in financial markets was not impaired. Nonetheless, policymakers, regulators, and others continued to maintain close surveillance of the system's performance.

Then, in the fall of 1985, the governor of FCA with subsequent agreement by the leadership of FCS concluded and announced that substantial Federal assistance could be needed in the next 18 to 24 months to keep the system solvent if farm financial conditions continued to deteriorate. After much debate, including concerns about the standing of the system's securities in the financial markets and equitable treatment for other troubled farm lenders, Federal legislation was passed that strengthened the regulatory authority of the FCA, strengthened the system's capacity for handling problem and loss loans, and essentially provided a contingent line of credit from the Federal Government if the system's own reserves proved inadequate to deal with continuing financial problems in agriculture. While further regulatory changes likely will occur in the future, these developments should enable FCS to come through the stress times in reasonably good condition.

Moreover, over the long term, FCS is clearly taking actions to perform more effectively in a more competitive, deregulated financial environment. One such action during 1983-85 has been the initiation of a significant self-study (called Project 1995) of the system's future missions and directions in all phases of its activities (agricultural financing, financial markets, government affairs, personnel, and management). Other actions have in general reflected the emergence of FCS as a vigorous commercial entity seeking to achieve high performance for its member borrowers. Among these actions have been a stronger emphasis on the development and marketing of new products and services; the continuing trend toward centralization and unification of territorial boundaries, management, service provisions, and other functions; the formalization of government affairs activities through trade association arrangements; and a moderately paced expansion of international activities.

From a policy perspective, FCS has also been caught up in the swift and significant changes in regulation and competition affecting the U.S. financial system. The effects have been less direct than on depository institutions but, over a longer term, basically involve the trade-offs between: 1) the needs by the U.S. agricultural sector for a specialized, reliable, nationally oriented credit system with special privileges in the financial markets; and 2) the trend toward greater openness in financial markets, with less emphasis on regulatory preferences in funding and mandated specialization in asset allocations.

These issues began to emerge during the debate preceding the passage of the Farm Credit Act Amendments of 1980. Much concern arose about the concept of a "level playing field" in the regulatory environment for commercial banks, FCS, and other types of lenders. Included in the debate were differences between institutions in their access to financial markets (the agency status issue), geographic restrictions, tax obligations, legal reserve requirements and lending limits, stringency of regulation and supervision, and the range of financial services and borrower clientele for these types of institutions. None of these issues affecting FCS were fully resolved in the debate on the 1980 Act, although the legislation that was finally passed did reflect responses to some of the concerns raised by commercial bankers and others.

Since 1980, much attention in policy circles has focused on the "agency status" of the securities that FCS sells in the financial markets. While FCS is privately owned and operated, the securities it sells still have some special regulatory privileges, giving rise to the "agency status" label. To some extent, agency status is a vestige of earlier times when FCS had significant Government involvement and formal backing. However, the system's securities have a set of regulatory exemptions and preferences that have continued since FCS reverted largely to a private status in the late 1960s (Lins and Barry, 1984; Barry, 1984). This status helps the system achieve a very large volume of security sales at interest costs that are just above those of the U.S. Government and below those of the largest, most creditworthy corporate issuers.

Several groups have studied the possible effects of removing agency status. While these effects are difficult to measure precisely, the general consensus is that loss of agency status would increase the interest cost on farm credit securities to the interest rate levels of high-grade corporate bonds or commercial paper. This might be an increase of 0.5 to 1 percent, or even more. In addition, the volume of marketable securities could decline significantly, since the past volume of these sales far exceeds the annual volumes of the largest corporate issuers. A contrary view, however, is that even without agency status the financial markets are efficient and deep enough and the Farm Credit securities have a favorable enough record that the entire funding needs of the system could still be met, although at higher interest rates.

The agency status issue will eventually be resolved by the political process that, in the mid-1980s, has favored continuation of agency privileges for FCS, especially in light of the financial stresses affecting agriculture. But it seems likely that attempts to remove agency status will continue, as has been the case for some of the housing agencies whose securities also have agency status. In its own self-study, FCS states that Government-sponsored agencies can probably retain agency status in some form through the mid-1990s. However, political pressures toward privatization will continue and will bring higher costs for the agencies involved, as well as perhaps greater interest in broadening their authorizations in funding methods and asset allocations as various agency attributes are diminished. Indeed, having a reliable source of funding is essential if FCS is to retain its mandate to provide credit in all regions of the United States and through all phases of the economic cycle. Thus, the agency status issue has important policy implications that affect the financial markets in general, the farm credit markets in particular, and especially the costs and availability of credit from FCS. In turn, these effects will have important implications for the structure and performance of the agricultural sector.

PUBLIC CREDIT PROGRAMS

Public credit programs currently administered through FmHA and CCC at the Federal level, and in numerous State governments as well, have long been important in achieving social objectives for the U.S. agricultural sector. These programs help channel funds to selected geographic areas and types of borrowers; they help foster a smaller scale, pluralistic structure for the farm sector; they provide financing opportunities for beginning and limited resource farmers; they provide valuable liquidity for emergency situations; and, in the case of CCC, they contribute valuable inventory financing to promote orderly marketing of farm commodities. In addition, from a policymaker's standpoint, credit programs are a popular, politically expedient policy instrument. They are relatively easy to administer; they are highly visible to constituents; they can be quickly developed for responding to ad hoc crises; and they do not directly influence commodity and resource markets, even though the secondary effects on asset value, income, and risk can be significant. Moreover, the administrative and risk-bearing costs of such programs are difficult to measure and are effectively hidden from taxpayers.

The growth in FmHA lending has been substantial since the late 1970s, especially through various emergency loan programs. This lending helped considerably in softening the impacts of high interest rates and weak farm income on some farmers and relieved commercial lenders of many problem loans. But this liberal lending may have helped worsen some farmers' financial conditions. Some observers have suggested that part of the financial stress of farmers is due to excessive public sector lending and that more credit will only worsen the conditions of highly leveraged farmers and will needlessly delay the departure of some farmers from the industry. Similar observations over a longer term perspective suggest that strong Government lending may have overfinanced the farm sector, accelerated the adoption of capital-intensive technology, shifted too much risk bearing to the Government, and capitalized the effects of easy financing terms into higher values of land and other assets.

Much concern has surfaced about the role of special credit treatment in agriculture, the proper balance between private and public sectors, which farmers are served, the level and form of subsidies, and the resulting tax burden. These are sensitive issues in the public arena. On the one hand, the stresses of the early 1980s have brought increasing pressure from farmers, farm groups, and others to provide additional public assistance to solve these problems. Yet, at the same time, the liberal, high-cost, public programs of the recent past have fostered growing dissatisfaction and closer scrutiny by nonfarm groups as well as by those farmers with stronger financial positions and less indebtedness.

In terms of regulatory change, the public programs have not, of course, experienced the same considerations of deregulation as those that affect lenders in the private sector. Nonetheless, these public programs must still operate under various regulations and practices affecting interest rates, lending limits, credit decisions, eligibility of borrowers, disaster declaration authorities, and relationships with other lenders. In general, the interest rates on public loan programs now reflect the level and frequency of changes in the Government's costs of funds. Thus interest rates on public credit follow market interest rates much more closely, and while rate levels are higher than in the past, they are still more favorable than commercial loan rates. An exception occurs in the case of various emergency loan programs in which significant concessions in interest rates may occur for the affected borrowers.

Lending limits on various loan programs in general are still set by law rather than by individual credit factors. These limits provide controls on the magnitude of appropriations, and impose an administered allocation of loan funds among eligible borrowers. The limits tend to adjust upward over time to reflect the effects of inflation and the costs of establishing and operating viable farm businesses. However, the adjustments occur at sporadic intervals with no formal indexing to other measures. A continuing dilemma in setting loan limits involves the choice between the levels of credit needed by individual borrowers to move in an orderly way toward eventual graduation to commercial financing versus the preference to spread an allocation of funds that is fixed in the short run among the greatest possible number of borrowers.

Closely related to lending limits for individual borrowers are the issues associated with the allocation of funds among various States and regions and over the various loan programs. It is not unusual for funds in some uses and locations to be fully allocated part way through a budget year so that otherwise eligible latecomers may find that loan funds are depleted. This process may then trigger the need for new appropriations, rechanneling of funds from other uses, discretionary rationing, or other responses. Thus, lending limits at the agency, program, and borrower levels may introduce considerable uncertainty about the availability of credit.

Another administrative issue in FmHA lending involves the form of credit programs—that is, the choice between direct (insured) loans and guarantees of loans made by commercial lenders. To date, nearly all FmHA lending to farmers has occurred through direct loans, even though both programs are available. In concept, direct loans and guaranteed loans have similar effects in that the bulk of the credit risk is still carried by the Government. However, the guarantee approach is considered to involve lower degrees of subsidy and to involve more formally the commerical lender in the credit decision and loan servicing. Thus loan guarantees can be a more efficient method of program design that has less disruptive effects on credit markets. Some of these possible benefits have been offset, however, by the commercial lenders' perception of the costly process of using the guarantee program, and by the greater effectiveness of direct loans in emergency situations. In response, FmHA has sought to simplify procedures for using guarantees through a "preferred lender" program that expedites the private lenders' use of the program.

Another administered change has involved the centralization of decision authority for declaring disaster conditions in various geographic areas. In the past the location of these authorities at the State level gave too much incentive to the parties involved to declare emergencies in their respective States in order to qualify for low-cost emergency loans. It is believed that centralizing this decision authority allows the allocation of emergency funds to be more objective.

FUTURE ROLE OF STATE CREDIT PROGRAMS

At the State level, a number of States have developed farm credit programs with a heavy emphasis on financing the aquisition of farmland and other capital assets by younger farmers (Lowenberg-DeBoer and Boehlje, 1983). These programs vary considerably, but tend, like FmHA, to have a set of regulations affecting borrower eligibility, loan purposes, loan limits, budget limits, interest rates, and so on. Heavy emphasis in many of these programs has been placed on lending financed by tax-exempt bonds and on various types of tax incentives affecting land purchases and leasing by young farmers. The tax-exempt bond programs appear to be less cost-effective compared with other program methods, since they essentially involve the Federal Government in sharing the State program costs. Recently, the Federal authorizations for States to offer tax-exempt bond programs have been curtailed, with further limitations anticipated for the future. In the future, the general importance of State credit programs could increase, especially if Federal credit programs are cut back. However, the scope, missions, and instruments used in these programs will likely receive careful review and revisions to assure that the programs are formulated in the best public interest of the States involved.

FUTURE ROLE OF FmHA

Central to the debate on FmHA's future role in the process of technological change is the question of adoption constraints.[4] Given the characteristics of the average FmHA borrower, it would seem that several barriers would have to be removed or diminished if this group is to be a major beneficiary of the emerging technologies. A number of options are available to FmHA if it undertakes the task of removing or reducing these barriers.

[4]This section on FmHAs role is based on a paper by David Trechter and Ronald Meekhof, "The Role of Federal Credit Assistance Programs in the Process of Technological Change," prepared for the Office of Technology Assessment, Washington, DC, August 1985.

Many of the technologies that will influence agriculture in the coming period will not require major capital investments in order to be adopted by the majority of farmers. However, even today many of FmHA's clients control too few resources to compete effectively. Some of the technologies that will be developed between now and year 2000 will only exacerbate this situation. One option for FmHA would be to change the type of clients it serves. However, it makes little political or economic sense to change the focus of FmHA to the larger, more economically viable farms.

A second option would be for FmHA to help its clients attain a more economically viable size. If FmHA increases its lending activities so that a specific subsection of the farm population can acquire a new technology and the resources that go with it, serious equity considerations are raised. Even in the best of times, the special benefits given FmHA farmers pose equity questions. When times in farming are difficult, the rumblings of farmers who cannot or have not taken advantage of FmHA loans grow louder. Selectively providing the means to acquire and use new technologies, particularly when this is accompanied by significant increases in the assets controlled by FmHA farmers, would be expected to increase the controversy surrounding the agency. In addition, providing FmHA's clients with more resources does not ensure success unless the management skills necessary to use them fully are also available.

A third option for FmHA is to alter its operations in an attempt to fill an empty market niche—the development of human capital. FmHA and other lenders are presently operated to facilitate the acquisition of physical assets. Most lenders are very reluctant to provide credit for the acquisition of human capital because payoffs are typically long-term in nature, repayment risks are substantial, and little collateral is available. The preference for financing physical capital acquisition is understandable from the individual bank's point of view but may result in suboptimal outcomes for society. For example, society might prefer that a farmer use a loan to buy training in integrated pest management techniques rather than more lethal pesticides.

FmHA could play a particularly important role in the acquisition of human capital, given the nature of most of its clients. FmHA farmers are relatively richly endowed with one resource —labor. Since it is impractical to expand its clients' base of physical capital, a fruitful role for FmHA could be in facilitating the acquisition of human capital. One means of implementing this would be to expand the training component that is attached to existing FmHA loan activities. A hallmark of early FmHA operations was a substantial farm management/advisory role for loan officers. An increased emphasis on this type of operation would entail a significant expansion of the number of personnel in FmHA, greater coordination with public advisory services such as the extension service, or increased use of private farm management firms. A second option would be to develop a loan program to finance human capital acquisition. Such loans could be used by the farmer to acquire training directly or to purchase the services of farm financial managers.

Technologies that might be especially appropriate for this loan category are those that lack congruence. Investments in human capital to learn how to use these technologies could be used by FmHA farmers to improve the management of their farms. In addition, these farmers might be capitalized by selling their expertise to other farms. Finally, these training investments would facilitate the transition out of agriculture for those who decide to leave the sector.

IMPLICATIONS FOR TECHNOLOGY ADOPTION AND STRUCTURAL CHANGE IN AGRICULTURE

The discussion in the preceding sections has indicated that the financing consequences of new technologies in agricultural production will likely depend on the relationships between three important factors: 1) the financing characteristics of the new technologies, 2) the creditworthiness of individual borrowers, and 3) the changing forces in financial markets that affect the cost and availability of financial capital. To review these factors briefly, the financing characteristics suggest that most of the new technologies should largely be financed with short- and intermediate-term loans that are part of the normal financing procedures for agricultural businesses. That is, the basic criteria of structuring loans to match loan maturities with anticipated payoff periods and to provide adequate loan security should not change in any fundamental way, although the risks associated with obsolescence and collateral values will need careful consideration. However, the technical characteristics of the technologies, together with the factors constituting the creditworthiness of individual borrowers, suggest that much greater emphasis in credit evaluations will be placed on the management capacity of the agricultural production units, on the ability of farm operators to demonstrate appropriate technical competence in using the new technologies, and on building human capital, where appropriate. In some cases—particularly for FmHA borrowers—significant investments in human capital, with related financing requirements, may accompany the adoption of new technologies. This is consistent with the more conservative responses by lenders to the agricultural stress conditions of the early 1980s. In turn, the lending institutions themselves must have sufficient technical knowledge and expertise to evaluate these management and credit factors, along with the other sources of business and financial risks in agriculture. Finally, some forms of new technology involving large investments and having long-run uncertain returns will likely rely more on equity capital for financing.

The changing regulatory and competitive forces in financial markets, including the preference for greater privatization of some credit institutions, means that the cost of borrowing for agricultural producers will likely remain higher and more volatile than in pre-1980 times and will follow market interest rates much more closely. Similarly, the continued geographic liberalization of banking and the emergence of more complex financial systems mean that the functions of marketing financial services, loan servicing, and credit decisions will become more distinct, with an increasing proportion of credit control and loan authority occurring subregionally and with regional money centers that are located away from the rural areas. This will continue to fragment and dichotomize the farm credit market so that commercial-scale agricultural borrowers are treated as part of a financial institution's commercial lending activities (although separate personnel for agricultural and commercial loans should still be prevalent) and so that smaller, part-time farmers are treated as part of consumer lending programs.

The competitive pressures on financial institutions and the risks involved will bring more emphasis on analyzing the profitability of various banking functions, including loan performance at the department level and individual customer level. Innovative lenders will strive more vigorously to differentiate their loan products and financial services, especially for more profitable borrowers, and will more precisely tailor financing programs to the specific needs of creditworthy borrowers. In turn, however, these agricultural borrowers must be highly skilled in the technical aspects of agricultural production and marketing as well as in financial accounting, management, and risk analysis as they compete for credit services.

In general, most forms of new technology in agricultural production should meet the tests of both economic and financial feasibility, although the structural characteristics of the adopting farm units will continue to evolve in response to managerial, economic, and market factors. The structural consequences of these factors are severalfold:

- a continuing push toward larger sizes of commercial-scale farm businesses, with greater skills in all aspects of business management;
- continuing evolution in the methods of entry into agriculture by young or new farmers, with greater emphasis on management skills and resource control, and less emphasis on land ownership;
- the continuing development of a marketing systems approach toward financing agriculture, with more sophisticated skills

in marketing analysis by farmers and higher degrees of coordination with commodity and resource markets;

- more formal management of financial leverage and credit by farmers, with greater diversity of funding sources by farmers and better developed markets for obtaining outside equity capital;
- further development in financial leasing and greater stability in leasing arrangements for real estate and non-real estate assets; and
- more complex business arrangements in production agriculture that accommodate various ways to package effectively debt and equity financing, leasing, management, accounting, and legal services for the farm business of the future.

Given the above consequences, FmHA clientele face severe challenges. The farmers served by FmHA have, with some notable exceptions, been drawn from the lower end of the economic spectrum. Given their resource endowments and the nature of many of the technologies that are emerging, these farmers are not the most likely adopters of new technologies, given the current institutional setting.

FmHA should consider a significant shift in how it serves this clientele. Historically, FmHA played an important role in human capital formation in agriculture. FmHA loan officers were actively involved in the management of their clients' farms, particularly the management of farm finances. Given the increasingly important role played by debt capital in agricultural finance and the volatility of agricultural markets, sound financial management of the farm business was never more important than it is today. FmHA might provide more farm financial management services.

At a more ambitious level, FmHA might consider the development of a special class of loans devoted to human capital formation. Loans used by farmers to acquire the skills necessary to take advantage of the emerging technologies that require major human capital development could have two beneficial effects: First, skills would be learned that would improve the management of these smaller farms. Given that many of these farms are at a competitive disadvantage in terms of the amount of resources they control, the management of their resources becomes of paramount importance. Second, the skills acquired by these farmers would have wide applicability in the farm and nonfarm sectors. It is possible that these skills could be sold to other farmers as a source of off-farm employment or could be used in the broader economy if the individual decided to leave the farm or seek off-farm employment. In short, human capital investments would be expected to increase the long-term economic viability of loan recipients, whether they remain in farming or make the transition to the nonfarm economy.

CHAPTER 7 REFERENCES

Barry, P.J., *Impacts of Financial Stress and Regulatory Forces on Financial Markets for Agriculture: Current Directions and Policy Issues,* Food and Agriculture Committee, National Planning Association, Washington, DC, 1984.

Farm Credit Administration, *Risk and the Structure of Agriculture,* Washington, DC, 1980.

Lee, J., Gabriel, S., and Boehlje, M., "Public Policy Toward Agricultural Credit," *Future Sources of Loanable Funds for Agricultural Banks,* Federal Reserve Bank of Kansas City, 1980.

Lins, D.A., and Barry, P.J., "Availability of Financial Capital as a Factor of Structural Changes in the U.S. Farm Production Sector," *Farm Structure,* U.S. Senate, Committee on Agriculture, Nutrition, and Forestry, Washington, DC, 1980.

Lins, D.A., and Barry, P.J., "Agency Status for the Farm Credit System," *American Journal of Agricultural Economics,* vol. 66, 1984.

Lowenberg-DeBoer, J., and Boehlje, M., "Evaluation of State Legislative Programs To Assist Beginning Farmers," *Agricultural Finance Review,* vol. 43, 1983, p. 9-20.

U.S. Department of Agriculture, Economic Research Service, *A Time to Choose,* Washington, DC, 1980.

Chapter 8

Emerging Technologies, Public Policy, and Various Size Crop Farms

Contents

Tables

Figure

Chapter 8

Emerging Technologies, Public Policy, and Various Size Crop Farms

The size, and therefore the survival, of farms is affected by several factors. Clearly, there are economies of size in many of the crop areas covered by farm policy. These economies motivate further concentration of resources. In addition, present farm policy, more than any other policy tool, makes major impacts on farm size and survival. Although very large farms can survive without these farm programs, moderate farms are very dependent on them for their survival.

Given these realities, an important question arises: What combination of policy and technology advances will encourage each size farm to grow, or at least maintain itself? To answer this question, this chapter and the next will present findings of an analysis of selected regions in the United States that represent significant agricultural production regions for dairy, corn, cotton, soybeans, rice, and wheat. The analysis is presented in two parts. First, the results of an analysis of size economies is presented for each commodity. This is followed by the findings from an analysis of the economic impacts of emerging technologies and selected public policies for crop farms of various sizes.

SIZE ECONOMIES AND COMPARATIVE ADVANTAGE IN CROP PRODUCTION IN VARIOUS AREAS OF THE UNITED STATES

A major question asked throughout this study is what is the impact of resource concentration in U.S. agriculture. Very little is known about the process of resource concentration. Willard Cochrane states that:

> . . . we are almost totally ignorant regarding the shape of the long-run planning or cost curve at very large volumes of output. Thus, we don't know whether we are working with one economic force that is so powerful that there is little or no possibility of controlling it with public policies or whether the force has largely expended itself and with some intelligent policy action we could slow the process of resource concentration to a walk (Cochrane, 1983).

The intent of this section is to provide policymakers with an overview of variation in cost of production by enterprise size and geographic location for corn, soybeans, wheat, rice, and cotton. This information should improve understanding of the process of resource concentration in American field crops.[1]

The concept of economies of size is defined as the relationship between enterprise size (measured in acres) in combination with other productive services, and the rate of output of the enterprise. Three enterprise size categories are used: very large, large, and moderate. If enterprises are arranged in increasing order by planted acreage, very large enterprises are those enterprises in the 90th percentile. Large enterprises comprise the 70th and 80th percentiles, and moderate enterprises comprise the 40th through the 60th percentiles. The evidence for size economies is based on a weighted average of measures outlined in appendix D. These measures include production cost, use of harvesting machinery, and the Herfindahl indices. In addition, a structural elasticity measure of internal size economies is applied to the percentage increase in enterprise size in acres relative to a 1-percent decrease in production costs. The geographic locations of the selected pro-

[1]Information presented in this section is based on the OTA paper "Size Economies and Comparative Advantage in the Production of Corn, Soybeans, Wheat, Rice, and Cotton in Various Areas of the United States," prepared by Stephen C. Cooke, 1985.

ducing areas for the five field crops are identified in table 8-1 and shown in figure 8-1. These production areas have been designated by the U.S. Department of Agriculture (USDA) on the basis of soil types and/or levels of rainfall.

Corn

In corn production, very large enterprises of about 1,000 acres have an 11-percent cost advantage over 250-acre, medium-size enterprises (table 8-2). The change in relative production and resource concentration between 1978 and 1982 is positive for very large corn enterprises and is increasing at a rate of about 10 percent a year. The structural elasticity measure is −15. This indicates that each 15-percent increase in enterprise acreage (between the 250- and 1,000-acre range) results in a 1-percent decrease in corn production costs. There is strong evidence to argue for the existence of increasing internal economies of size in corn production across the selected producing areas.

Table 8-1.—Homogeneous Production Areas for the Various Commodities

Description	Abbreviations and FEDS[a] designation
Corn:	
East Central Illinois	IL 300
Central Indiana	IN 101
North Central Iowa	IA 201
South Central Nebraska	NE 400
Soybeans:	
East Central Illinois	IL 300
North Central Iowa	IA 201
Mississippi Delta	MS 100
Western Ohio	OH 101
Wheat:	
Western Kansas	KS 100
Northeast Montana	MT 200
Central North Dakota	ND 200
Eastern Washington	WA 400
Rice:	
North Central California	CA 400
Texas Upper Gulf Coast	TX 100
Mississippi and Arkansas Delta	DLT 100 & 300
Northeast Arkansas	AR 200
Cotton:	
Northern Alabama	AL 600
South Central California	CA 500
Mississippi Delta	MS 100
Texas High Plains	TX 200

[a]Firm Enterprise Data System.

SOURCE: Office of Technology Assessment.

In general, the sources of internal size economies among the selected corn-producing areas are as follows. Corn yields are from 3 to 10 percent higher on very large enterprises than on medium-size enterprises. Very large corn enterprises spend less on fertilizer, fuel, lubrication, repairs, and custom harvesting than do medium-size enterprises. Finally, very large corn enterprises consistently use some combination of fewer and/or more efficient machines and tractors that go over the field fewer times.

Of the areas studied, Iowa has a comparative advantage in corn production (table 8-3), regardless of whether the return to land is assumed to be 10 or 5 percent of its 1982 value. Iowa also has the smallest average corn enterprise size, the lowest level of resource concentration, and the lowest percentage of change in resource concentration of the selected corn-producing areas studied. The trends in relative yield and land values indicate that the comparative advantage in corn production in Iowa will continue in the future, other things being equal. There are also size economies that can be exploited. Therefore, the data suggest that corn producers in Iowa are not operating at the least-cost production point. The quality of initial resource endowment in Iowa provides producers with the ability to remain highly competitive in corn production without fully exploiting size economies.

The comparative disadvantage for Nebraska increases from 8 to 29 percent relative to Iowa, associated with a 10- and 5-percent return to land, respectively. This 21-percent difference results in Nebraska becoming a very marginal corn-producing area in the presence of lower interest rates. This is because Nebraska's comparative advantage is largely associated with inexpensive land relative to that of other areas. A lower rate of return to land effectively discounts the source of comparative advantage in corn production in this area.

Soybeans

In soybean production, very large enterprises of about 800 acres have a 1-percent cost advantage over 300-acre, medium-size enterprises (table 8-2). The change in relative production

Figure 8-1.—Firm Enterprise Data System Production Areas

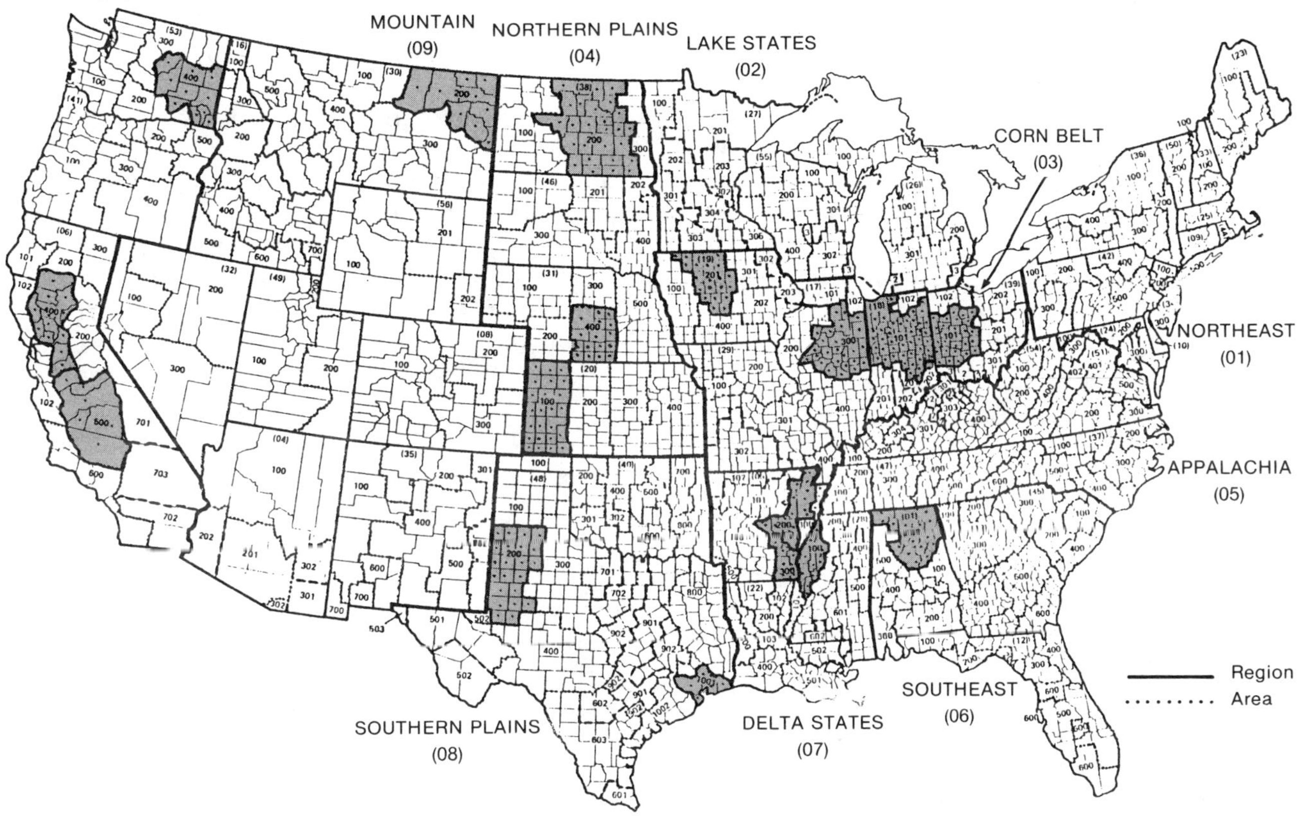

SOURCE: U.S. Department of Agriculture.

Table 8-2.—Summary of the Measures of Internal Economies of Size in Selected Corn, Soybean, and Wheat Producing Areas, 1983

Commodity, State, and area	Economies of size	Structural elasticity[a] (%)	Enterprise size (acres)		Enterprise cost ($/unit)	
			Medium	Very large	Medium	Very large
Corn:						
IL 300	Very large	−7	246	1,113	$119	$100
IN 101	Very large	−22	271	903	105	100
IA 201	Very large	−22	170	576	105	100
NE 400[b]	Very large	−12	266	1,715	113	100
Overall	Increasing	−15	234	1,003	111	100
Soybeans:						
IL 300	None	−44	270	684	102	100
IA 201	Very large	−14	210	707	108	100
MS 100	None	45	795	1,262	99	100
OH 101	None	8	244	897	86	100
Overall	Constant	−179	302	790	101	100
Wheat:						
KS 100	Very large	−8	774	3,909	118	100
MT 200	Very large	−12	421	577	110	100
ND 200	None	−38	338	1,283	103	100
WA 400	None	6	753	2,388	85	100
Overall	Constant	−36	647	2,661	103	100

[a]The percentage change in enterprise size in acres for a 1-percent change in production costs relative to the medium and very large enterprise size range.
[b]Irrigated.

SOURCE: Office of Technology Assessment.

Table 8-3.—Summary of the Measures of Interregional Comparative Advantage Among Selected Corn, Soybean, and Wheat Producing Areas, 1983

Commodity, State, and area	Economies of size	Comparative advantage A[a] (%)	Comparative advantage B[b] (%)	Enterprise size (acres)
Corn:				
IL 300	Very large	114	116	663
IN 101	Very large	110	114	586
IA 201	Very large	100	100	335
NE 400[c]	Very large	108	129	1,095
Total	Increasing	109	113	610
Soybeans:				
IL 300	None	103	102	464
IA 201	Very large	100	100	362
MS 100	None	124	138	1,126
OH 101	None	105	109	570
Total	Constant	105	106	521
Wheat:				
KS 100	Very large	100	100	2,152
MT 200	Very large	103	112	1,352
ND 200	None	109	133	893
WA 400	None	100	104	1,613
Total	Constant	102	109	1,647

[a]Assumes a 10-percent rate of return on 1982 land values.
[b]Assumes a 5-percent rate of return on 1982 land values.
[c]Irrigated.

SOURCE: Office of Technology Assessment.

and resource concentration is positive for very large enterprises and is increasing at a rate of about 13 percent per year. The structural elasticity is −179, indicating that a 179-percent increase in (or nearly tripling from 300 to 500 acres) an average soybean enterprise acreage results in a 1-percent decrease in production cost. There is evidence to argue for the existence of constant internal economies of size in soybean production across the selected producing areas.

In general, internal size economies do not exist (except for Iowa) among the selected soybean areas because soybean yields are only from 1 to 6 percent higher on very large enterprises relative to medium-size enterprises. Very large enterprises have higher expenditures on fertilizer, herbicides, and insecticides in proportion to yield. And very large enterprises use relatively more efficient machines and tractors, particularly in land preparation, relative to medium-size enterprises.

Iowa also has a comparative advantage in soybean production among the selected producing areas (table 8-3), regardless of whether the return to land is assumed to be 5 or 10 percent of its 1982 value. This area has the smallest average soybean enterprise size, the lowest level of resource concentration, and one of the lowest percentages of change in resource concentration of the selected producing areas. The trends in relative yield and land prices indicate that the comparative advantage of Iowa in soybean production will decrease in the future, particularly in relation to Illinois. Iowa is unique among the soybean areas studied in that it has substantial internal economies of size, particularly for custom harvesting of very large enterprises, that can be exploited. The data suggest that soybean producers in Iowa are not operating at the least cost of production. The quality of the initial resource endowment in Iowa provides producers with the ability to remain highly competitive in soybean production without having to exploit size economies fully.

In Mississippi the comparative disadvantage in soybean production increases from 24 to 38 percent relative to Iowa, associated with a 10- and 5-percent return to land, respectively. This 14-percent difference results in Mississippi becoming an even more marginal soybean-producing area as interest rates decrease. This is because Mississippi's "comparative advantage" is associated with less expensive land. A lower rate of return to land effectively discounts the source of comparative advantage in soybean production in this area.

Wheat

In wheat production, very large enterprises of about 2,600 acres have a 3-percent cost advantage over 640-acre, medium-size enterprises. The change in relative production and resource concentration is positive for very large enterprises and increases at a rate of about 5 percent per year. The structural elasticity is −36, which indicates that each 36-percent increase in an average wheat enterprise acreage (between the 640- and 2,600-acre range) results in a 1-percent decrease in production cost. There is evidence to argue for the existence of constant internal economies of size in wheat production across the selected producing areas overall.

In general, the absence of internal size economies in wheat production is the result of statistical averaging. Only North Dakota exhibits constant returns to scale in wheat production. Kansas and Montana both have increasing returns to scale, while Washington has a decreasing return to scale. In Kansas, economies associated with custom harvesting are an important source of size economies for very large wheat enterprises. In Montana, size economies for very large wheat enterprises are the result of the combination of slightly higher yield and the economies associated with custom harvesting. In North Dakota, the lack of internal economies of size for very large wheat enterprises is the result of slightly higher yields offset by higher harvesting costs relative to those of medium-size enterprises. In Washington, size economies do not exist for very large wheat enterprises relative to that of medium enterprises because of the substantial diseconomies related to yield and the slightly higher capital ownership and harvesting costs.

Kansas has a comparative advantage in wheat production among the selected areas, maintaining this advantage regardless of whether the return to land is assumed to be 5 or 10 percent of its 1982 value. This area has the largest average enterprise size, the lowest resource concentration, and one of the lowest percentages of change in resource concentration of the selected wheat-producing areas studied. The trends in relative wheat yield and land prices indicate that the comparative advantage of this area may decrease slightly in the future. Kansas has substantial size economies that can be exploited. The data suggest that wheat producers in Kansas are not operating at the least cost of production. The quality of the initial resource endowment in Kansas provides producers with the ability to remain highly competitive without having to exploit size economies fully.

In North Dakota the comparative disadvantage in wheat production increases from 9 to 35 percent relative to Kansas, associated with a 10- and 5-percent return to land, respectively. This 24-percent difference results in North Dakota becoming a marginal wheat-producing area as interest rates decrease. This is because North Dakota's comparative advantage is associated with less expensive land. A lower rate of return effectively discounts the source of comparative advantage in wheat production in this area.

Rice

In rice production, very large enterprises of about 2,400 acres have a 4-percent cost disadvantage over 600-acre, medium-size enterprises (table 8-4). The change in relative production and resource concentration is positive for very large enterprises and increases at a rate of 1 percent per year. The structural elasticity is 27, indicating that each 27-percent decrease in the average acreage for a rice enterprise (between 600 and 2,400 acres) results in a 1-percent decrease in production cost. There is evidence to argue for the existence of decreasing internal economies of size in rice production across the selected producing areas.

In general, the presence of internal diseconomies of size among the selected rice-producing areas is a result of the following: Rice yields are from about 0 to 10 percent lower on very large enterprises than that of medium-size enterprises. Rice yield diseconomies are related to the ability to maintain a uniform distribution of flood irrigation water on the field. The uniform distribution of water is affected by the extent to which the fields have been leveled and the water levels maintained. The data indicate that the extent of leveling is approximately 100 percent across enterprise size among the rice-producing areas. Therefore, yield diseconomies necessarily relate to the timeliness of water, fertilizer, herbicide, insecticide, and fungicide application. Rice producers on medium enterprises are better able to manage these applications than those on very large enterprises. Finally, in Texas there are diseconomies associated with purchased canal water used for irrigation.

California has a comparative advantage in rice production among the selected rice-producing areas (table 8-5), maintaining this advantage regardless of whether the return to land is assumed to be 5 or 10 percent of its 1982 value. California has the largest average rice enterprise

Table 8-4.—Summary of the Measures of Internal Economies of Size in Selected Rice and Cotton Producing Areas, 1979 and 1982

Commodity, State, and area	Economies of size	Structural elasticity[a] (%)	Enterprise size (acres)		Enterprise cost ($/unit)	
			Medium	Very large	Medium	Very large
Rice:						
CA 400[b]	None	40	850	3,575	$ 97	$100
TX 1001[b]	None	33	691	2,068	97	100
DLT 100 & 300[b]	None	14	509	1,904	92	100
AR 200[b]	None	30	377	1,619	96	100
Overall	Decreasing	27	625	2,397	96	100
Cotton:						
AL 600	VL	−14	568	1,842	108	100
CA 500[b]	None	64	614	2,833	98	100
MS 100	None	Infinite	754	2,868	100	100
TX 200[b]	None	118	436	1,707	99	100
TX 200	VL	−9	972	5,920	117	100
Overall	Constant	−77	653	3,040	102	100

[a]The percentage change in enterprise size in acres for a 1-percent change in production costs relative to the medium and very large enterprise size range.
[b]Irrigated.

SOURCE: Office of Technology Assessment.

Table 8-5.—Summary of the Measures of Interregional Comparative Advantage Among Selected Rice and Cotton Producing Areas, 1979 and 1982

Commodity, State, and area	Economies of size	Comparative advantage A[a] (%)	Comparative advantage B[b] (%)	Enterprise size (acres)
Rice:				
CA 400[c]	None	100	100	1,071
TX 1001[c]	None	124	134	802
DLT 100 & 300[c]	None	114	120	694
AR 200[c]	None	108	114	45
Total	Decreasing	100	115	776
Cotton:				
AL 600	VL	123	125	1,225
CA 500[c]	None	109	107	2,088
MS 100	None	100	101	1,787
TX 200[c]	None	101	103	1,224
TX 200	VL	104	100	3,283
Total	Constant	106	105	2,020

[a]Assumes a 10-percent rate of return on 1982 land values (1978 values for rice).
[b]Assumes a 5-percent rate of return on 1982 land values (1978 values for rice).
[c]Irrigated.

SOURCE: Office of Technology Assessment.

size and the second highest level of production or resource concentration, after Mississippi, of the selected rice-producing areas. Production and resources are becoming less concentrated in California, at a rate of −2 percent per year. The trends in relative yield and land prices indicate that comparative advantage in this area will remain unchanged. Internal size economies have been fully exploited in this area. The data suggest that the minimum average cost point in rice production is reached at the medium enterprise size. The data also suggest that the support price for rice is such that the average revenue curve is substantially above the minimum point on the average cost curve. In this case firms can achieve a higher profit by extending output beyond the minimum point, even though they experience higher average total costs than smaller firms operating at a low point on the average cost curve.

In Texas the comparative disadvantage in rice production increases from 24 to 34 percent relative to that of California, associated with a 10- to 5-percent return to land, respectively. This 10-percent difference results in Texas becoming an even more marginal rice-producing area as interest rates decrease, because Texas' comparative advantage is associated with less expensive land. A lower rate of return effectively discounts the source of comparative advantage in rice production in this area.

Cotton

In cotton production very large enterprises of about 3,000 acres have a 2-percent cost advantage over 650-acre, medium-size enterprises. The change in relative production and resource concentration is positive and increases at a rate of 10 percent per year for very large cotton enterprises. The structural elasticity is −77, which indicates that each 77-percent increase in the average acreage for a cotton enterprise (between 650 and 3,000 acres) results in a 1-percent decrease in production cost. There is evidence to argue for the existence of constant internal economies of size in cotton production across the selected producing areas.

In general, the absence of internal size economies in cotton production is the result again of statistical averaging. Only three of the five selected cotton-producing areas have constant returns to scale. These areas include California, Mississippi, and Texas (irrigated). On the other hand, Alabama and Texas (dryland) both have increasing returns to scale in cotton production. Size economies exist for Alabama's very large cotton enterprises because these enterprises incur lower machinery and tractor-related expenses for a given field operation and still manage to obtain a slightly higher yield. In California, size diseconomies are primarily related to the diseconomies of purchased irrigation water for cotton production. In Mississippi the lack of size economies in cotton production for very large enterprises relates to similar preharvest and ownership costs without significantly higher yields. In Texas (irrigation), the lack of size economies relates to the combination of size diseconomies in harvesting and cultivation, along with slightly higher yields enjoyed by the medium enterprise in this area. In Texas (dryland), size economies for very large enterprises relate to the substantial preharvest and ownership cost advantage associated with lower machinery and tractor expenses for a given field operation without substantial loss in yield.

In terms of comparative advantage, there is a three-way tie for comparative advantage in the production of cotton among the selected areas. These three areas are Mississippi and Texas (both irrigated and dryland). These three areas maintain their comparative advantage (within 2 to 3 percentage points), regardless of whether returns to land are assumed to be 5 or 10 percent of its 1982 value. These three areas have the lowest resource concentration and an average enterprise size of between 1,200 to 3,300 acres. The percentage of change in resource concentration in these areas ranges from 2 to 20 percent per year. The trend in relative yield and land prices indicates that Mississippi will have a comparative advantage over Texas in cotton production. On the other hand, Mississippi loses comparative advantage in cotton production to Texas as interest rates decrease. Even though land prices are higher in Mississippi, a lower rate of return on land works to the advantage of Texas because land there is a larger percentage of total cost. Also, in Texas more of the cotton is planted in strip rows to conserve soil moisture, and fewer herbicides or insecticides are used than is the case in Mississippi. The data also suggest that cotton producers in Texas (dryland) are not operating at the least cost of production. The quality of the initial resource endowment in Texas (dryland) provides producers with the ability to remain competitive without having to exploit size economies fully. As a result, Mississippi and Texas have nearly equal comparative advantages in cotton production.

In Alabama the comparative disadvantage in cotton production increases from 23 to 25 percent relative to that of Mississippi and Texas, associated with a 10- and 5-percent return to land, respectively. A decrease in the interest rate does not substantially alter the marginal position of Alabama in cotton production. The comparative disadvantage in cotton production in Alabama is due to low relative yields, combined with high fertilizer, lime, insecticide, and harvesting costs. Aside from more fully exploiting size economies, Alabama has few options for increasing its comparative advantage in cotton production in the future.

Summary and Conclusions

The data suggest the following conclusion regarding internal economies of size for the selected commodities (table 8-6): The evidence for increasing returns to scale for corn and decreasing returns to scale for rice, given the configurations for the enterprises for these commodities in 1983 and 1979, respectively, are the strongest. Each of the selected corn areas has substantial internal economies as enterprise size increases from medium to very large. On the

Table 8-6.—Economies of Size and Structural Elasticities for Selected Commodities and Areas

Commodity	Size economies	Structural elasticity	Internal economies of size: Increasing	Constant	Decreasing
Corn	Increasing	−15	IL, IN, IA, NE		
Wheat	Constant	−26	KS, MT	ND	WA
Cotton	Constant	−77	AL, TX	CA[a], MS, TX[a]	
Soybeans	Constant	−179	IA	IL, MS	OH
Rice	Decreasing	27			CA[a], TX[a], DLT[a], AR[a]

[a]Irrigated.

SOURCE: Office of Technology Assessment.

other hand, each of the selected rice areas has internal diseconomies as enterprise size increases from medium to very large.

The strongest case for constant returns to scale can be made for cotton. Three of the five selected cotton-producing areas have constant size economies, given the configuration of cotton enterprises in 1982. Only Alabama and Texas (dryland) have increasing economies of size. Alabama is a marginal cotton-producing area, however, as indicated by the data on comparative advantage.

Texas (dryland) is one of the most competitive cotton-producing areas studied. The average enterprise size in Texas is the largest of the selected areas, while the percentage change in resource concentration between 1978 and 1982 was the lowest. This seems to indicate that producers of dryland cotton in Texas are willing to forego the potential gain associated with expanding enterprise size to avoid the additional exposure to risk and uncertainty. The capital investment in land and machinery required as cotton enterprises in this area expand from 1,000 to 6,000 acres substantially increases indebtedness, which threatens survivability of the firm. In particular, the uncertainty associated with cotton yields in this area of Texas make indebtedness unattractive.

The arguments for constant returns to scale in the cases of wheat and soybeans are somewhat more ambiguous. Selected areas having increasing, constant, and decreasing returns to scale are present for both of these commodities. In the case of wheat, Kansas has substantial size economies as enterprises increase from 800 to 4,000 acres. This area has the largest average wheat enterprise size of the selected areas and the lowest change in resource concentration between 1978 and 1982. This seems to indicate that producers of wheat in Kansas are willing to forego the potential gains in order to avoid additional exposure to risk and uncertainty associated with expanding enterprise size. A similar case can be made for wheat production in Montana. In the case of soybeans, enterprises in Iowa have size economies as they increase from 200 to 700 acres. This area has the lowest average enterprise size and the lowest percentage of change in resource concentration between 1978 and 1982 of the selected soybean areas.

Therefore, size economies for corn and size diseconomies for rice are constant and broadly based results. Constant size economies for soybeans, wheat, and cotton exist in general, but important exceptions exist for increasing (and decreasing) economies of scale in each of these commodities. Two of the important exceptions of increasing returns are Texas dryland cotton and Kansas wheat. The data suggest in both of these areas for their respective commodities that producers are willing to forego increasing return to scale to avoid additional risk and uncertainty associated with expanding enterprise size, which potentially could also threaten the survival of the firm.

It should be noted that the analysis focused on technical efficiencies in determining economies of size. Pecuniary economies (e.g., discounts on input supplies or services purchased in volume) that are important for very large enterprises were not taken into account. If they had been considered, pecuniary economies would have provided more economies of size for the very large operations. For wheat, soybeans, and cotton this could have changed the above overall analysis from constant to increasing economies of size.

This information strongly suggests that resource concentration for most American field crops will probably continue for some time. Powerful forces at work in the farm economy will lead to fewer and larger farms.

ECONOMIC IMPACTS OF EMERGING TECHNOLOGIES AND SELECTED POLICIES FOR VARIOUS SIZE CROP FARMS

The analysis of size economies did not take into account various policies and their impacts on resource concentration. This section will present the findings of just such an analysis of four selected regions in the United States that represent significant agricultural production in the commodities considered in farm policy: corn, cotton, soybeans, rice, and wheat. Within each production region, representative moderate, large, and very large commercial farms were identified and analyzed.[2] It was assumed for the analysis that the technology development and adoption conditions in existence would be those of the most likely environment, outlined in chapter 2.

Two techniques were used to analyze the effects of selected policy provisions and technology on farms within each region. Information was obtained on resource characteristics, acreage devoted to specific crops, and historic projected yields of crops eligible for farm program provisions. These data were used to develop resource characteristics of the three different farm sizes. Then a simulation model was used to analyze the economic viability and growth potential of each representative farm for selected policy and technology advance scenarios (appendix E contains a detailed discussion of the model).

The following presents the representative farms and major findings for the production areas analyzed. Obviously, more areas could have been analyzed, but neither time nor the resources allocated to this study would permit their inclusion. It is expected that the results will apply in broad principle to the major production region of which each area is a part. It is important to remember that the results of this analysis are mainly illustrative. Thus the relative results for the several farm sizes and for the several alternative policy and technology scenarios are probably more important than any specific numbers generated by the analysis.

[2]Small and part-time farms were not included because these farm operators in general depend on off-farm employment for their primary source of income.

Crop Farms Analyzed

Corn-Soybean Farms in the Corn Belt

The North Central Region of the United States produces approximately 50 percent of the U.S. total production of corn and soybeans. Representative farms for this region are three farms from the corn-soybean cash grain area of east central Illinois and three farms from the irrigated row crop area of south central Nebraska.[3]

The representative farm situations developed and used in this analysis were constructed from two basic data sources: 1) national cost-of-production surveys by USDA in 1978 and 1983, and 2) farm record data collected and analyzed by the Universities of Illinois and Nebraska. The size of the representative farms and the acreages of owned and rented cropland were developed from the size distributions in the USDA cost-of-production surveys. The very large farms approximate the largest 10 percent of farms in the surveys; the large farms, the 70th to 90th percentiles; and the moderate farms, the 40th to 70th percentiles.

[3]These representative farms were developed and analyzed in the OTA paper "Economic Impacts of Selected Farm Policies, Income Tax Provisions, and Production Technology on the Economic Viability of Corn-Soybean Farms in East Central Illinois and Irrigated Row Crop Farms in South Central Nebraska," prepared by W.B. Sundquist, 1985.

Financial status, as measured by net worth, debt load (both intermediate-term and long-term), and leverage ratio, differs dramatically from farmer to farmer. Data from the most recent USDA Agricultural Finance Survey were used to depict the beginning financial characteristics for the six representative farms (tables 8-7 and 8-8).

All of the representative farms are well-mechanized production units ranging from 640 to 2,085 acres of cropland, and all farms include a combination of owned and rented land. Of the six representative farms, only the very large units in each area employ full-time workers. The other farms operate with a combination of family and part-time workers. The Illinois farms have all of their cropland devoted to cash crop production of corn and soybeans. The Nebraska farms are cash crop operations that combine both gravity and sprinkler technologies to irrigate corn and a small acreage of soybeans. In addition, these farms produce a substantial acreage of grain sorghum under a nonirrigated (dryland) regime. Production on this dryland acreage tends to be somewhat riskier than that for the irrigated component of the farming operations, but irrigated farming still has some year-to-year yield variability, owing to weather. Although a number of these irrigated corn farms also produce some wheat and/or corn silage, those enterprises have not been included in the analysis.

The crop mix for the Nebraska farms is identical for all three farm sizes: irrigated corn (58.3 percent of cropland acres), irrigated soybeans (6 percent), dryland sorghum (35.7 percent). On the Illinois farms, the proportion of corn to soybeans varies only slightly for the three representative farms, with corn planted on 52 to 55 percent of the cropland acreage and soybeans on the balance.

For the Illinois farms, all cropland has the same per-acre value, whereas the price of cropland on the Nebraska farms reflects the differentials for four categories of land: 1) gravity ir-

Table 8-7.—Financial Characteristics of Three Representative Corn-Soybean Farms in East Central Illinois[a]

Characteristics	Farm size (acres) Moderate	Large	Very large
Cropland acres	640	982	1,630
Acres owned	260	429	458
Acres leased	380	553	1,172
Value of owned real estate ($1,000[b])	900.5	1,480.6	1,538.4
Value of machinery ($1,000)	92.2	104.8	129.0
Long-term debt ($1,000)	126.1	557.4	579.4
Intermediate-term debt ($1,000)	55.3	62.9	83.8
Initial net worth ($1,000[c])	855.4	1,027.6	1,106.4
Leverage ratio (fraction)	0.21	0.61	0.60
Long-term debt/asset ratio (fraction)	0.14	0.38	0.38
Intermediate-term debt/asset ratio (fraction)	0.60	0.60	0.65
Equity ratio (fraction)	0.82	0.62	0.63
Off-farm income ($1,000)	8.2	7.4	7.6
Minimum family living expenses ($1,000)	18.0	20.0	24.0
Maximum family living expenses ($1,000)	36.0	40.0	48.0
Marginal propensity to consume (fraction)	0.20	0.20	0.20

[a]A family size of four persons was assumed for the purposes of estimating family labor supply and determining appropriate income tax rates.
[b]Includes land and building.
[c]May include assets other than land, buildings, and machinery.

SOURCE: Office of Technology Assessment.

Table 8-8.—Financial Characteristics of Three Representative Irrigated Corn Farms in South Central Nebraska[a]

Characteristics	Farm size (acres) Moderate	Large	Very large
Cropland acres	672	920	2,085
Acres owned	302	530	1,042
Acres leased	370	390	1,043
Value of owned real estate ($1,000[b])	477.7	838.4	1,648.3
Value of machinery ($1,000)	102.7	112.1	183.9
Long-term debt ($1,000)	123.2	102.0	291.1
Intermediate-term debt ($1,000)	40.1	53.7	98.0
Initial net worth ($1,000[c])	448.3	839.0	1,463.1
Leverage ratio (fraction)	0.39	0.20	0.27
Long-term debt/asset ratio (fraction)	0.26	0.12	0.18
Intermediate-term debt/asset ratio (fraction)	0.39	0.48	0.53
Equity ratio (fraction)	0.72	0.84	0.79
Off-farm income ($1,000)	8.2	8.2	9.7
Minimum family living expenses ($1,000)	18.0	18.0	24.0
Maximum family living expenses ($1,000)	36.0	36.0	48.0
Marginal propensity to consume (fraction)	0.20	0.20	0.20

[a]A family size of four persons was assumed for the purposes of estimating family labor supply and determining appropriate income tax rates.
[b]Include land and building.
[c]May include assets other than land, buildings, and machinery.

SOURCE: Office of Technology Assessment.

rigated, 2) sprinkler irrigated, 3) dryland with irrigation potential, and 4) dryland without irrigation potential. Each of the three Nebraska farms, however, has the same proportions of gravity irrigation, sprinkler irrigation, and dryland acres.

Wheat Farms in the Southern Plains

Approximately 65 percent of U.S. wheat production is produced in the Great Plains. For the analysis of representative wheat farms, farms were selected from the Southern Plains region. They are representative of wheat farms in western Kansas, eastern Colorado, and the Oklahoma and Texas Panhandle.[4]

The three farms selected for the analysis are a typical moderate farm in the region (1,280 acres), a large farm (1,900 acres), and a very large farm (3,200 acres). The initial financial characteristics for the three representative farms are summarized in table 8-9. The proportion of cropland owned by each farm was obtained from the most recent Agricultural Finance Survey, summarized for wheat farmers in western Kansas, eastern Colorado, the Oklahoma Panhandle, and the Northern High Plains of Texas.

Average long- and intermediate-term debt-to-asset ratios from the Agricultural Finance Survey were used to estimate initial values for long- and intermediate-term debts. All three wheat farms had about the same beginning equity levels (75 percent) (table 8-9). Minimum family living expenses were based on values obtained from a Texas A&M survey that asked for the minimum annual cash expenditure for family living. The Agricultural Finance Survey was used to obtain values of off-farm income for the three representative farm operators.

A typical cropping pattern in the Southern Plains is to irrigate 50 percent of all cropland and to raise wheat on one-half of this irrigated land. Grain sorghum is typically raised on the other half of the irrigated cropland. Wheat is generally also raised on the portion of the cropland that is not irrigated. This cropping pattern was assumed for all three farms.

Table 8-9.—Financial Characteristics of Three Representative Wheat Farms in the Southern Plains

Characteristics	Farm size (acres)		
	Moderate	Large	Very large
Cropland acres	1,280	1,920	3,200
Acres owned	640	840	1,400
Acres leased	640	1,080	1,800
Acres of pastureland owned	120	220	360
Value of owned cropland ($1,000)	296.0	388.5	647.5
Value of owned pastureland	29.4	53.9	88.2
Value of machinery ($1,000)	241.9	352.2	477.2
Value of off-farm investments ($1,000)	37.3	49.0	53.5
Beginning cash reserve ($1,000)	10.0	12.0	20.0
Long-term debt ($1,000)	60.2	86.3	143.5
Intermediate-term debt ($1,000)	83.2	126.5	171.3
Initial net worth ($1,000)	470.3	642.3	970.7
Equity ratio (fraction)	0.77	0.75	0.75
Leverage ratio (fraction)	0.31	0.33	0.33
Long-term debt/asset ratio (fraction)	0.19	0.20	0.20
Intermediate-term debt/asset ratio (fraction)	0.34	0.36	0.36
Off-farm income ($1,000)	12.4	9.8	9.0
Minimum family living expenses ($1,000)	18.0	20.0	23.0
Maximum family living expenses ($1,000)	40.0	50.0	50.0
Marginal propensity to consume (fraction)	0.25	0.25	0.25

SOURCE: Office of Technology Assessment.

Numerous crop share arrangements prevail in the region for leased land. However, these arrangements generally involve the producer paying the landlord about 25 percent of the crop and the landlord paying none of the production and harvesting costs. This crop share arrangement was assumed for all leased cropland.

General Crop Farms in the Delta Region of Mississippi

The Mississippi Delta is an excellent region for analysis of general crop farms.[5] Farms in this area can produce a variety of crops not possible in other parts of the United States. The rep-

[4]These representative farms were developed and analyzed in the OTA paper "Economic Impacts of Selected Policies and Technology on the Economic Viability of Three Representative Wheat Farms in the High Plains," prepared by James W. Richardson, 1985.

[5]These representative farms were developed and analyzed in the OTA paper "Economic Effects of Selected Policies and Technology on the Economic Viability of General Crops Farms in the Delta Region of Mississippi," prepared by B.R. Eddleman, 1985.

resentative farms in this region produce cotton, rice, soybeans, and wheat (or other small grains).

The three representative farms developed for this study are a moderate farm (1,443 acres), a large farm (3,119 acres), and a very large farm (6,184 acres). Table 8-10 provides a summary of the financial and resource characteristics for the three representative farms. The long-term and intermediate-term debt-to-asset ratios for the 1,443-acre farm and the 3,119-acre farm were obtained from the Agricultural Finance Survey and adjusted to reflect the equity levels as reported from a 1983 mail survey of farms in the delta. These debt ratios are the average for part-owner general crops farms in the Mississippi Delta region that had debt on real estate in 1979. Financial ratios for the largest farm were developed by extending the ratios on a per-acre basis for a 3,457-acre farm, as reported in the most recent Agricultural Finance Survey, and were adjusted by the equity levels reported for the largest farm-size group.

The mix of acreages planted in each crop changed by farm size. In general, the acreage planted in cotton and soybeans increased relative to the acreage planted in rice and wheat as farm size increased. The moderate farm planted 73 percent of tillable cropland in cotton and soybeans, while the large and the very large farms planted 89 and 82 percent, respectively, of tillable cropland in cotton and soybeans. In the analysis, as the farm was allowed to grow in size to the next largest farm size, the proportion of cropland planted in each crop was changed to reflect these differences in crop mix.

Table 8-10.—Financial Characteristics of Three Representative General Crops Farms in the Delta of Mississippi

	Farm size (acres)		
Characteristics	Moderate	Large	Very large
Cropland acres	1,443	3,119	6,184
Acres owned	533	1,419	3,064
Acres leased	910	1,700	3,120
Acreage of principal crops in 1983:			
Cotton	395	1,088	2,250
Rice	305	574	871
Soybeans	640	1,190	2,539
Wheat (or other small grains)	82	247	180
Value of owned cropland ($1,000)	799.5	2,128.5	4,596
Value of farm machinery ($1,000)	378.9	786.7	1,209.8
Value of off-farm investments ($1,000)	129.1	210.3	358.7
Beginning cash reserve ($1,000)	31.9	71.1	141.6
Long-term debt (1,000)	331.4	840.8	1,640.8
Intermediate-term debt ($1,000)	243.8	413.0	574.7
Net worth ($1,000)	748.6	1,921.5	4,047.5
Total equity to assets (fraction)	0.56	0.60	0.64
Long-term debt/asset ratio (fraction)	0.41	0.40	0.36
Intermediate-term debt/asset ratio (fraction)	0.64	0.52	0.48
Off-farm income ($1,000)	18.3	18.2	36.0
Minimum family living expense ($1,000)	18.0	24.0	30.0
Maximum family living expense ($1,000)	27.0	36.0	45.0
Marginal propensity to consume (fraction)	0.25	0.25	0.25

SOURCE: Office of Technology Assessment.

Cotton Farms in the Texas Southern High Plains

Cotton is an important commodity in the United States, and over one-half of the cotton produced can be found in the Southern High Plains of Texas. The three farms selected for analysis are a typical moderate farm in the region (1,088 acres), a large farm (3,383 acres), and a very large farm (5,570 acres).[6] These size farms accounted for 31 percent of the farms and 62 percent of the cotton lint produced in the Texas Southern High Plains.

Table 8-11 provides a summary of the demographic and financial characteristics for the three representative cotton farms used in the present study. The long- and intermediate-term debt-to-asset ratios for the moderate farm were obtained from the Agricultural Finance Survey. These debt ratios are the average for part-owner cotton farmers in the Texas High Plains who had debt on real estate in 1979. Financial ratios reported by Smith (1982) for the two larger farms were used because the Agricultural Finance Survey did not provide information for farms in these categories.

A special survey of farmers identified average annual off-farm income and minimum fam-

[6]These representative farms were developed and analyzed in the OTA paper "Economic Impacts of Selected Policies and Technology on the Economic Viability of Three Representative Cotton Farms in the Texas Southern High Plains," prepared by James W. Richardson, 1985.

Table 8-11.—Financial Characteristics of Three Representative Cotton Farms in the Texas Southern High Plains

Characteristics	Farm size (acres)		
	Moderate	Large	Very large
Cropland acres	1,088	3,383	5,570
Acres owned	381	1,048	3,453
Acres leased	707	2,335	2,117
Value of owned real estate ($1,000)	222.4	611.7	2,015.4
Value of machinery ($1,000)	144.5	420.8	713.9
Value of off-farm investments ($1,000)	59.0	110.0	213.7
Beginning cash reserve ($1,000)	16.7	52.0	85.5
Long-term debt ($1,000)	61.1	120.9	488.7
Intermediate-term debt ($1,000)	98.3	203.6	475.4
Initial net worth ($1,000[c])	275.0	854.8	2,032.3
Leverage ratio (fraction)	0.62	0.72	0.67
Long-term debt/asset ratio (fraction)	0.61	0.40	0.49
Intermediate-term debt/asset ratio (fraction)	0.27	0.20	0.24
Equity ratio (fraction)	0.68	0.48	0.67
Off-farm income ($1,000)	16.0	0.0	0.0
Minimum family living expenses ($1,000)	15.2	29.1	38.0
Maximum family living expenses ($1,000)	50.0	50.0	60.0
Marginal propensity to consume (fraction)	0.25	0.25	0.25

SOURCE: Office of Technology Assessment.

ily living expenses by farm size (Smith, 1982). Maximum annual family living expenses were assumed to be $50,000 to $60,000, depending on farm size. The model assumes the family will use 25 cents of every additional dollar of disposable income, over and above the minimum requirement, for family living. In no instance, however, will family living expenses exceed the maximum indicated in table 8-11.

Cotton production costs for the three farms were estimated based on Smith's (1982) study. The two larger farms had a 13 percent lower total cost of production, per pound of cotton lint, than the moderate-size farm. The mix of irrigated and nonirrigated cotton changed across farm size. The moderate farm irrigated 32 percent of its available cotton acreage, while the two larger farms irrigated only 23 percent. In the simulation analysis, as the moderate farm grew in size, its proportion of irrigated cropland was decreased to 23 percent.

Farm Policy, Tax Policy, and Technology Scenarios

The three representative farms for each production region were analyzed for the period 1983-92 under alternative policy scenarios.[7] Six farm policy scenarios (including a continuation of the 1981 farm bill), an income tax provision scenario, two financial stress scenarios, a technology option, and a new entrant scenario were analyzed for each farm. All assumptions and policy values associated with each scenario were held constant across farm sizes to allow direct comparison of their impacts on different size farms. Appendix E contains a summary of the analysis for each farm size by region.

Current Policy

The current policy scenario involves continuing through 1992 the current income tax provisions and the price supports, income support, and supply control programs of the 1981 farm bill. In addition, it was assumed that annual mean crop yields for the three representative farms in each of the four production regions will increase as new technologies are introduced and adopted by farmers in the most likely technology environment. For this policy scenario it was assumed that the following farm policies were in effect:

- The Commodity Credit Corporation (CCC) loan program is available to producers for corn, cotton, rice, sorghum, soybeans, and wheat.
- A 3-year, indirect, farmer-owned reserve (FOR) is available for feed grains and wheat.[8]
- An acreage diversion/set-aside program is in effect for 1983 to 1985, using the actual acreage reduction levels and diversion payment rates specified for these years.

[7]The current version of the Firm Level Income Tax and Farm Policy Simulator (FLIPSIM V), developed by James W. Richardson and Clair J. Nixon, was used to simulate the three representative farms in each region.

[8]The 1977 farm bill established FOR as a 3-year extension of the CCC loan after grain had been in the regular loan for 9 months. Stocks remain in the farm operator's control until the Secretary of Agriculture authorizes release.

- A target price-deficiency payment program is available for corn, cotton, rice, sorghum, and wheat in all years.
- The $50,000-payment limitation for deficiency and diversion payments is in effect and is effective on the farm as specified.
- Farms of all sizes are eligible to participate in these farm program provisions.

Values for loan rates, target prices, diversion rates, and diversion payment rates for 1983 and 1984 are set at their actual values, expressed in 1982 dollars. Values for these variables for 1985 are set at their respective levels announced on or before September 14, 1984, by Secretary of Agriculture Block. Loan rates and target prices for 1985 are held constant through 1992. No acreage reduction program is assumed to be in effect after 1985.

It was assumed that the following options for depreciating machinery and calculating income taxes are used for the current policy scenario:

- Machinery, livestock, and buildings placed in use prior to 1981 are depreciated using the double declining balance method.
- Machinery, livestock, and buildings placed in use after 1980 are depreciated using the accelerated cost recovery method.
- The operator elects to claim first-year expensing for all depreciable items placed into use after 1980.
- The operator elects to take maximum investment tax credit (ITC) and thus reduces the basis for all depreciable assets placed into service after 1980.
- The operator adjusts crop sales across tax years to reduce current-year taxes.
- The operator may use either the regular income tax computation or income averaging to calculate Federal income tax liabilities.
- There is no maximum interest deduction for calculating taxable income.
- The actual self-employment tax rates and maximum income levels subject to this tax for 1983 and 1984 are used. Announced values for these variables in 1985-86 are used, and the 1986 values are held constant through 1992.
- The operator elects to trade in old machinery on new replacements at the end of each item's economic life.

Results Expected

Since this policy includes price supports, income supports, and supply control programs to maintain and stabilize prices and farm income at a reasonable level and reduce the price and income risks, it is anticipated that all farms under this program will have a higher probability of remaining solvent over the 10-year planning horizon, will have higher net farm incomes, and will have stronger financial positions.

Results Obtained

- Except for Texas cotton farms, all farms in the other four regions had a 100-percent probability of remaining solvent over the 10-year period. For Texas cotton farms, the probability of survival ranged from 92 percent for the moderate farms to 94 percent for very large farms.

- All farms in four of the five regions increased their absolute net worth by the end of the period, with very large farms increasing more than the moderate farms. The two smaller farms in Illinois experienced a loss in net worth over the period, while the largest farm experienced a 14.5-percent increase in real net worth.

- On average, all three farms were able to grow by purchasing and leasing cropland. Moderate farms grew faster than the very large farms. The moderate and large grain farms grew at approximately the same rate.

- Average annual net farm incomes for all farms substantially benefited by the presence of price and income supports in the current policy. Removal of these program provisions resulted in negative average annual net farm incomes for farms in all regions except Illinois. (Illinois net farm incomes did not fall below zero because a large portion of cropland was devoted to soybeans, which do not receive a deficiency payment.)

- Ratios of net farm income to total Government payments reveal that, across all regions, the moderate farms were more dependent on Government payments to maintain their incomes than were the very large farms.

Price Supports

The price supports program is designed to prevent prices from falling below a certain level and to stabilize prices through the CCC nonrecourse loans at established loan rates to farmers. Such loans, plus interest and storage costs, can be repaid within 9 to 12 months when the commodity is sold on the cash market. If the market is not favorable for a farmer to sell the commodity and repay his loan, CCC accepts the commodity in full payment of the loan.

CCC releases its stock to the market when prices are high and withdraws stocks from the market when prices are low. Thus the program also stabilizes prices.

Results Expected

- Since price supports stabilize prices and prevent prices from falling below the loan rate, this program should increase farm income and reduce the price risk for farmers.
- All farms should have a higher probability of survival, greater net present value,[9] and higher net farm incomes than they would have had without the program.

Results Obtained

- Price supports increased the probability of survival for all three representative farms in all regions.
- Net farm incomes for these farms also increased with the price supports program. In all regions, the larger the farm, the greater the increase in net farm income.
- With increased farm incomes and reduced price risk, all three farms in all regions experienced increases in real net worth.
- Average ending farm sizes were not significantly different because of the price support program.

Income Supports

Income supports are accomplished through deficiency payments and the target price. Deficiency payments are paid to farmers to make up the difference between a price determined to achieve a politically acceptable income level (target price) and the average market price. Deficiency payments are made on each farm's base acres and farm program yield. The farm program yield is based on each farm's yield history. Target prices were set initially to reflect an average cost of production.

Deficiency payments were initiated to raise and stabilize farmer incomes to the level of the nonfarm population while allowing farm prices to be competitive in the export market. Total annual Government payments (deficiency and diversion) were limited to $50,000.

Results Expected

- The major impact of deficiency payments should be to increase the income level of producers who participate in the farm program. Since the payments are based on the quantity of eligible production, large-scale producers benefit more than small-scale producers, up to the $50,000-payment limitation.
- Deficiency payments also reduce income risk for producers, increase their ability to obtain financing, and thus increase the probability of all farms remaining solvent.

Results Obtained

- The deficiency payment program increased the probability of survival more for moderate Texas cotton farms than for the very large Texas farm. For farms of other regions, the

[9]The concept of present value is used to help measure the profit potential of an investment decision. Simply put, a dollar today is worth more than a dollar in the future because today's dollar can be invested and can accrue interest. Thus the present value of a specified amount of money payable at a specified future date is the amount of money that one would have to invest now in order to have that future amount by that future date. In analyzing an investment over several periods, a positive present value would indicate an economically attractive decision; a negative present value would not.

probability of survival was 100 percent, with or without income supports.

- Income supports increased net farm incomes substantially for all farms, often moving net farm incomes from negative to positive.
- Income supports enhanced net farm incomes of all farms more than the price support program.
- The presence of the $50,000-payment limitation causes the income support program to benefit moderate farms relatively more than very large farms. In contrast, the price support program results in a greater advantage for large and very large farms.
- With reduced income risk and greater farm incomes under the income support program, all farms improved real wealth, and average after-tax net present value increased for all farms.
- Income supports increased the average ending farm size for all farms. Average ending farm size increased at a faster rate for moderate farms than for very large farms.
- Removal of the $50,000 limitation on deficiency payments benefited larger farms more than smaller farms. Big winners of this program were big farms in Texas and Mississippi. In Texas, for example, when the $50,000-payment limitation was removed, average annual net farm income increased $3,600, $50,000, and $104,000 for moderate, large, and very large farms, respectively.
- Increased farm income strengthened the financial positions of larger farms, increasing their ability to obtain more financing. All three representative farms, especially the very large farms, had increased net worth at the end of the 10-year period. For example, removal of the $50,000 limitation increased the ending net worth of the moderate Texas cotton farm by $37,000, of the large Texas farm by $441,000, and of the very large Texas farm by $1,019,000.

Supply Control Policy (Acreage Reduction Program)

The objective of acreage reduction programs is to reduce the quantity produced, and thus the supply, of a given commodity. Acreage reduction consists of an acreage set-aside and/or acreage diversion that is generally voluntary. Acreage set-aside programs require participating farmers to idle a percentage of their crop base acres so that they are eligible for other program benefits. Acreage diversion programs pay producers a given amount per acre to idle a percentage of their base acres. A farmer's base acres are determined by the production history of the crop.

For this analysis the provisions of the current policy were modified by adding a 15-percent set-aside with a 5-percent diversion for corn, cotton, rice, sorghum, and wheat in 1986-92. Normal slippage[10] (30 percent for corn and 70 percent for all other crops) and program participation rates were used to estimate the resulting real increase in mean prices for these crops in 1986-92. All other provisions of the current policy were used without change.

Results Expected

- To the extent that acreage reduction programs reduce production, they reduce supply and stocks and increase prices domestically for those commodities. Higher prices will result in higher total and net incomes for all farm sizes. Farms that participate in diversion payments also benefit from the program through increased cash receipts, up to the $50,000 limit.
- Slippage in the programs reduces the programs' effectiveness, increases the farms' net present value, and increases farm size.

[10]Slippage is the difference between the percent of production decrease and the percent of acreage reduced. These two percentages are different because farmers tend to set aside marginal lands in Government programs or intensify the cultivation of remaining land.

- Higher incomes lead to more disposable income for debt repayment and retained earnings for accelerating farm growth.
- Farm operators' average net present value should increase.
- Faster rates of growth should be experienced by the farms because of increased cash accumulation, repayment capacity, and equity in existing land assets.

Results Obtained

- Imposing a 20-percent acreage reduction program increased the average net present value and ending net worth for all three farms in all regions except for the large farm in Illinois.
- Imposing a 20-percent acreage reduction to existing farm programs resulted in an increase of 20 to 300 percent in net farm income for almost all farms.
- Average ending farm size for all three farm sizes increased relative to the initial farm size.
- Imposing additional supply controls to existing farm programs does not substantially change the rate of growth or ending farm size of all farms. Moderate farms continued to grow faster than larger farms.
- Eliminating slippage reduced the rate of growth relative to that in the current policy for all three farm sizes.
- The less slippage in an acreage reduction program, the smaller the increase in average net present value for all three farm sizes.

No Farm Program

In the no-farm-program scenario, all farm programs outlined for the current policy were eliminated for all 10 years of the planning horizon. In this essentially free market environment, farm prices and income are very unstable because: 1) production varies, owing to weather and biological factors; and 2) demand for farm products changes. The inelastic nature of supply and demand for farm products makes farm prices particularly unstable. The variability in prices and incomes has both favorable and unfavorable aspects. From a favorable perspective, the movement in prices reflects changes in supply and demand conditions and is a signal for production regarding market needs. However, when prices become highly unstable, the signals may be misinterpreted, and mistakes may be made in production and marketing decisions. The result frequently is misallocation of resources. In addition, variability in price and income increases the risk and uncertainty to the farm business.

Results Expected

- Average farm incomes will be less with no loans or price supports because the floor on prices received for these commodities has been removed, allowing prices to fluctuate freely.
- Net present value will be lower and more unstable than with price and income supports.
- Net worth of farms will decline because the market value of cropland will be less, since there are no benefits from the programs to be capitalized into the land.
- Farms will have less probability of survival because of increased instability in prices for crops. The impact will be more pronounced for highly leveraged farms that cannot survive without price and/or income support and for smaller farms that cannot survive with high price risk.

Results Obtained

- Removing all farm programs reduced the probability of survival for all three farm sizes in cotton and wheat regions, relative to the base policy. The probability of survival fell more for the moderate farms in these regions than for the very large farms. For example, in cotton the moderate farm's chance of remaining solvent for 10 years decreased from 92 to 42 percent. The chance for the solvency of very large farms decreased from 94 to 78 percent.

- The probability of having a positive after-tax net present value declined significantly for all farm sizes in all regions except for those farms in the Mississippi Delta. For example, in the Southern Plains the probability of a positive net present value for the moderate farm declined to about 10 percent. In most cases the very large farms had a higher probability of positive net present value than did the moderate farms. The probability of a positive net present value was 100 percent in the Mississippi Delta without the farm program, owing primarily to diversification of crop production and the reduced relative yield variability in the Delta compared with that of the other regions.
- Ending net worth declined for all three farm sizes in all regions. In most regions the absolute decline in net worth was greater for the large and very large farms than for the moderate farms. For example, the large and very large Texas cotton farms experienced a decline of $743,000 and $1,100,800 in net worth, respectively, from that of the current policy, while the moderate farms' net worth declined $396,800. The ending net worth of the Mississippi Delta farms declined the least of all regions because a significant portion of crop acreage was devoted to soybeans.
- In the absence of farm programs, all three farm sizes continued to grow in all regions, but at a much slower rate than under the current policy. For example, farms in the Southern Plains declined from the current policy an average of about 20 percent in ending farm size.

Target Farm Program Benefits

For the target farm program benefits scenario, all farm program and income tax provisions of the current policy were used except that large farms were not eligible to participate in farm program provisions. Farms producing more than $300,000 worth of program commodities (corn, cotton, rice, sorghum, soybeans, and rice) valued at their localized loan rate were not permitted to participate directly in the program provisions (CCC loan, FOR, target price/deficiency payments, and diversions/set-asides). Mean prices and relative variability in prices were not adjusted because it was assumed that a sufficient number of "small" farms participated in the farm program for the price support actions of the CCC loan and FOR to function normally.

Results Expected

- Findings for moderate farms will be the same as the findings for the current policy.
- Large and very large farms exempted from the programs will receive indirect benefits from other farms participating in the programs.
- Compared to the no-farm-program scenario, the following should be observed for large and very large farms:
  —Net present value will be higher and more stable.
  —Net worth of these farms will be greater.
  —Farms will have a greater probability of survival because of the increased stability in prices.
  —Farms will be larger because of increased income and large repayment capacity.

Results Obtained

- Moderate farms consistently producing less than $300,000 in program crops exhibited the same growth rates, net farm incomes, and ending financial positions as they do under the current policy.
- Farms that grew beyond or were initially larger than the $300,000 threshold level of sales experienced lower average Government payments, net farm incomes, average net present values, and net worths than under the current policy, owing to targeting program benefits.
- The larger the farm, the greater the reduction in average ending acres from the current policy for farms in the Southern Plains, Nebraska, and Illinois. Moderate grain farms in these regions experienced no real change in average ending farm size, because their level of total sales was less than $300,000.

- Growth rates for the very large farms in Texas and the Delta were similar to those experienced under the no-farm-program option. The moderate and large farms in the Delta experienced reduced rates of growth relative to that of the very large farms. A similar relationship was observed between the large and very large cotton farms in Texas. The reason for these different rates of growth is that the very large farms in these regions depend less on farm programs than smaller size farms do.

Reduced Income Tax Benefits and Current Farm Program

The Federal income tax provisions in place for the current policy were made more restrictive in the reduced income tax benefits and base farm program scenario. All farm policy provisions of the current policy were left unchanged. The more restrictive Federal income tax provisions included the following:

- Machinery, livestock, and buildings were depreciated using the straight-line cost recovery method.
- First-year expensing provisions were eliminated for all depreciable items.
- Maximum ITC provisions were eliminated.
- The maximum annual interest expense that could be used to reduce taxable income was $15,600.
- The operator was required to sell obsolete machinery upon disposition rather than trading it in on new replacements, thus forcing recapture of excess depreciation deductions.

Results Expected

- Making Federal income tax policies less favorable tends to increase income tax payments by reducing tax deductions. Net cash farm income is not affected directly in the first 4 to 6 years. After that, interest income usually becomes a factor, and higher tax payments in the first 4 years reduce cash available for interest income in later years.
- The farm operator will have lower tax deductions and tax credits when machinery is replaced. The length of time machinery is kept will not likely be shortened from the current policy because machinery was replaced based on its normal economic life, not its depreciation life.
- Reducing tax deductions and tax credits will mean greater annual income tax payments, resulting in greater cash flow requirements and reduced ending cash reserves. Net present value will likely be reduced because of lower retained earnings and the slower accumulation of wealth.

Results Obtained

- Adoption of a more restrictive set of Federal income tax provisions had little impact on farm survival.
- Increasing the Federal tax burden on farmers reduced the average annual rate of growth in farm size about the same for all sizes of farms in each region. Average ending farm size was about 8 percent less than that for the current policy for large and very large farms and about 4 percent less for moderate farms.
- The more restrictive income tax provisions reduced the propensity to grow through purchasing cropland and increased the propensity to lease cropland for growth. For example, in the Mississippi Delta the growth rate in owned cropland for the moderate farm was reduced to 4 percent, and the growth rate in leased cropland increased by 49 percent.
- The changes in the tax provisions resulted in reduced annual net farm incomes for all sizes of farms in all regions. The reduction in net farm income was greater for the very large farm relative to the moderate farm because the very large farm had more depreciable items affected by changes in depreciation rules, ITC s, and capital gains treatment of sales of used machinery.

Technology Scenarios

To determine the impact of technology on structure, selected farm policy scenarios were simulated, assuming increases in mean yields of crops only from the use of existing technologies. A comparison of these simulated results with those of the previous farm policy scenarios, which included increases in mean yields from emerging technologies, indicates the impact of new technology on structure. Three policy alternatives were analyzed under these conditions. They were the base farm policy, which continues all provisions of the 1981 farm bill, the elimination of income support provisions, and the elimination of all farm program provisions.

Results Expected

- The longer the technology is in use for each farm, the greater should be the benefit to wealth accumulation, net income, and rate of growth in acres controlled.
- The greater the increase in productivity, the greater should be the increases in wealth, net income, and rate of growth in acres controlled.

Results Obtained

- Farm policies had more effect on the final amount of acres controlled than did technology, across all sizes of farms in all regions.
- Technology had a greater impact on the final amount of acres controlled for the very large farms in all regions (except Nebraska) than for the moderate and large farms. Yield-enhancing benefits from emerging technologies increased average final farm size 1 to 2 percent in the Delta, Illinois, and Texas and 10 percent in the Southern Plains. The greatest increase in farm size occurred in the Southern Plains, because these farms are principally wheat producers. The greatest increases in yields were predicted by OTA to occur for wheat.
- Small increases in final farm size for the other regions can be explained by the relatively smaller increases in yields (based on the results of OTA workshops for corn, soybeans, cotton, and rice).
- Flows of new technology for all commodities in all regions were found to increase annual net farm incomes for each size of farm. Net farm income was increased relatively more for the very large farms than for the moderate farms, across all farm policies evaluated.

Summary and Conclusions

- Farm programs have major impacts on rates of growth in farm size, wealth, and incomes of commercial farmers.
- Most farm program benefits are capitalized into land values and net worth. Very large farms increase their net worth significantly more than moderate farms under current farm programs and account for a very large share of the program payments.
- Moderate farms depend more than very large farms on farm programs to maintain their incomes.
- Income supports provide significantly greater benefits to moderate farms than to very large farms. (In contrast, price supports provide more wealth and growth benefits to very large farms.) Targeting of income supports to moderate farms is an effective policy for prolonging those farms' survival.
- Very large farms can survive without income supports.
- Adoption of a more restrictive set of Federal income tax provisions had little impact on farm survival.
- Farm policies had more effect on the final amount of acres controlled than did technology, across all sizes of farms in all regions. However, in a relative context, technology had a greater impact on the final amount of acres controlled for the very large farm than for the moderate and large farms.
- Flow of new technology will increase annual net farm income for all sizes of farms. How-

ever, net farm income increased more for the very large farms than for the moderate farms.

Financial Stress and New Entrants Scenarios

Financial Stress Scenario

The financial position of many farmers is currently under severe stress. The situation is serious and may not improve for some time. Policymakers are considering various solutions to this problem. Two of the most discussed alternatives are interest rate subsidy and debt restructuring.

An *interest rate subsidy* is a loan at below-market interest rates. For example, if the Government's cost of money is 11 percent and the Farmers Home Administration makes loans at 5 percent, there is a 6-percent direct interest rate subsidy. To analyze the effects of such a credit policy, the financial positions of the three representative farms in each of the five regions were modified to depict highly leveraged farms. The long-term debt-to-asset ratio for each farm was increased to 55 percent, the intermediate-term debt-to-asset ratios were set equal to 60 percent, and annual interest rates on old loans were increased to their average values for 1980 to 1983.

The object of an interest rate subsidy is to reduce the cash expenses for interest costs, thus increasing total net cash farm income. The total cash requirements are reduced, thereby benefiting all farms. The total saving is greater for larger farms because of the total debt being larger on these farms.

Debt restructuring refers to the rescheduling of loan commitments. Debt may be restructured by rewriting short- or intermediate-term debt to a long-term basis if the collateral justifies such change. The amount paid per year is then reduced. Without sufficient additional long-term collateral, debt restructuring is limited to rescheduling each class of loans—short-, intermediate-, and long-term—over a longer repayment period. Also, if the debt is on a fixed interest rate basis and interest rates have declined, the debt might be rescheduled in part to take advantage of lower interest rates and to obtain a longer repayment period.

Restructuring debt has the same type of expected effects as interest rate subsidy; however, the methods differ. Debt restructuring does not reduce the annual interest payments in the initial period unless long-term interest rates are less than intermediate-term interest rates. Annual principal payments are reduced, thus reducing cash flow needs of the farm operator.

Results Expected for Interest Rate Subsidy and Restructuring Debt

- Higher probability of survival.
- Higher land values, net worth, and average net present value.
- An increase in the equity ratio because current debts are paid and longer-term debts are reduced, allowing greater opportunity for the farm to grow in size because of the increased ability to leverage existing equity.

Results Obtained From Financial Stress Scenarios

- Restructuring initial debt for highly leveraged farms failed to increase appreciably the probability of survival for each size of farm in any region except for moderate and large wheat farms in the Southern Plains.
- In all regions, the interest rate subsidy strategy substantially increased the average net farm income more than did the restructuring of farms' debts.
- Both debt restructuring and interest rate subsidy policies resulted in increased growth in farm size and real wealth (i.e., ending net worth) on the very large farms in all regions. In all regions but Texas, very large farms with high debts are not as dependent on financial bailout strategies for survival as moderate farms are.

- Both alternatives increased growth in farm size. Debt restructuring resulted in more rapid rates of growth than did interest rate subsidies.

New Entrants Into Farming Scenario

All previous simulations of the effects from the farm commodity policy alternatives were based on representative farms operated by *established* farm producers. These simulations provide indications of the short-run effects of the alternative farm commodity policy provisions on economic survival and growth characteristics of established farm operations. They do not provide information on the survivability and economic viability of potentially new entrants into farming. To gain some general notions of the effects of selected farm commodity policies on newly established farming operations, the smallest farm in each region was simulated under the condition that the farm operator was a new entrant.

In this scenario the entering farm operator was allowed to have only minimum equity in owned farmland (30 percent) and farm machinery (35 percent). All farm machinery was considered to have a new machinery cost, and annual interest rates on long- and intermediate-term loans were equal to the 1980-83 averages. The operator was not allowed to have any off-farm investments. Because the farm operator was paying the full cost of all inputs (land, capital, machinery, and labor), these simulations provide an indication of long-run survivability and profitability of the representative farms. Three policy alternatives were analyzed under these conditions for the new entrant. They were the base farm policy, which continues all provisions of the 1981 farm bill, the elimination of the target price/deficiency payments provision of the program (no income support provisions), and the elimination of all farm program provisions.

Results Expected

- New entrants would be expected to face lower probabilities of survival, slower rates of real wealth accumulation, and slower rates of growth in farm size than would current operators on the representative farms in each region under existing farm legislation. Because both depreciation adjustments on machinery and annual cash requirements for debt repayment on real estate and machinery loans are based on new 1982 costs and current (1980-83) interest rates, annual net farm incomes will be lower for new entrants than for current operators, under existing policy.
- Elimination of income support provisions of the 1981 farm bill will be expected to reduce the probability of survival, rate of growth in real net worth and farm size, and annual net farm incomes of new entrants in each region. The greatest impacts would be expected for specialized crop farms producing commodities eligible for target prices and deficiency payments. Elimination of all farm program provisions would be expected to reduce further the rate of growth in real wealth and farm size. Annual net farm incomes for new entrants would be expected to be even lower, particularly on representative farms producing commodities eligible for set-asides and paid diversion provisions.

Results Obtained

- New entrants exhibited considerably lower probabilities of survival under the base farm policy than did current operators for all specialized crop farms. The diversified crop farms in Nebraska and the Mississippi Delta exhibited relatively high probabilities of survival for new entrants.
- New entrants experienced much lower rates of real wealth accumulation than did current operators under current policy. In two of the regions—High Plains wheat farm and Nebraska and Illinois crop farms—real net worth after 10 years was lower than initial net worth on the farms, indicating that the new entrant operator had to sell owned cropland to remain solvent. Net farm incomes were negative for all farms, with the High Plains wheat farm experiencing the largest relative decline in annual net income.

- New entrant farm operators in the High Plains wheat and Nebraska and Illinois crop regions were unable to increase farm size over the 10-year period under current farm policy. The Texas cotton farm and the Mississippi Delta crops farms experienced considerable growth, 20 and 33 percent, respectively.
- Eliminating the target price/deficiency payments provision of current legislation substantially decreased the probability of survival and ending net worth on all farms. Only the Texas cotton farms exhibited any appreciable growth in farm average (about 10 percent).
- Under the policy alternative of no farm programs, none of the farms exhibited reasonable potentials for remaining solvent over the 10 years. Farms in the Texas High Plains, Southern Plains, and Corn Belt had less than a 10-percent probability of survival. Mississippi Delta farms had only a 60-percent chance for remaining solvent over the 10 years.
- Under the current farm program only the Nebraska and Mississippi Delta crop farms had sufficient returns for new farmers to enter agriculture with a reasonable chance of remaining solvent and making a reasonable return on their investment.
- Elimination of income support, price support, and supply control provisions of current farm policy resulted in new entrant farmers in all five regions facing little chance of surviving and becoming an economically viable farming operation.
- Other sources of income, economic assistance, or wealth accumulation will be required for these new entrants to survive economically in an open market farm policy environment.

Summary and Conclusions

- Restructuring of debt for highly leveraged farms does not appreciably increase their probability of survival.
- Interest rate subsidy substantially increases average net farm income more than debt restructuring. It is, therefore, a more effective strategy for easing financial stress.
- Very large farms with high debts do not depend on these financial programs for survival as moderate farms do. Under these programs, very large farms will grow significantly in farm size and real wealth.
- New entrants into agriculture will not likely survive even with current farm programs. Other sources of income, economic assistance, or wealth accumulation will be required.

CHAPTER 8 REFERENCES

Cochrane, Willard W., "Farm Commodity Policy, Technological Advance and Structural Change," paper prepared for the Policy, Technology, and Structure of Agriculture Workshop, Office of Technology Assessment, 1983.

Cooke, Stephen C., "Size Economies and Comparative Advantage in the Production of Corn, Soybeans, Wheat, Rice, and Cotton in Various Areas of the United States," prepared for the Office of Technology Assessment, Washington, DC, 1985.

Smith, E.G., "Economic Impact of Current and Alternative Farm Programs on Farm Structure on the Southern High Plains of Texas," unpublished doctoral dissertation, Texas A&M University, 1982.

Chapter 9

Emerging Technologies, Public Policy, and Various Size Dairy Farms

Contents

Tables

Figures

Chapter 9

Emerging Technologies, Public Policy, and Various Size Dairy Farms

One of the most controversial areas involving agricultural technology, policy, and structural change in the United States is in the dairy sector. In 1983 the large amount of surplus milk production cost taxpayers approximately $2.6 billion. Emerging technologies promise to dramatically increase milk production per cow by year 2000, from a national average of 12,000 pounds in 1982 to over 24,000 pounds in 2000. As discussed in chapter 3 a reduction of approximately 30 percent in cow numbers will be needed by year 2000 to counteract the effect of the emerging technologies and the static demand for milk and milk products. Thus the impact of these technologies and policy on the dairy industry will be dramatic. This chapter attempts to provide the foundation for understanding these changes and for analyzing various policies to cope with the dynamic interaction between policy and technological advance.

One of the changes will be a major regional shift in milk production: the Midwest and the Northeast will lose their comparative advantage to the Southwest. During the 1970s milk production increased 41 percent in the Southwest region of the United States, while U.S. milk production increased only 11 percent (figure 9-1). Much of the increased production came from dairies with more than 500 cows, with herds of 1,500 to 2,000 cows being common. Although 303,710 farms in the United States reported having milk cows in 1983, all the milk that sold that year could have been produced by less than 5,000 well-managed dairies with 1,500 cows each.

This chapter examines important economic factors that will affect the trend to fewer and larger dairies and the regional shift in milk production. The first part of the chapter estimates size economies and comparative advantage of milk production for moderate, large, and very large dairy operations in five major U.S. dairy areas. These comparisons provide an indication of the most competitive farm sizes and regions. They are based on returns on investment after all costs are paid, including the regional

Figure 9-1.—How the Dairying Picture Has Changed
(percent change in milk production in various regions from 1970-71 to 1980-81)

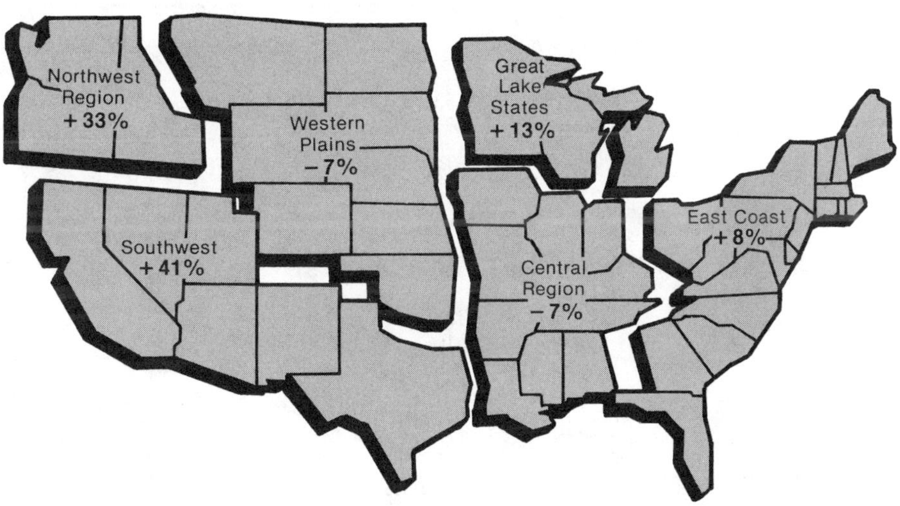

SOURCE: U.S. Department of Agriculture.

replacement of depreciable assets needed to maintain the long-term productive capacity and viability of the farm.

The second part of the chapter develops a beginning financial situation for eight dairy operations in three regions. The ability of these operations to remain solvent and increase net worth over a 10-year planning horizon is simulated under conditions of risk and under alternative policy and technology scenarios. These results provide an indication of how alternative policies affect individual dairy farm operations.[1]

[1]The representative farms were developed and analyzed in the OTA paper "Economic, Policy, and Technology Factors Affecting Herd Size and Regional Location of U.S. Milk Production," prepared by Boyd M. Buxton.

SIZE ECONOMIES AND COMPARATIVE ADVANTAGE IN MILK PRODUCTION

Dairy Operations Analyzed

Herd size, technologies employed, and practices used in milk production vary considerably throughout the United States. In May 1983 the average herd size for 120,655 producers selling milk to plants regulated by Federal milk marketing orders was 63 cows per farm (table 9-1). However, the average herd size in each State varied from 49 cows in Pennsylvania, to 532 cows in Florida.

The variation in herd size within each State was even more dramatic. Although the average herd size in Florida was 532 cows, the average herd size for the largest 10 percent of the herds in that State was 1,861 cows (table 9-1). Similarly, the average herd size for the largest 10 percent of herds regionally was about 1,700 cows in the Southwest, but only 125 cows in the Great Lake States region. Generally, dairy herds are much larger in the Southwest, Southeast, and Northwest than in the Great Lake States and the Northeast.

From the herd size information in table 9-1, 22 dairies were selected to represent existing herd sizes in five major dairy areas (table 9-2).

Table 9-1.—Total Producers and Size Distribution of Herds Selling Milk to Plants Regulated by Federal Milk Marketing Orders, May 1983[a]

		Average herd size (milk cows)				
Region (State)	Number of total producers	All farms	Largest 10%	70 to 89%	40 to 69%	Smallest 40%
Great Lake States:						
Minnesota	9,968	53	116	74	49	30
Wisconsin	24,400	54	133	68	52	28
Northeast:						
Pennsylvania	12,928	49	127	66	44	25
New York	13,374	59	162	81	53	27
Southeast:						
Georgia	962	127	343	181	117	54
Florida	352	532	1,861	931	355	133
Southwest:						
New Mexico	176	333	1,832	433	169	32
Arizona	160	510	1,733	714	433	160
California	13	400	1,640	580	253	110
Northwest:						
Idaho	574	135	607	169	90	34
Washington	1,647	127	418	171	108	46
United States	120,655	63	202	82	54	26

[a]The 120,655 farms accounted for about 69 percent of all milk produced in May 1983, but excluded most farms in California and other States where there is no Federal milk order.

SOURCE: Boyd M. Buxton and John P. Rourke, "Size Distribution of Dairy Farms Marketing Milk Under Federal Milk Orders," unpublished report, U.S. Department of Agriculture, Economic Research Service, April 1984.

Table 9-2.—Representative Dairies Selected for Preparation of Whole Farm Budgets, by Region and Herd Size

Region/State	Herd size (cows)	Cropland (acres)	Housing facilities[a] (type)	Sun shades	Feed produced	Silage storage (type)	Total labor (W/e)[b]
Great Lake States:							
Minnesota	52	188	Stanchion	No	Most	Upright	2.03
Minnesota	125	449	Free stall	No	Most	Upright	3.30
Northeast:							
Pennsylvania	52	156	Stanchion	No	Forage	Trench	2.2
Pennsylvania	125	375	Free stall	No	Forage	Trench	3.8
Pennsylvania	200	600	Free stall	No	Forage	Trench	5.54
New York	52	156	Stanchion	No	Forage	Trench	2.21
New York	200	600	Free stall	No	Forage	Trench	5.54
New York	600	1,800	Free stall	No	Forage	Trench	14.36
Southeast:							
Georgia	200	400	Free stall	Yes	Forage	Trench	4.5
Georgia	350	700	Free stall	Yes	Forage	Trench	7.84
Florida	350	0	Open field	Yes	None	NA	7
Florida	600	0	Open field	Yes	None	NA	11
Florida	1,436	0	Open field	Yes	None	NA	18
Southwest:							
New Mexico	900	0	Corral	Yes	None	NA	13
Arizona	359	0	Corral	Yes	None	NA	7
Arizona	834	0	Corral	Yes	None	NA	12
Arizona	1,436	0	Corral	Yes	None	NA	16
California	550	0	Corral	Yes	None	NA	9
California	1,436	0	Corral	Yes	None	NA	16
Northwest:							
Washington	140	51	Free stall	No	Silage	Trench	2.96
Idaho	200	400	Corral	No	Most	Trench	5.0
Idaho	550	0	Corral	No	None	NA	10.5

[a]Housing types are:
- *Stanchion* — A conventional barn with locking stanchions in which cows are milked and fed.
- *Free stall* — A covered barn with individual stalls in which cows freely enter and exit.
- *Open field* — A field where cows are kept that is large enough to maintain plant cover.
- *Corral* — A drylot open pen where cows are kept and fed at a fenceline feeder.

[b]Labor in worker equivalents of 2,500 hours annually.
NA—not applicable.

SOURCE: Office of Technology Assessment.

The 200-cow Pennsylvania and 600-cow New York dairies exceed the average size of the largest 10 percent of dairies in those States. However, such larger sized dairies exist in these States and will become more prevalent in the near future.

Technologies and Practices

The technologies and practices assumed for each of the 22 dairy operations were based on discussions with dairy producers, university and Government employees, and equipment representatives. The objective of these discussions was to describe efficiently organized dairy operations that use proven technologies and practices for each specified herd size. Therefore, the dairy operations in this analysis are not the average of what now exists, but rather approximate modern sizes and types of operations.

The 52-cow dairies in Minnesota, Pennsylvania, and New York use the conventional stanchion barns for housing and milking cows (table 9-2). For larger herds in the Great Lake States, the Northeast, Washington, and Georgia, free-stall housing and milking parlors are assumed.

Cows are kept in open corrals throughout the Southwest and on larger Idaho dairies. Sun shades in the corrals are assumed for farms in New Mexico, Arizona, and California (Southwest), but not in Idaho. Cows are milked twice a day in milking parlors and fed at fenceline bunks from a feed wagon or truck.

Open fields with sun shades are assumed for farms in Florida. One-half acre per cow is provided, allowing fields to remain grass-covered to minimize mud problems. Cows are milked twice a day in a milking parlor. After leaving the milking parlor, they are fed concentrates in a feed barn before being released back to the field. Roughage is fed loose in the open fields.

The source of feed follows the common practice existing in the various States. For New Mexico, Arizona, California, and Florida, most feed is purchased from off the dairy operation. The same is assumed for the 550-cow Idaho dairy. Dairy operations in Pennsylvania, New York, and Georgia purchase most of the concentrates but produce most of the forage used by their dairy herds. All feed is assumed to be produced on farm for the Minnesota and the 200-cow Idaho dairies.

Costs and Returns

The specialized dairy operations considered in this chapter receive all revenue from the dairy enterprise. Milk sales are the single largest source of revenue, but the sale of cull cows, bull calves, and replacement heifers are also important. The prices received for milk delivered to plants vary from one State to another, largely reflecting the classified pricing policy of Federal and State milk orders and the proportion of milk used as fluid in the various States.

Costs are divided into operating and ownership costs. Operating costs include purchased feed and a wide range of expenses such as farm repairs, hired and operator labor, utilities and fuel for the dairy herd, and veterinary and breeding fees. Annual ownership costs include depreciation, property taxes, and insurance premiums.

Based on the above, the estimated costs per cow for assets required on the 22 dairies are illustrated in figure 9-2. These costs reflect an amount sufficient to replace wornout assets when needed and thus reflect an amount needed to maintain the long-term viability of the operation.

In calculating relative rate of return for these dairies, milk prices received by dairy operators were assumed to be those prices received in

Figure 9-2.—Total Cost per Hundredweight of Milk by Herd Size and State, 1982

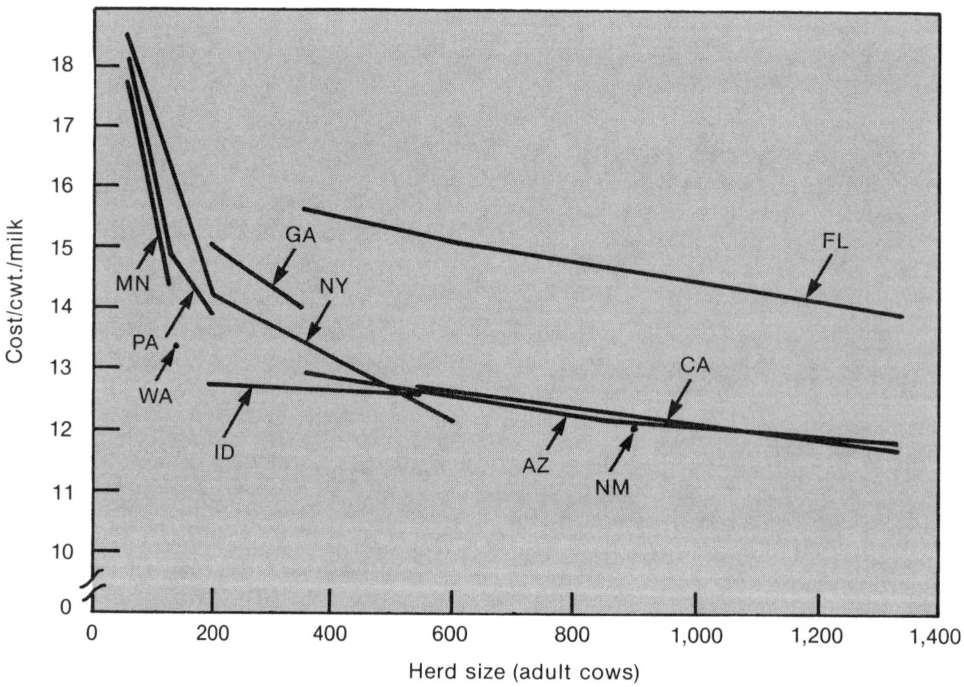

SOURCE: Office of Technology Assessment.

1982. The price level varies from $12.70 per hundredweight (cwt) in Idaho to $16.40 in Florida. The difference in price between States is due in large part to pricing policies under Federal and State milk marketing orders. States with relatively high prices are areas where milk used as a fluid beverage is priced relatively high and is a relatively large share of total sales.

Given the above assumptions, costs and returns for the 22 operations were calculated (figure 9-3). The rate of return ranged from −2.15 percent on the 52-cow New York dairy to about 15.72 percent for the 1,436-cow Florida and 900-cow New Mexico dairies.[2] The differences are due mostly to herd size. The differences between New Mexico, Arizona, and Florida (relatively high return) and California (relatively low return) in part reflect differences in milk prices.

[2]New Mexico, Arizona, California, Idaho, and Washington costs and returns were based on the current subsidized irrigation costs for water to produce alfalfa hay. If the irrigated water were priced to reflect actual costs more closely, which are about three times the subsidized costs, the rate of return would be 2 to 3 percentage points below the rates shown in figure 9-3 for these States. For details of this analysis see Boyd M. Buxton, "Economic, Policy, and Technology Factors Affecting Herd Size and Regional Location of U.S. Milk Production," paper prepared for the U.S. Congress, Office of Technology Assessment, 1985.

Note that the rate of return for the 600-cow New York dairy was favorable compared with that for herds of similar sizes in other States. The differences between the 600-cow Florida, 600-cow New York, 550-cow California, and 550-cow Idaho dairies are in part related to differences in milk prices.

Summary and Conclusions

- Investment or replacement cost per cow is less on larger farms.
- For herd sizes that characterize dairy farming in each region, investment per cow is less for the large dairy operations in the Southwest, Northwest, and Southwest than for the Great Lake States and Northeast regions.
- The larger dairies with 500 cows or more are more profitable than smaller dairies. Dairies in New Mexico, Arizona, and Florida are

Figure 9-3.—Long-Term Rate of Return to Investment by Herd Size and State

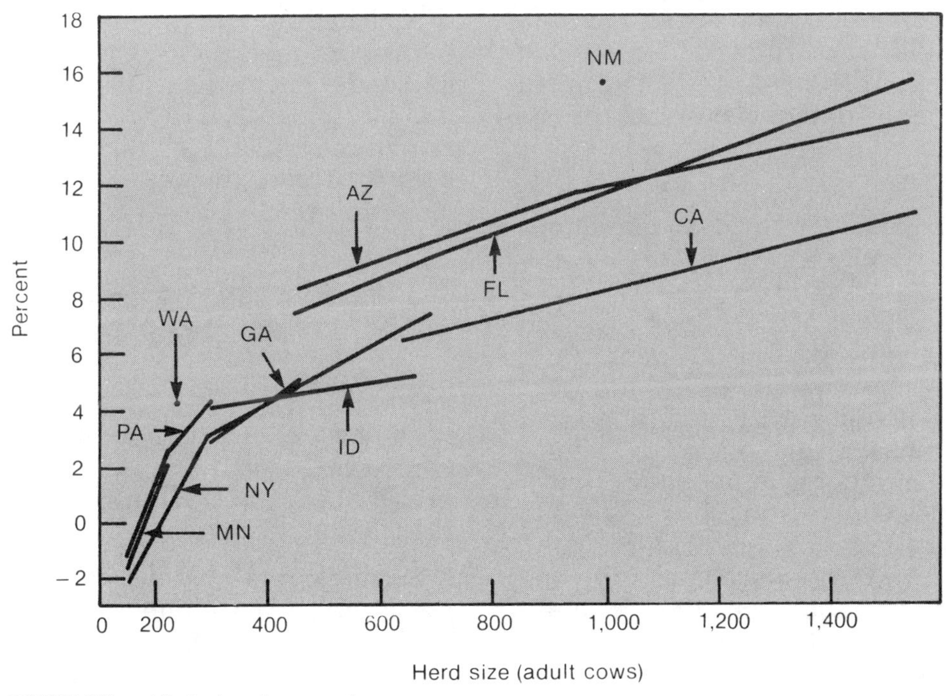

SOURCE: Office of Technology Assessment.

more profitable than their counterparts in Minnesota and the Northeast.

- Although costs are highest in Florida, the relatively high price received for milk provides a competitive return. The profitability of California and Idaho dairies is adversely affected by lower milk prices than those of New Mexico, Arizona, and Florida dairies.
- Strong economic pressure exists for herds to grow larger in all regions. This will continue the trend to fewer and larger dairies.
- The relatively favorable rate of return of large-scale dairy farming in the Southwest, Southeast, and Northwest regions will likely result, over the long term, in a continued shift in milk production to those areas. Those areas will likely increase their relative share of total U.S. milk production, placing increased competitive pressure on the traditional Great Lake States and Northeast dairy areas.

ECONOMIC IMPACTS OF EMERGING TECHNOLOGIES AND SELECTED POLICIES FOR VARIOUS SIZE DAIRY FARMS

The preceding section considered the long-term relative rates of return of different size dairies in five regions. Implicit in the analysis is that the real production costs will remain constant and the support price cannot be permanently above or below the price level that will balance supply and demand. Under the present purchase-type price support program, decisions to set support prices above the long-term market clearing level would, in the long run, have to be modified. Otherwise, the Government expenditures would grow to a level unacceptable to policymakers. This means that alternative support levels must reflect market conditions in the long run.

This section considers the economic impacts of selected policy decisions on dairy operations over a 10-year period. Panelists at the OTA animal technology workshop discussed in chapter 2 identified likely new technologies that would be available over the next 10 years, their adoption rate by the industry, and their impact on milk production per cow. They found that the adoption of new technology would reduce the real cost of producing milk. In turn, lower real costs of production would be reflected in lower milk prices. Eight of the 22 dairy operations analyzed in the previous section were selected for analysis about the impact of alternative technologies and policies. Three of the five regions are reprsented: Great Lake States, Southeast, and Southwest.

A Firm Level Income Tax and Farm Policy Simulator used in the previous crop farms analysis, and discussed in detail in appendix E, was used to simulate the eight dairy operations for selected policy and technology scenarios for 10 years, beginning in 1983. The planning horizon was simulated 50 times (iterations) using a different set of: 1) random milk, cull cow, and replacement cow prices; 2) feed costs; and 3) milk production per cow for each iteration. At the end of each iteration, values for present value of net returns (revenue minus cost over 10-year period) and ending net worth, long-term and intermediate-term debt, equity-to-asset ratio, internal rate of return, and net farm income were calculated. The results of OTA's analysis are discussed in further detail in appendix F.

Farm Policy, Tax Policy, and Technology Scenarios

Base Scenario

The base scenario assumptions were those considered most likely over the 10-year planning horizon and are summarized in the following sections.

Technology.—The impact of productivity gains achieved through new technologies and management practices are largely reflected in increased milk production per cow and reductions in the real cost of producing milk. In the longer run, lower production costs are largely passed on to consumers through lower milk prices. Dairy farmers who adopt technology and achieve productivity gains are able to compete and remain financially solvent, whereas those farmers that cannot will likely become insolvent.

The pooled knowledge and judgment of the above-mentioned panel at the OTA workshop identified the most likely new technologies and adoption rates and productivity gains over the 1983-92 period. Although milk production per cow is expected to increase for all herd sizes in all regions, the panel expected operators of larger herds within each region to adopt new technology more rapidly than operators of smaller herds. While the 125-cow dairy is considered very large in Minnesota, it would be considered very small in California, Arizona, or Florida. However, operators of the very large 125-cow Minnesota dairies are expected to adopt new technology as rapidly as operators of the very large 1,436-cow dairies in California and Florida (table 9-3).

Milk production per cow is expected to continue a long-term trend by increasing 1 percent annually for all herd sizes (table 9-4). New technology likely to increase milk production above this long-term trend was grouped into three main categories: 1) information and nutrition, including such technologies as computer management and feeding systems, communication and information systems, improved environmental management, and feed additives;[3] 2) bovine growth hormone (bGH); and 3) other biotechnologies, including embryo transplants and sexing, genetic engineering, and pest and disease control.

Information and nutrition technologies are expected to increase milk production per cow an additional 1.8 percent annually, starting in 1983 for very large dairies, in 1985 for large dairies, and in 1987 for medium dairies (table 9-4).

[3]Information and nutrition technologies were grouped together because their availability to dairy farmers will come at about the same time. Information technologies will account for a 1.2-percent annual increase and nutrition technologies 0.6-percent annual increase for a total of a 1.8-percent annual increase.

Table 9-3.—Very Large, Large, and Moderate Herd Sizes, by State

State	Herd size (milk cows)[a]		
	Very large	Large	Moderate
Minnesota	125	b	52
California	1,436	550	b
Arizona	b	b	359
Florida	1,436	600	350

[a]Size groups based on market order data as found in Boyd M. Buxton and John P. Rourke, "Size Distribution of Dairy Farms Marketing Milk Under Federal Milk Orders," unpublished report, U.S. Department of Agriculture, Economic Research Service, April 1984.
[b]Dairies for these herd size groups were not simulated.

SOURCE: Office of Technology Assessment.

Bovine growth hormone is expected to increase milk per cow 25.6 percent when adopted. This jump in milk production is expected to continue after the adoption of bGH in 1987 by operators of very large dairies, in 1988 by operators of large dairies, and in 1989 by operators of medium dairies. The favorable economic incentives suggest a more rapid adoption than other technology groups once bGH is available.

Increased feed costs per hundredweight of milk due to bGH is estimated at $4.49 for California, Arizona, and Florida dairies and $3.59 for Minnesota dairies. About 90 percent of the increased feed cost would be for concentrates, and 10 percent for forage. The cost is less in Minnesota because concentrate prices are lower there than in other States (Kalter, 1984).

The other biotechnology group, which includes embryo transplants, genetic engineering, and pest and disease control, is expected to increase milk production per cow an additional 0.5 percent annually, starting in 1987 for very large dairies, in 1989 for large dairies, and in 1991 for moderate dairies (table 9-4).

Table 9-5 summarizes the expected milk production per cow for various size herds in three regions, given the above technology assumptions.

Milk Prices.—The base scenario assumes the support price specified in the Dairy and Tobacco Production and Stabilization Act of 1983. The 1984 price likely will be 32 cents lower than the 1983 price, reflecting a 50-cent reduction in the support price on December 1, 1983. The Government purchases of surplus dairy products

Table 9-4.—Year-to-Year Percentage Increase in Milk Production per Cow for Three Technology Groups, by Herd Size, 1983-92

		Information and nutrition			Bovine growth hormone			Other biotechnologies		
Year	Trend	Very large	Large	Medium	Very large	Large	Medium	Very large	Large	Medium
1983	1.0%	1.8%	0	0	0	0	0	0	0	0
1984	1.0	1.8	0	0	0	0	0	0	0	0
1985	1.0	1.8	1.8%	0	0	0	0	0	0	0
1986	1.0	1.8	1.8	0	0	0	0	0	0	0
1987	1.0	1.8	1.8	1.8%	25.6%	0	0	0.5%	0	0
1988	1.0	1.8	1.8	1.8	0	25.6%	0	0.5	0	0
1989	1.0	1.8	1.8	1.8	0	0	25.6%	0.5	0.5%	0
1990	1.0	1.8	1.8	1.8	0	0	0	0.5	0.5	0
1991	1.0	1.8	1.8	1.8	0	0	0	0.5	0.5	0.5%
1992	1.0	1.8	1.8	1.8	0	0	0	0.5	0.5	0.5

[a]Percentage increases are for specified year and are maintained in all subsequent years. Percentage increases are above 1982 production per cow levels of 147.1 cwt in Minnesota, 165.7 cwt in California and Arizona, and 131.1 cwt in Florida.

SOURCE: Office of Technology Assessment, Animal Technology Workshop, Washington, DC, April 1984; and Robert J. Kalter, et al., *Biotechnology and the Dairy Industry: Production Costs and Commercial Potential of the Bovine Growth Hormone*, AE Research 84-22 (Ithaca, NY: Cornell University Department of Agricultural Economics, December 1984).

Table 9-5.—Milk Production per Cow for Most Likely Technology Scenario, 1982-92 (hundredweight)

	Minnesota		California and Arizona			Florida		
Year	Very large	Medium	Very large	Large	Medium	Very large	Large	Medium
1982	147.1	147.1	165.7	165.7	165.7	131.1	131.1	131.1
1983	151.2	148.6	170.3	167.3	167.3	134.8	132.4	132.4
1984	155.5	150.0	175.1	169.0	169.0	138.6	133.7	133.7
1985	159.9	151.5	180.1	173.0	170.0	142.5	137.5	135.0
1986	164.3	153.1	185.1	178.6	172.0	146.4	141.3	136.5
1987	169.0	157.2	190.4	183.6	177.1	150.6	145.3	140.1
1988	225.4	161.7	253.8	188.7	182.1	200.8	149.3	144.1
1989	232.9	166.4	262.3	252.0	187.4	207.5	199.4	148.3
1990	240.7	221.9	271.0	260.0	260.0	214.5	206.1	197.8
1991	248.6	229.3	280.0	269.1	258.3	221.6	212.9	204.4
1992	256.8	237.1	289.3	278.0	267.1	228.9	219.9	211.3

SOURCE: Office of Technology Assessment.

are assumed to be high enough through 1985 and 1986 to trigger a 50-cent drop in support price on April 1, 1985, and again on July 1, 1985, as specified in the above-mentioned 1983 Act. This price is projected through 1986.

From 1987 to 1992 the dairy support price is expected to be reduced 50 cents per year as long as the estimated variable milk production costs, given assumed technological changes and associated declines in real costs, are less than market prices in the previous year. It is expected that the 50-cent-per-hundredweight declines will occur through 1992.[4]

Financial Characteristics.—The likelihood of a particular dairy remaining solvent under alternative policies is directly affected by its initial financial characteristics. The characteristics of most importance include the value of assets, cash reserves, debt, net worth, equity, and family consumption needs. A policy change can have quite different implications for the operator of a dairy with a high level of debt than one with a low level of debt.

The average financial situation that exists on the eight dairies of the size and location selected are shown in table 9-6. The averages were approximated from a U.S. Department of Agriculture farm financial survey.[5] Equity ranged from 69 to 76 percent of total assets. In contrast to

[4]These assumptions approximate the actual policy for dairy as specified in the Food Security Act of 1985.

[5]Summary of financial characteristics of dairy farms were estimated from farm financial summary data provided by Neil Peterson, Economic Research Service, USDA.

Table 9-6.—Financial Characteristics Assumed for Eight Dairy Operations in Four States

	Minnesota		Arizona	California		Florida		
Financial characteristics	52	125	359	550	1,436	350	600	1,436
Value of:								
Cropland and farmstead ($1,000)	293.4	679.1	39.4	160.0	312.0	262.5	450.0	1,074.0
Buildings ($1,000)	92.7	176.7	192.8	284.4	512.6	87.9	108.9	211.7
Farm machinery ($1,000)	104.1	159.0	120.3	183.1	303.0	114.6	180.0	260.7
All livestock ($1,000)	77.9	181.4	599.6	960.7	2,505.0	525.5	981.4	2,344.3
Off-farm investments ($1,000)	5.5	13.1	0	0	0	0	0	0
Beginning cash reserves ($1,000)	12.0	62.5	89.8	137.5	35.9	70.0	212.0	505.5
Debt ($1,000)	268.3	302.4	297.7	464.3	1,130.0	303.7	461.9	944.6
Initial net worth ($1,000)	417.1	969.4	744.2	1,261.3	2,537.5	756.9	1,464.7	3,343.0
Equity ratio (fraction)	0.71	0.76	0.71	0.73	0.69	0.71	0.76	0.76
Family living								
Minimum ($1,000)	20.0	25.0	25	27	30	25	27	30
Maximum ($1,000)	32.0	35.0	30	38	40	30	38	40
Marginal propensity (fraction)	0.3	0.4	0.3	0.4	0.4	0.35	0.4	0.4
Off-farm income ($1,000)	0	0	0	0	0	0	0	0

SOURCE: Office of Technology Assessment.

replacement values used in the previous section, the value of buildings and machinery are market values; it was assumed that each asset was about half depreciated.

Results Expected.—Under the base scenario, it was expected that a well-managed dairy of average size would about break even after paying expenses and farm overhead and making withdrawals for family living. It was also expected that well-managed dairies in all regions should be able to survive under a continuation of the current program. Farms that were not in a position to realize most of the economies of size in dairying would be gradually forced out of business. In other words, an extension of current policy would force dairies to compete on the basis of cost and efficiency.

Results Obtained:

- All dairies except the 52-cow Minnesota operation were able to increase their real net worth over the 10-year planning horizon (table 9-7). The 52-cow dairy experienced a 42-percent reduction in net worth.
- The larger the dairy, the greater its financial success. Dairies in Florida and the Southwest were more profitable than dairies in Minnesota. The Florida dairy benefited greatly from higher milk prices.
- The 52-cow dairy had the lowest probability of survival (74 percent), owing to having the highest unit cost of production. It lost an average of $22,000 annually in net farm income.

A Crop Acreage Reduction Program

The present feed grain program was assumed through 1985. From 1986 to 1992 a 15-percent set-aside with a 5-percent diversion for corn, cotton, rice, sorghum, and wheat was assumed. This program results in dairy feed prices being 9-percent higher than those under the base scenario.

Results Expected.—Feed cost would represent about 50 to 60 percent of total costs per cow. A crop program that results in a 9-percent higher feed cost is roughly equal to a 5-percent reduction in the price of milk. This would have an adverse impact on a dairy's ability to increase net worth, reduce debts, and achieve as high an internal rate of return as under current policy. In the short run, dairies that raise most of their feed would be less directly affected. The probability of survival would probably be reduced for dairies operating at or below the break-even point under the current policy because they would be unable to absorb the higher feed costs.

Table 9-7.—Comparison of Continuation of Present Policy (Base Scenario) on Dairy Farms From Various Regions

Dairy herds	Probability of survival	Beginning net worth	Ending net worth	Average net cash income	Average net income
Minnesota:					
52 cows	74%	$ 417,000	$ 240,000	$−7,000	$−22,000
125 cows	100	969,000	1,120,000	49,000	20,000
California:					
550 cows	96	1,261,000	2,055,000	101,000	10,000
1,436 cows	98	2,538,000	7,332,000	628,000	449,000
Arizona:					
359 cows	96	744,000	1,296,000	77,000	14,000
Florida:					
350 cows	96	757,000	1,004,000	41,000	−6,000
600 cows	100	1,465,000	2,453,000	153,000	83,000
1,436 cows	100	3,343,000	9,257,000	759,000	635,000

SOURCE: Office of Technology Assessment.

Results Obtained:

- The associated higher feed prices had the greatest adverse financial impact on dairies that purchased most of the feed from off the farm. For example, compared with that of the current policy, the average annual net farm income of the 1,436-cow California dairy declined 62 percent, from $449,000 to $171,000.
- The probability of survival was reduced for all dairies except the 1,436-cow Florida dairy and the 125-cow Minnesota dairy.
- There was relatively little impact on Minnesota dairies, where most feed is raised at the dairies.

No Crop Programs

There is much discussion of a desire to move to more market-oriented crop programs. Removing all price supports and income supports would increase the variability of feed prices, subjecting the dairyman who purchases feed to greater risk. For this scenario the Commodity Credit Corporation (CCC) loan, farmer-owned reserve (FOR), and target price provisions were eliminated for all years in the planning horizon (1983-92). This increased the variability in feed costs facing dairy operations. The impact of this variability was evaluated.

Results Expected.—Feed prices paid by dairies would be higher in some years but lower in other years. Over time, high and low price years would be expected to balance out, leaving a surviving dairy about as prosperous as it is under the current policy. However, the cost associated with possible borrowing to tide a dairy over periods of high feed costs might be expected to affect somewhat adversely its ability to retire debt and increase net worth. Dairies under tight financial conditions under current policy would be expected to have a lower probability of survival without crop programs because they would be less able to absorb the effects of periods of relatively high feed prices. This would be less a problem for dairies in a relatively strong financial position under current policy because they would be better able to absorb these shocks.

Results Obtained:

- The increased variability in feed prices, associated with eliminating all crop programs, had little financial impact on all dairies compared with the results under the current policy. Average net present value declined less than 2 percent for all dairies.
- Increased price risk did not reduce the probability of survival for any of the farms.

Fifty Cents Reduction in Price

All the assumptions of the current policy were retained except that mean milk prices were assumed to be reduced 50 cents per hundredweight and the variability of milk prices was assumed to have increased. This scenario was included in the analysis because of the current high level of Government stocks and program costs.

Results Expected.—Lower support prices would be expected to affect adversely the dairies' net incomes as well as their survival and growth. The dairies most adversely affected would be those that were already in financial difficulties under the base policy.

Results Obtained:

- All farms, compared with results for the current policy, were negatively affected. All farms experienced losses in net farm income, net present value, and net worth compared with results under current policy.
- The largest dairies in each region experienced little reduction in the probability of survival.
- The greatest adverse impact was on the smallest Minnesota dairy, where the probability of survival declined from 74 to 58 percent and the probability of a positive net present value declined from 26 to 18 percent. Other dairies that were adversely affected included the smaller Florida and California dairies. Therefore, reduced price supports would force many small dairies out of business.

No Dairy Program

All assumptions of the base scenario were retained except that milk price variability was assumed to have increased. Milk price was expected to fall to the estimated average variable cost for the most efficient dairies until 1990. Price was then expected to recover in 1991 and 1992 until in 1992 the price would be equal to the average total cost for the most efficient operations. However, with no price support program, the actual price may be either above or below the average price. The model randomly selects milk prices from a distribution that may be as much as 20 percent above or 25 percent below the mean price.[6]

Results Expected.—Without a dairy price support program there would be no guaranteed price floor. In some years milk prices would be higher, while in other years they would be lower than under current policy. However, they would still fluctuate about the long-term equilibrium price. Over time, favorable and unfavorable prices should balance out, meaning that the ability of a dairy to increase net worth, repay debt, and achieve a favorable internal rate of return would not be seriously affected. However, the probability of survival for dairies in tight financial situations would be adversely affected.

Results Obtained:

- The probability of survival fell for all farms, with the greatest reduction experienced by the moderate farms analyzed. The lowest probability of survival was 22 percent for the 52-cow Minnesota dairy (table 9-8).
- Ending net worth declined significantly on all farms except for the very large farms in California and Florida. For example, net worth declined 73 percent for the 52-cow Minnesota dairy and 37 percent for the 550-cow California dairy.
- Average net income was negative for all farm sizes except for the very large dairies in California and Florida.
- Very large farms were the only farms able to survive under no price support program.

Supply Control

All assumptions of the base current policy were retained except that mandatory quotas were assumed to be imposed on dairies. Quotas equal to 96.5 percent of a producer's normal production would, over time, be expected to maintain milk prices $1 above those under current policy. Herd size would be reduced about 4 percent in order to reduce milk production 3.5 percent, assuming that poorer-than-average cows would be culled in complying with the quota.

Results Expected.—The financial performance of all dairies would likely be improved as a result of permanently higher milk prices, despite those dairies having to reduce total milk produced within the designated quota. The probability of survival would increase along with a greater ability to reduce debt and increase net worth for dairies existing at the time the pro-

[6]The variation of milk prices without a dairy price support program was developed from the following study: Cameron S. Thraen and Jerome W. Hammond, *Price Supports, Risk Aversion and U.S. Dairy: An Alternative Perspective of the Long-Term Impacts*, Economic Report ER83-9, Department of Agricultural and Applied Economics, University of Minnesota, June 1983.

Table 9-8.—Comparison of No Milk Price Support Program on Dairy Farms From Various Regions

Dairy herds	Probability of survival	Beginning net worth	Ending net worth	Average net cash income	Average net income
Minnesota:					
52 cows	22%	$ 417,000	$ 114,000	$ −19,000	$ −38,000
125 cows	98	969,000	835,000	6,000	−21,000
California:					
550 cows	62	1,261,000	800,000	−72,000	−166,000
1,436 cows	96	2,538,000	4,418,000	187,000	7,000
Arizona:					
359 cows	42	744,000	276,000	−55,000	−121,000
Florida:					
350 cows	36	757,000	317,000	−49,000	−97,000
600 cows	72	1,465,000	1,268,000	−23,000	−97,000
1,436 cows	100	3,343,000	6,625,000	366,000	242,000

SOURCE: Office of Technology Assessment.

gram is implemented. However, this economic advantage could be capitalized into the quota value, thereby eroding the advantage for new entrants or producers that would have to purchase quotas to expand milk production.

Results Obtained:

- Probability of survival was increased for all farms of all regions (table 9-9). The 52-cow Minnesota dairy experienced the largest increase in the probability of survival, from 74 percent under the base scenario to 92 percent.
- Average net present value increased for all dairy farms. The 52-cow Minnesota dairy increased from −$61,000 to $13,000.
- Ending net worth was increased for all dairies, owing to retained earnings and repayment of debt.
- Net farm income for Minnesota dairies was increased by at least $8,000 compared to the base scenario. These dairies previously had the lowest income.

Income Tax Changes

All assumptions of the base scenario were retained except that more restrictive Federal income tax provisions were included, such as the following:

- Machinery, livestock, and buildings were depreciated using the straight-line cost recovery method.
- First-year expensing provisions were eliminated for all depreciable items.
- Maximum investment tax credit provisions were eliminated.
- The maximum annual interest expense that

Table 9-9.—Comparison of Supply Control Program on Dairy Farms From Various Regions

Dairy herds	Probability of survival	Beginning net worth	Ending net worth	Average net cash income	Average net income
Minnesota:					
52 cows	92%	$ 417,000	$ 310,000	$−2,000	$−14,000
125 cows	100	969,000	1,190,000	59,000	33,000
California:					
550 cows	96	1,261,000	2,349,000	161,000	76,000
1,436 cows	100	2,538,000	8,543,000	812,000	653,000
Arizona:					
359 cows	96	744,000	1,486,000	112,000	54,000
Florida:					
350 cows	98	757,000	1,164,000	67,000	25,000
600 cows	100	1,465,000	2,681,000	201,000	137,000
1,436 cows	100	3,343,000	10,038,000	877,000	769,000

SOURCE: Office of Technology Assessment.

could be used to reduce taxable income was $15,600.
- The operator must sell obsolete machinery on disposition rather than trading it in on new replacements, thus forcing recapture of excess depreciation deductions.

Results Expected.—These tax policy changes would have an adverse impact on the ability of a dairy to reduce debt, increase net worth, and, if in a tight financial situation, reduce the probability of survival. All tax changes would increase the tax liability, reducing the net income of the operation and leaving less for debt retirement and increases in net worth.

Results Obtained:

- Eliminating the tax benefits increased tax liabilities and reduced the net present value and net worth for all farms. These reductions, however, were relatively small—in the range of 1 to 10 percent.
- The increased tax liabilities were not large enough to reduce significantly the probability of survival.

Technology Scenarios

The milk price assumption of the base scenario was retained for the two technology scenarios discussed below. It should be recognized that milk prices would be expected to be higher than the base scenario prices if productivity gains from the designated technologies did not materialize. Therefore, the adverse effect of these technology scenarios is overstated.

No Information and Nutrition Technology.—The 1.8-percent annual increase in production per cow attributable to information and nutrition technology was excluded from the base assumption for this scenario. The financial performance of all dairies would be adversely affected under this scenario. For the very large farms the per-cow increase in 1982 milk production by 1992 was only two-thirds as much as under the base scenario.

No Bovine Growth Hormone.—The 25.6-percent jump in milk production when bgh is adopted was excluded from the base assumption for this scenario. The financial performance of all dairies was expected to be adversely affected under this scenario compared with the base scenario. For the very large dairies, the increase in 1982 milk production per cow by 1992 was assumed to be only 40 percent as much as under the base scenario.

Results Expected.—The expected impact of not adopting these technologies was to affect significantly the financial performance of the dairies. The probability of survival and all measures of financial performance would decline compared with the base scenario.

Results Obtained.—Large decreases in net farm income, net present value, and ending net worth were experienced for all dairies compared with results from the base scenario.

Financial Stress Scenarios

The assumed beginning financial conditions for four of the eight dairies were changed to reflect high-debt operators and new entrants. Debt load was doubled to reflect high-debt situations. For new entrants all equipment was assumed to be new, which increased both the initial value of the machinery and the total debt load.

Two policies were considered for high-debt dairies. One was to subsidize interest rates on all debt so that the effective rate for all loans paid would be 8 percent rather than the higher rates used in the current policy. The second was to restructure the debt by converting a portion of intermediate debt into long-term loans and/or to extend the length of intermediate-term loans. In the second case, interest rates, total debt loads, and other assumptions of the high-debt dairies remained the same as in the base scenario.

The impact of higher feed costs and of eliminating the dairy price support program was evaluated for new entrants with a high-debt position. The results obtained included the following:

- The probability of survival for any dairy depends greatly on its initial financial position. Dairies with high debt and new entrants with high debt had significantly lower probabilities of surviving than dairies with initial financial situations assumed under current policy.

- Neither interest subsidies nor opportunities for debt restructuring greatly improved the chances of high-debt dairy farms remaining solvent.
- The probability of survival for both Minnesota dairies was zero for all policy scenarios. The implication is that high-debt producers in this region cannot survive, even under the current dairy policy.

Summary and Conclusions

- Policies and technologies that are favorable for dairy provide greater financial opportunities for large rather than small dairies.
- Policies that adversely affect the dairy industry such as higher feed costs, fewer income tax benefits, and no dairy price support program will negatively affect small dairies more than larger dairies.
- The major advantage enjoyed by larger dairies is more related to the efficiency of operation than to specific dairy policies.
- There will be a continued trend to fewer and larger dairies in all regions. Milk production can be expected to continue to increase in the lower cost regions of the Southwest and West.
- Traditional dairy regions will continue to experience increased competitive pressure from larger scale, more efficient producers in other parts of the United States. Substantial restructuring of dairies in the Great Lake States and the Northeast will be required for those dairies to compete.
- Emerging technologies need to be transferred to moderate-size dairy farms at a much earlier time in the technology adoption process for these farms to survive.
- Dairy price supports must be sufficiently flexible to adjust to the increased production and lower costs spurred by technological change. This could be accomplished either by adjusting the price support level to changes in production costs per unit of output or by adjusting the level of CCC purchases.
- Current geographic price alignment systems in Federal milk marketing orders are becoming increasingly outdated. A comprehensive study is needed of changes required to modernize the Federal order system in light of technological changes.

CHAPTER 9 REFERENCE

Kalter, Robert J., et al., *Biotechnology and the Dairy Industry: Production Costs and Commercial Potential of the Bovine Growth Hormone,* A.E. Research 84-22 (Ithaca, NY: Cornell University Department of Agricultural Economics, December 1984).

Chapter 10

Impacts on the Environment and Natural Resources

Contents

Tables

Chapter 10

Impacts on the Environment and Natural Resources

Overall, the emerging technologies are expected to reduce land and water requirements for agricultural use. They are also expected to reduce certain adverse environmental impacts associated with land and water use, such as soil erosion, threats to wildlife, and pollution from the use of farm chemicals. What impacts will result from biotechnologies, however, are more uncertain, for there are no good predictive ecological models or systems in existence that could help evaluate the potential impacts of a release of genetically altered organisms into the environment.

This chapter evaluates the implications for the environmental and natural resource use of emerging agricultural technologies. It is divided into four parts: 1) impacts of technology, which includes the methodology for identifying the emerging technologies and evaluating their environmental/resource impacts, the evaluations of the technologies, and some limitations associated with the evaluations; 2) the relationship between the size and structure of farms and the adoption of new technologies; 3) environmental concerns of emerging technologies; and 4) policies for mitigating adverse consequences and for enhancing the favorable aspects of the new technologies.

IMPACTS OF EMERGING TECHNOLOGIES

Methodological Approach

One problem that arises in identifying the technologies to be evaluated is the establishment of a time horizon. Although the emerging technologies studied are expected to be adopted before the end of the century, some of the technologies identified are not expected to be introduced for commercial adoption until very late in the century. Even then, adoption may be rather limited, and widespread adoption may not occur until the first decade of the next century. Thus the environmental and resource impacts of such technologies may not manifest themselves until early in the next century.

Since policy concerns focus on technologies that should be discouraged or encouraged because of expected environmental/resource consequences, holding rigidly to the end of the century for evaluating the emerging technologies is too limiting. Therefore, the convention used in determining which technologies were to be evaluated was that the technology be *available* for adoption by the end of the century, even though the environmental and resource impacts from use of that technology might not occur until later.

The Delphi approach was used to facilitate consensus in the evaluation of the impacts of emerging technologies (Coates, in Teich, 1981; Gordon and Ament, in Teich, 1981). While the technique does not provide for a high level of scientific rigor, the difficulties inherent in foreseeing the myriad possible consequences of a new technology in a complex socioeconomic system require subjective evaluation from a well-informed, multidisciplinary team. The Delphi technique lends itself to identification of consensus.

A team of 11 experts was assembled for a 2-day workshop to perform the evaluation.[1] These

[1]This chapter is based largely on the results of the workshop as analyzed by James Hite in the OTA paper "Environmental and Natural Resource Impacts of Emerging Technologies in American Agriculture" and reviewed by the workshop participants.

experts represented a broad range of backgrounds and regions within the United States. (The names, affiliations, and disciplinary specialty of each member of the team is included in appendix C.) The first task of the team was to group the technologies in a way that was meaningful for evaluation of their environmental/resource impacts. The first division was between animal and plant agriculture.

Four general types of technologies related to animal agriculture were identified: 1) genetic engineering, growth and development, reproduction, and nutrition; 2) animal disease, pest control, environment, and behavior; 3) animal waste and crop residues; and 4) aquaculture.

The nine general areas of technologies related to plant agriculture included: 1) genetic engineering in plants; 2) photosynthesis; 3) nitrogen fixation; 4) plant growth regulators; 5) organic farming; 6) multiple cropping; 7) water and soil-water-plant relationships; 8) soil erosion and land management; 9) disease, insect, and weed control.

Each general area of technology was then evaluated relative to eight types of impacts: 1) water quality, 2) water quantity, 3) soil erosion, 4) soil productivity, 5) air quality, 6) wildlife, 7) solid waste, and 8) human health.

The evaluation was performed on a 10-point scale. A technology with a strongly favorable impact on the environment and/or natural resources would receive a rating of 10.0. A technology with a strongly adverse impact would receive a rating of zero. If the impact were judged to be neutral, the rating would be 5.0. A computer-driven device, a Consensor, was used to tabulate the ratings assigned by each expert. In addition, the device allowed each expert to weight his or her rating according to the degree of confidence he or she had in the rating. That level of confidence could be set at zero, 25, 50, 75, or 100 percent.

The Consensor provided an immediate video screen readout of the rating distribution, the weighted average rating, and the average degree of confidence. If the first vote showed a very wide distribution of ratings, those experts with outlying ratings were asked to explain their reasons for their ratings. After additional discussion, another vote was taken. Since lack of a consensus after such discussion is, in itself, an indication of considerable uncertainty about the impacts of new technology, no attempt was made to force a consensus beyond a second vote.

The "with and without test" was adopted as a basic guide in making the judgments necessary to assign a rating. Simply, the test involves evaluating what the environmental/resource situation would be with and without the technology. The rating, therefore, is based on an assessment of the net effect of the emerging technology. A rating that suggests that a particular technology will result in environmental improvement cannot be taken to mean that the environment will be better after adoption of that technology. Rather, such a rating means that the group's judgment was that the environment will be better with the new technology than it would be if the old technology were continued into the future. The converse is also true.

Evaluation Results

Technology in Animal Agriculture

Genetic Engineering, Growth and Development, Reproduction, and Nutrition.—Emerging technologies in the broad area of animal growth and development center on recombinant deoxyribonucleic acid (rDNA), monoclonal antibodies, estrous-cycle regulation, and embryo transfer. All of these technologies are expected to lead to production of increased output with fewer animals and reduced input of feed. That means that a given future demand can be met with fewer animals, less land devoted to production of feed grains and to pasture, and, in general, less demand on natural resources than would otherwise be the case.

On the other hand is the effect these new technologies might have on the structure of animal agriculture. If the new technologies encourage fewer but larger herds and greater geographic concentrations of animal agriculture, localized environmental problems might intensify. For example, disposal of manure and increased use of antibiotics, hormones, and other chemicals could result.

The rating results indicate that only two of the nine environmental impacts—water quality and human health—were judged relevant to this group of technologies (table 10-1). Given fewer animals, some marginal improvement in water quality would result. The possibilities for using genetic engineering techniques to reduce unsaturated fats in red meats would have a marginally beneficial effect on human health.

Animal Disease, Pest Control, Environment, and Behavior.—The emerging technologies in the area of animal disease, pests, environment, and behavior combined biotechnology and computer systems. In addition, it is expected that increased use will be made of existing technologies for diagnostic testing, slow-release insecticides and vaccines, and photoregulation.

Technologies in this area are viewed as similar in their effects on production to those associated with the previous area of animal growth and development, reproduction, nutrition, and genetic engineering, as shown in the rating. The use of these technologies will result in increased output of animal products with fewer inputs of natural resources. The environmental/resource consequences were also judged to be essentially the same as those in the previous area. With fewer animals needed to meet a given future demand, less natural resources would be required for feed production. Thus reduced pressure would be exerted on the environment and natural resources. However, as indicated above, concentration of animals could cause environmental problems.

Animal Waste and Crop Residue.—The basic features of emerging technologies in the handling of animal waste and crop residues center on chemical and biological conversion, recycling, and fuel production. All of these technologies are already being used to varying degrees. The new features involve increased adoption and application of the technologies to specific crops. One example is the use of corn cobs as fuel in thermal gasifiers for drying.

In general, economic factors will prevent widespread use of biomass for fuel or the conversion of animal waste to methane for the foreseeable future. Incorporating crop residues into the soil is expensive and sometimes creates disease and insect problems. Only if the field burning of crop residues is banned by law, thus raising the cost of conventional methods of managing these residues, would many new technologies in this area be widely adopted.

Assuming that the new technologies are adopted, increased use of animal waste for energy would reduce some water pollution and would marginally improve water quality. If crop residues were removed from the fields in large quantities, some additional soil erosion and loss of soil quality would result from the reduction of humus, but it is thought that the new technologies would not have a large effect on residues left in the field. Any movement toward less burning of crop residues, however, would produce a marginal improvement in air quality in selected localities.

Aquaculture.—Aquacultural activities have considerable potential for adverse impacts on water quality and quantity. In some parts of the country, aquacultural enterprises remove large quantities of groundwater from aquifers. In addition, some potential exists for wastewater from such enterprises to pose a water quality

Table 10-1.—Impacts of Animal Technologies on the Environment and Natural Resources

Technology group	Water quality	Water quantity	Soil erosion	Soil productivity	Air quality	Wildlife	Solid waste	Human health
Genetic engineering, growth, reproduction, and nutrition	5.9	NR	NR	NR	NR	NR	NR	5.5
Animal disease, pest control, environment, and behavior	5.9	NR	NR	NR	NR	NR	NR	5.5
Animal waste and crop residues	5.4	NR	5.0	NR	5.3	NR	NR	NR

Rating system: 10 = strongly favorable impact; 5 = neutral; 0 = strongly adverse impact; and NR = not relevant.

SOURCE: Office of Technology Assessment.

problem. Yet there are important economic questions about the potential growth in markets for aquacultural products. Until these questions are answered, it is not possible to make meaningful comments about potential environmental/resource impacts.

Technology in Plant Agriculture

Genetic Engineering.—While the technology of genetic engineering in plants offers dramatic possibilities for agriculture, the scientists working in the area believe that actual adoption of new technologies on the farm is some years away. Basic work in developing gene maps for plants is somewhat behind that for animals.

The technology involves rDNA, cell culture, cell fusion, and monoclonal antibodies. Much of the effort will focus on moisture and drought stress in plants, suggesting a reduced need for irrigation water. There should also be reductions in the use of chemicals as resistances become engineered into plants, with favorable results for water quality. To produce a given level of output, increased yields will allow retirement of some marginal, erosion-prone land, increasing the amount of habitat available for wildlife. Possibilities for using genetic engineering techniques to improve soil microbes was thought especially promising for soil productivity. On balance, therefore, these genetic engineering techniques in plant agriculture would enhance the environment rather strongly (see table 10-2).

Enhanced Photosynthesis.—Technologies that enhance photosynthesis address the plant's central productive process, increasing yields per unit of land. With these technologies, marginal lands could be retired, water needs could be held down, and erosion would be reduced.

In general, the technologies would be environmentally helpful. One possible exception concerns soil productivity. For example, with enhanced photosynthesis, crops on land left in production will draw out soil nutrients faster than would otherwise be the case. Therefore, the effect of the technology on those lands would be to reduce the natural productivity of the soils. However, with this technology there would be less land in production. On those lands not in production, soil productivity would be restored, or at least maintained. On balance, the effect of enhanced photosynthesis technologies on soil productivity would be about neutral.

Nitrogen Fixation.—Legumes have the ability to fix nitrogen from the atmosphere and transform it into plant food. However, cereal plants generally lack this ability. Breeding cereal plants with nitrogen-fixing abilities has long been the Holy Grail of agricultural geneticists, but many difficult problems have been encountered in its pursuit. Advances in genetic engineering, however, have opened up new avenues for plant breeders, and renewed hope exists of developing cereal plants with nitrogen-fixing capabilities. Even though the possibilities for significant breakthroughs prior to the end of the century are considered remote, some incremental advances are expected.

If such nitrogen-fixing technologies develop, the environmental/resource implications would be significant and positive. These technologies

Table 10-2.—Impacts of Plant Technologies on the Environment and Natural Resources

Technology group	Water quality	Water quantity	Soil erosion	Soil productivity	Air quality	Wildlife	Solid waste	Human health
Genetic engineering	6.4	6.9	6.5	7.4	5.9	6.3	5.4	6.1
Photosynthesis	6.2	6.2	6.3	5.0	NR	5.6	NR	NR
Nitrogen fixation	7.1	NR	NR	5.6	5.4	6.3	NR	5.9
Plant growth regulators	6.2	6.2	6.3	5.0	NR	5.6	NR	NR
Organic farming	5.7	5.1	5.6	5.5	NR	5.5	5.5	5.9
Multiple cropping	6.4	4.8	6.8	5.0	NR	4.8	NR	NR
Water and soil-water-plant relationships	6.2	7.5	7.1	5.8	NR	5.0	NR	NR
Soil erosion and land management	6.3	6.6	9.1	7.7	6.6	6.7	NR	5.1
Disease, insect, and weed control	6.9	5.3	7.0	5.7	5.7	7.1	6.1	7.4

Rating system: 10 = strongly favorable impact; 5 = neutral; 0 = strongly adverse impact; and NR = not relevant.

SOURCE: Office of Technology Assessment.

would allow substantial reductions in the use of nitrogen fertilizers, resulting in a decrease in nitrogen runoff into surface waters and percolation into groundwater, with beneficial effects on water quality. With less nitrogen being manufactured, fewer people would be exposed to health risks in fertilizer plants and on the farm. Improved air quality would result from reduced fertilizer manufacturing, and wildlife, especially aquatic life, would also benefit from reduction of nitrogen runoff into surface waters.

Growth Regulation.—Plant growth regulators are typically organic chemical compounds sprayed on the surface of plants. They increase yield by affecting the way the plant uses its nutrients. Chemical concentration in the sprays is usually quite low. The compounds are rather quickly metabolized by the soil and usually present few environmental problems because the compounds themselves tend to break down quickly. The environmental/resource impacts of new technologies in plant growth regulation are likely to be quite similar to those determined for enhanced photosynthesis.

Organic Farming.—In some sense, organic farming represents an old and traditional set of technologies. However, in recent years, the concept of organic farming has undergone some changes. At the heart of organic farming are technologies concerned with nutrient self-reliance and recycling and minimum use of, but not necessarily total elimination of, chemicals. New organic farming in particular could be expected to make use of advances flowing from genetic engineering, enhanced photosynthesis, simultaneous cropping, and several other technologies discussed in this chapter.

Assessing the environmental/resource implications of organic farming presented more problems than any other single set of technologies. While there was general agreement that organic farming approaches would require more land to meet expected demand, there was skepticism about the extent to which organic farming technologies would be adopted. If widely adopted, organic farming would disperse animal agriculture geographically, since there would be a greater need to keep animals on many farms to produce manures. As a consequence, farm energy consumption would be reduced. If energy prices rise substantially, organic farming techniques might be adopted rather widely; but barring such an increase, the panel thought it unlikely that organic farming would account for more than a small percentage of the Nation's farm output.

Organic farming could have adverse impacts on environmental resources if it were widely adopted. Increased pressures would be brought on marginal and erodible lands, and widespread use of animal manures could have some negative consequences for water and air quality. While wildlife might be less threatened by the use of fewer chemical compounds, wildlife habitats could be threatened by the need for more land for crops.

Perhaps the strongest positive impact was thought to be in the area of human health. Organic farming would reduce human exposure to agricultural chemicals and could result in food products that have higher nutritional value and less chemical contamination.

Multiple Cropping.—The concept of multiple cropping involves two separate types of practices. The first, called simultaneous cropping, involves growing two or more crops in the same field at about the same time. The second type involves growing a second crop closely behind the harvest of another in the same field in the same year. The two types merge in some cases where one crop is begun before the other is harvested.

The latter type of multiple cropping has been increasing rather rapidly in the Southeast and in California. Because the land is covered with a crop for longer periods during the year, runoff and erosion are reduced. The result is improved water quality and soil conservation. In certain instances, increased irrigation water is required. Multiple cropping can be either beneficial or harmful to soil productivity, depending on the crops grown. In the wheat-soybean systems of the Southeast, for instance, multiple cropping might improve soil productivity marginally, but other systems would intensify the removal of soil nutrients.

On balance, multiple cropping would probably have some adverse consequences for wildlife. Machinery would be in the fields more often, disturbing nesting areas and wildlife generally. Also, there would be less stubble and other crop residues available for cover and feed. On the other hand, however, multiple cropping has the potential to reduce the land needs of agriculture and, as a result, to protect habitat and soil resources.

Water and Soil-Water-Plant Relationships.—Technologies that affect water and soil-water-plant relationships include certain genetic engineering approaches. Improvements in plants' capabilities to close leaf pore openings (stomata) for longer periods to retain moisture are possibilities. The ratings, however, were assigned primarily on perceptions of still-to-be-applied irrigation technologies, particularly improvements in onfarm irrigation technologies.

Movement toward improved irrigation efficiency is considered environmentally benign. Less water applied means less return flow and, thus, less threat to water quality. It also means substantial savings in water and reduced soil erosion. The effects on soil productivity are mixed, however. While improved technologies will allow plants to make better use of existing soil nutrients, they will also allow those nutrients to be used up faster. So, improved irrigation technologies would have a marginally beneficial effect on soil productivity. Similarly, effects on wildlife are likely to be mixed. The concentration of salts in runoff water might be higher, and with less water, there might be fewer reservoirs and other habitats. On the other hand, with less water being drawn away from irrigation, more clean water might be available elsewhere. Thus the impact on wildlife would probably be neutral.

Soil Erosion/Land Management.—One major technological change in agriculture in the 1970s was the growth of what is called conservation tillage. Conservation tillage implies limited tillage. It has several forms: in some cases, corn and other grains are actually planted into grass or stubble along with an herbicide applied in a very narrow strip where the seed is injected. A newer innovation is called prescribed tillage, a practice that uses computer technology to monitor soil conditions. This technique integrates such information with weather forecasts, for example, to determine when and how to undertake tillage.

Conservation tillage has enormous possibilities for reducing soil erosion. The major environmental problem is the increased use of herbicides. Another problem is the reduced crop yield that results from conservation tillage. To meet given production demands additional acreage must be cultivated. Although the health impact on humans is likely to be negative because of an increaseed threat to groundwater from agricultural chemicals, reduced tillage could reduce mechanical energy consumption and incidences of farm accidents associated with tillage activities.

Disease, Insect, and Weed Control.—The emerging technologies in plant disease, insect, and weed control begin with integrated pest management (IPM), an approach to pest control that does not eliminate use of pesticides but does attempt to minimize those pesticides by making maximum use of predators, by attempting to protect beneficial insects, and by applying pesticides in limited quantities only after no other control mechanism is deemed feasible. IPM has the potential to reduce pesticide use by as much as 50 percent.

Integrated weed management is similar in concept. In this practice changes in cultivation practices are integrated into reduced use of herbicides.

These new technologies, combined with a new generation of agricultural chemicals expected to appear on the market by late in the decade, will tend to cause considerable environmental improvements over existing technologies, if properly applied. However, some concerns were expressed about application in the field. Application machinery often is not precise, and knowing that, farmers sometimes deliberately use a greater application rate than that called for in farm-chemical instructions. Unlike many of the other technologies examined, these disease, insect, and weed control technologies will reduce the amount of land needed for crops. Most of the environmental improvements are expected to be associated with reduced use of agricultural chemicals, which will clearly be environmentally beneficial to water quality, soil quality, human and animal health, air pollution, and energy requirements.

EFFECTS OF FARM STRUCTURE ON THE ENVIRONMENT

To evaluate the impact of farm structure on the environment and natural resources, three scenarios related to three different structures of production agriculture were postulated:

1. a continuation of current policy, which could be expected to result, by the end of the century, in a notably dual distribution of farms by size—many small farms with sales of less than $20,000 annually and many large farms with annual sales of $500,000 or more each year;
2. policies that accelerate the trend toward a dual distribution and a significant reduction in the number of moderate farms; and
3. policies that would slow down the move toward the bipolar distribution, maintaining the number of moderate farms at the expense of larger farms.

The question posed was: what effect, if any, would these structural scenarios have on the environment and on natural resource use? The scale used in assigning ratings was the same as that used in evaluating the various sets of technologies, that is, a rating of 5.0 meant that the scenario was expected to make no perceptible difference. Ratings higher than 5.0 suggested environmental improvement, all things being equal, whereas ratings below 5.0 suggested some environmental degradation.

Considerable evidence suggested that large farms were more likely than small farms to adopt new technologies. Small farms are constrained by time limitations in the use of technologies that require intensive management. They may also be constrained by access to financing. In this context, both considerations are especially important, since many of the new technologies are management-intensive and will require substantial front-end outlays.

On the other hand, it was noted that some large farms may currently be more heavily leveraged financially than the moderate farms. Moreover, since almost all of the small farms are operated by persons or families with some outside income, the small farms may be less constrained financially to use technologies that save labor and do not require enormous front-end outlays that necessitate borrowing. Organic farming is an example of a technology that may have more appeal to small than to large farmers.

It is also important to note that many large farms making use of hired labor may concentrate on minimizing labor costs. The new technologies, in the main, are not primarily labor-saving. The principal savings to be had from these new technologies are in a reduction in land and environmental degradation. Since the environmental effects are usually offsite and external to the farm firm's accounts, there may be only modest incentives for some of the larger farms to adopt the new technologies unless the farms are under strong regulatory pressures from environmental agencies.

On balance, however, the technologies would favor large-farm operators. Because the technologies, in general, tend to be environmentally enhancing, it follows that movement toward a greater concentration of production in the hands of large operators would have beneficial environmental effects. However, several panelists insisted on the caveat that the major factors influencing adoption of the new technologies are access to front-end capital and managerial capability. Thus the technologies are not confined exclusively to large farms; many moderate farms are also in a position to adopt the technologies. In the areas of animal agriculture, particularly, many of the new technologies are being adopted first by seed-stock producers, who tend to be moderate operators.

These reservations are important as background for the interpretation of the ratings. Scenario 1 represents a continuation of recent trends and, given the "with and without" rule used in evaluating the technologies, must be assigned a rating of 5.0 (environmentally neutral) in all impact areas. Movement toward greater concentration of production on large farms would, in the panel's judgment, have a general tendency to enhance the environment because the larger farms (as a class) will be more likely to adopt the new technologies. The panel emphasizes, however, in strong terms, that movement toward greater concentration of large farms is not a necessary condition for realization of environmental improvement flowing from the new technologies. Public policies that improve the access of small and moderate farms to the new technologies would accomplish the same end.

Scenario 3 represents public policy designed to improve the survival rates of moderate farms. Such policy, taken alone, was judged to have unfavorable environmental consequences in five of the nine impact areas addressed. If such farms survive, but do not prosper, they will have few resources available to use in adopting the new technologies. Policies that improve the opportunities of moderate farms to prosper and survive, however, would allow such farms to avail themselves of new technologies that are environmentally enhancing. Indeed, under such conditions, moderate farms might well adopt these new technologies more rapidly than the larger farms for those reasons cited above.

ENVIRONMENTAL CONCERNS OF EMERGING TECHNOLOGIES

Looking at all the technologies assessed, the environmental/resource impacts were believed by the majority of the panel to be, at least marginally, environmentally enhancing. The panel noted particularly the potential for new tillage technologies to reduce soil erosion and improve soil productivity, for new irrigation technologies to conserve water, for nitrogen-fixing technologies to improve water quality by reducing nitrogen runoff, and for genetic engineering to improve agricultural productivity. The technologies should reduce land needs and thereby reduce threats to wildlife habitats. They should also reduce the use of chemicals and the resulting possible threats to human health.

In only a few cases were the new technologies thought likely to have unfavorable environmental or resource impacts. Those concern the impacts of multiple cropping on water quality and on wildlife. In both these cases, however, the unfavorable impacts were thought to be relatively mild. Concern also arose over possible problems associated with human error or machine malfunction in the application of chemicals used in conservation tillage.

Perhaps the chief cause for concern was the lack of knowledge about the potential effects on the environment of the release of genetically altered organisms. Most of the biotechnologies applicable to agriculture that are expected to be commercially adopted in the next few years involve release of new organisms into the environment. The question of "deliberate release" of engineered micro-organisms has already arisen in agriculture, however, in connection with the testing of genetically altered bacteria in potato fields to prevent freeze damage. Other cases are almost certain to arise. The potential for these genetically altered micro-organisms to interact with the environment in unpredictably harmful ways cannot be ignored. Considerable debate is occurring within the scientific community and within the Federal bureaucracy over proper controls on the testing and use of genetically altered organisms prior to deliberate release. The economic benefits from uses of biotechnology that require deliberate release of modified organisms probably are substantial, but the panel recognized the need to work out suitable regulatory safeguards controlling such releases (Doyle, 1985; Healy, 1985; Kendrick, 1985; Schatzow, 1985).

The safety issue of biotechnology was debated when the first gene was about to be inserted into a micro-organism. Concerned about potential hazards of new rDNA techniques, a group of the world's leading scientists, headed by Paul Berg, met 10 years ago for the second time at the Asilomar Conference Center in Pacific Grove, California. They agreed to strictly regulate those experiments using rDNA techniques until more data could be collected for assessing the potential hazards and until safety could be assured. One day after the second Asilomar conference, a National Institutes of Health (NIH) Recombinant DNA Advisory Committee, commonly known as RAC, held its first meeting and began drafting a set of safety guidelines for the rDNA experiments, guidelines that have governed rDNA research in the United States ever since (Tangley, 1985).

As scientists learned more about rDNA techniques, the guidelines were periodically revised. Each revision further relaxed the rules as scientists came to realize that the fears of hazards from rDNA research, although not groundless, were greatly overestimated a decade ago. During the last decade, hundreds of laboratories around the world have been cutting and splicing DNA in a multitude of combinations.

As new products of rDNA research approach field testing, clinical trials, and commercial introduction, safety and ethical issues have rekindled. Unlike 10 years ago, when concerns came exclusively from scientists, concerns today come from scientists, industry, social activists, and the public, all for different reasons. Some scientists, social activists, and members of the public are concerned about possible adverse impacts on human health and the environment, while some scientists and industry fear that public concerns may lead to overregulation.

One example illustrates the controversy over rDNA research in agriculture. Two years ago, Steven Lindow and Nicholas Panopoulos of the University of California, Berkeley, successfully constructed non-ice-nucleation bacteria that inhibit frost formation on potato plants. As these researchers readied to field test the new organisms to see if they could protect crops from frost damage, a coalition of public interest groups filed a lawsuit to postpone the field trials (Tangley, 1983). These groups believe that field tests of genetically modified organisms should not proceed until scientists develop a method for establishing the safety of such releases. In May 1984, 9 days before the scheduled release of the micro-organisms, a U.S. District Court issued a temporary injunction halting the first proposed release and prohibiting NIH from approving any more releases until the case was fully resolved (*Bioscience*, 1984). In February 1985 a U.S. Court of Appeals upheld part of the lower court's decision, stating that NIH was required to prepare an environmental assessment of the one field test in question but that the institute could go ahead and consider other release proposals. Recently, the Environmental Protection Agency approved the first two field tests of genetically altered organisms. In the first experiment, Agracetus of Middletown, Wisconsin, would test the effects of their new products, genetically modified disease resistant tobacco plants, on the natural environment. In the second experiment, Advanced Genetic Sciences of Oakland, California, would test bacteria that have been genetically altered to prevent frost formation on strawberry plants. This company would spray the modified bacteria on 2,400 blossoming strawberry plants on a one-fifth acre plot in Salinas Valley.

The central issue of these controversies is whether genetically engineered micro-organisms will disrupt the ecosystem into which they are released and will have adverse impacts on human health and the environment. In the case of non-ice-nucleation bacteria, scientists know virtually nothing about the normal role these bacteria play in the biosphere. Closely related bacteria are apparently ubiquitous, and some scientists suggest that they play a role in the moisture nucleation in clouds, and consequently in rain or snowfall (Feldberg, 1985). What happens, however, if these new strains really are effective in competing for the same ecological niche as the natural strains? What if they allow clouds to hold much more moisture before precipitation occurs?

The major point the proponents of biotechnology, mainly the biotechnology industry and some scientists, use to defend biotechnology is that genetic engineering techniques have been used in plant breeding and animal husbandry for centuries. During the last several decades, biotechnology has been used in chemical and food processing industries (Fraley, 1985). For example, antibiotics, amino acids, and other supplements produced by fermentation technology are routinely added to feeds to stimulate animal growth and prevent disease. Microbial seed inoculums are commonly used to increase crop yields. Immobilized cells and enzymes are being used extensively to catalyze biochemical conversions in the production of specialty chemicals and feedstock (Fraley, 1985). Genetic engineering methods for manipulating genes in micro-organisms such as bacteria and yeast have also existed for several years.

Speaking for the biotechnology industry, Hardy and Glass (1985) argue that genetically engineered organisms are similar either to genetically engineered organisms already in commercial use or to naturally occurring organisms indigenous to habitats where they would be introduced. The fact that similar organisms exist or have been previously introduced into the environment suggests that no adverse effects would occur from the introduction of novel micro-organisms. Hardy and Glass claim that after nearly 10 years of close scrutiny, risk assessment studies, and worldwide experience with molecular and cellular genetic manipulations, there is no evidence of any significant hazards associated with this technology. The risks remain only speculative. In fact, with the accumulation of knowledge, many of the fears voiced in the early days of molecular genetics have been shown to be unfounded. Therefore, there is no reason to believe that cellular and molecular genetic engineering should present any greater hazard than that posed by the whole-organism genetic manipulation that has been practiced for centuries. With millions of dollars invested in biotechnology research and development, industry is concerned with overregulation that could stifle future growth of the biotechnology industry and cause the U.S. industry to lose its competitive edge.

Opponents argue that the issue of deliberate release of genetically engineered micro-organisms into the environment is quite different from the genetic engineering methods, such as production of antibiotics and amino acids by fermentation technology, used in the past. McGarity (1985) points out several reasons why risks of large-scale release of micro-organisms are of much greater concern than the risks of fermentation biotechnology; for example: 1) a large-scale release of genetically engineered micro-organisms into the environment significantly reduces the degree of human control over the novel micro-organisms; 2) biological containment, by which strains of micro-organisms are weakened so that they cannot survive outside the laboratory environment, can no longer be used as a safeguard against potential hazards; and 3) it is difficult to assess potential risks to human health and the environment.

Alexander (1985) also expresses concerns that there is not enough information to predict the ecological consequences. The best model for predicting ecological consequences is the exotic species model, but this model has been criticized as inappropriate for predicting the potential ecological consequences of the deliberate release of novel micro-organisms because the model is based on outdated ecological thinking (Regal, 1985). Because there is no adequate model, Alexander (1985) suggests that history be used as a guide for the future. The history of the application of these emerging technologies provides lessons for assessing new technologies. Alexander asserts that no technology was without risk, and a risk-free technology probably does not exist now. Regal (1985) also indicates that there is no great power that is only good, that has no dangers, and that cannot be misused. Although the risks from deliberate release appear to be small, the consequences of an unlikely event could be disastrous. It is the fear that new micro-organisms or their genetic traits might survive and multiply unchecked and thus have adverse impacts on the ecosystem that make scientists worry about the release of the novel micro-organisms (Robbins and Freeman, 1985).

Now rDNA technology is entering a crucial stage in its development from laboratory to the marketplace. Goldberg (1985) demands that there be a social responsibility to ensure that rDNA technology is safe and suggests that all sectors of society must participate in decision-making about new developments such as the release of genetically modified organisms. In light of public concerns, proponents of biotechnology realize that without public confidence and support, this promising technology could falter as it moves from the laboratory to the marketplace. Those scientists who believe the risk of genetically modified micro-organisms is next to nothing envision that public concerns about such organisms must be addressed. Industry also recognizes the need for some regulation of the environmental applications of genetic engineering and suggests that better risk assessment tools be developed (Hardy and Glass, 1985).

Some biotechnology companies that are or will be introducing genetically engineered products believe that public perception translates into public policy and that commercialization of biotechnology may be in peril if ignorance engenders fear of biotechnological research and applications (Price, 1985). They call for public education about biotechnology research and applications.

Since deliberate release of genetically engineered micro-organisms is such an important and controversial issue, OTA and the National Science Foundation cosponsored a workshop late in 1985 to address this issue.

POLICY IMPLICATIONS

Although the discussion above addresses some policy matters obliquely, the purpose of this section is to discuss the way public policy might enhance the positive environmental effects or mitigate the negative effects of the emerging technologies.

The first point on which a consensus emerged was the observation that the net movement of new technologies is environmentally enhancing. Given the public pressure for environmental improvement and the increased regulatory activity by Government in the environmental area, that movement is consistent with economic theory. It follows, then, that increased Government expenditures on research and education would tend to have positive environmental effects. The assessments presented here suggest that there is a strong public interest in accelerating technological change in agriculture, and if that is the objective, the action required is increased research and education.

The second approach, which complements the research and education effort, is to develop more stringent environmental regulations and provide stronger enforcement of regulations. Such regulations tend to have the economic effect of raising the cost of using the environment in agricultural production. Economic theory suggests that as costs of particular inputs are increased, economic agents will find ways to reduce the use of the relatively higher priced inputs. So, increasing the costs of environmental inputs through stronger regulations and better enforcement will accelerate the adoption of the new technologies and give impetus to research that has further environmental benefits.

Finally, policies that reward farmers who adopt environmentally enhancing technologies have some precedent. Cost-sharing programs in the area of soil conservation are the best known example. Targeted cost-sharing and creative use of the relationship between environmentally enhancing farm practices and the price support program, such as the so-called cross-compliance proposal for using conservation cost-sharing funds, represents, at least in a generic way, a policy option likely to enhance the environmental benefits of the new technologies.

Turning to policies to mitigate the undesirable environmental impacts of new technologies, the panel moved quickly to the classic prescription of exacting a user charge. If the research base were available to determine the appropriate *T* value for soil erosion, a soil erosion tax could be levied on each ton of loss above that level. Ideally, the tax would be equal to the environmental damage caused by the erosion. Unfortunately, the research base for setting the *T* value is not sufficient. There would also be very difficult enforcement problems with such a tax. But the idea of making users of environmental inputs pay for them was considered fundamental to policy that mitigates undesirable environmental consequences.

More practical applications of this concept focus on raising the costs of agricultural chemicals by placing excise taxes on those chemicals. The more expensive the chemicals, the less they will be used and the greater care the user will take to be sure that the application rate is not excessive. Similarly, policies to raise the price of irrigation water might have some beneficial impacts on water quantity (although there are studies suggesting that the price elasticity on such water approximates unity).

SUMMARY AND CONCLUSIONS

In general, with a few notable exceptions, most of the emerging technologies are expected to reduce substantially the land and water requirements for meeting future agricultural needs. As a result, these technologies are also expected to reduce environmental problems associated with the use of land and water. The technologies were thought to have beneficial effects relative to soil erosion, to reduce threats to wildlife habitat, and to reduce dangers associated with the use of agricultural chemicals. New tillage technologies, however, may reduce erosion and threats to wildlife while increasing the dangers from the use of agricultural chemicals.

The panel concluded that the new technologies were most likely to receive first adoption by farmers who were well financed and were capable of providing the sophisticated management required to make profitable uses of the technologies. In the main, such farmers will tend to be those with relatively large operations. Hence, the technologies will tend to give additional economic advantages to large farm firms relative to moderate and smaller farms, accentuating the trend toward a bipolar or dual farm structure in the United States.

In addition, since the new technologies tend to be, at the margin, environmentally enhancing, there is public interest in research and education that leads to their rapid development and widespread adoption. That conclusion becomes even stronger if public policy is aimed at maintenance of the moderate farm. Larger farms, with their own access to research results and scientific expertise, may be able to advance the new technologies with relatively little publicly sponsored research. But moderate and small farms will have to depend on publicly sponsored research and extension education to obtain access to the new technologies and to adapt them to their individual situations.

The new technologies will require more stringent environmental regulations and stronger enforcement of regulations. The complexities of some of the emerging technologies will pose significant challenges for promulgation of wise environmental regulations. The economic benefits from the technologies cannot be passed by, but users may have little private incentive to make use of the technologies in ways that avoid unnecessary, adverse, third-party effects. The panel considered that economic incentives or disincentives, including the use of excise taxes to discourage overuse of potentially threatening materials, represented a more intelligent approach to protection of environmental values than did direct regulations. Yet the panel also concluded that: 1) some additional effort to enforce existing regulations would hasten the adoption of the new technologies that are, at least potentially, less environmentally threatening; and 2) new regulations will be required to deal with some aspects of the emerging technologies.

Perhaps the most revolutionary of the new technologies are those associated with rDNA. While specific applications of such technologies that are currently apparent would appear to reduce resource needs and threats to the environment arising from agricultural activities, the panel recognized possible dangers associated with the deliberate release of genetically altered micro-organisms. The very revolutionary nature of the new biotechnologies and the lack of a scientifically accepted predictive ecology prevented the panel from providing specific evaluations of resource/environmental impacts associated with the deliberate release of new forms of life.

Ten years ago, scientists concerned about the impact of rDNA agreed to regulate the rDNA experiments. Now many scientists see little danger in the applications of the planned rDNA technology. But as new products of rDNA research approach field testing and commercial introduction, safety and ethical issues have rekindled. Both sides of the issue agree that more research should be conducted to assess the potential benefits and risks.

CHAPTER 10 REFERENCES

Alexander, Martin, "Ecological Consequences: Reducing the Uncertainties," *Issues in Science and Technology,* spring 1985, pp. 57-68.

Bioscience, "Court Halts Genetic Engineering Field Test," 34:4, July/August 1984, p. 413.

Doyle, Jack, "The Regulation of Commercial Biotechnology," Testimony Before the Subcommittee on Investigations and Oversight of the Committee on Science and Technology, U.S. House of Representatives, Washington, DC, Apr. 17, 1985.

Feldberg, Ross, "Framing the Issue: The Social Implications of Biotechnology," *Science for the People* 17:3, May/June 1985, pp. 4-9 and 53.

Fraley, Robert T., "Impact of Genetic Engineering on Plant Agriculture," paper prepared for the Office of Technology Assessment, Washington, DC, 1985.

Goldberg, Terri, "Social Responsibility and Applications of Biotechnology," paper presented at the annual meetings of the American Association for the Advancement of Science, 1985.

Hardy, Ralph W.F., and Glass, David J., "Our Investment: What Is At State?" *Issues in Science and Technology,* spring 1985, pp. 69-82.

Healy, Bernadine, M.D., "Statement," Testimony Before the Subcommittee on Investigations and Oversight of the Committee on Science and Technology, U.S. House of Representatives, Washington, DC, Apr. 17, 1985.

Hite, James C., and Laurent, Eugene A., *Environmental Planning: An Economic Analysis* (New York: Praeger Publishers, 1972).

Kendrick, Edgar L., "Statement," Testimony Before the Subcommittee on Investigations and Oversight of the Committee on Science and Technology, U.S. House of Representatives, Washington, DC, Apr. 17, 1985.

McGarity, Thomas O., "Regulating Biotechnology," *Issues in Science and Technology,* spring 1985, pp. 40-56.

Price, Harvey S., "Biotechnology: The Need for Public Education," *BioScience* 35:4, April 1985, p. 211.

Regal, Philip J., "Biotechnology and a Safe Future: Mobilizing Scientific Expertise," paper presented at the annual meetings of the American Association for the Advancement of Science, 1985.

Robbins, Anthony, and Freeman, Phillis, "Biotechnology and the Public Purpose, or a Public Agenda for Biotechnology in the Congress," paper presented at the annual meetings of the American Association for the Advancement of Science, 1985.

Schatzow, Steven, "Statement," Testimony Before the Subcommittee on Investigations and Oversight of the Committee on Science and Technology, U.S. House of Representatives, Washington, DC, Apr. 17, 1985.

Tangley, Laura, "New Biology Enters a New Era," *BioScience* 35:5, May 1985, pp. 270-275.

Tangley, Laura, "Engineered Organisms in the Environment? Not Yet," *BioScience* 33:11, December 1983, pp. 681-682.

Teich, A.H. (ed.), *Technology and Man's Future,* 3d ed. (New York: St. Martin's Press, 1981).

Chapter 11

Impacts on Rural Communities

Contents

Tables

Figure

Chapter 11

Impacts on Rural Communities

The impacts of technological and structural change in agriculture do not end with the individuals who live and work on farms. A variety of other consequences are to be expected at the level of the rural[1] communities that have direct and indirect linkages to farms and farmers. As with individual farmers, some communities are likely to benefit from change, while others are likely to be affected adversely. Much depends on the type of overall labor force in the community and on the opportunities for labor to move to other areas of employment.

Hard-hit communities may need technical assistance to attract new businesses to their areas, to develop labor retraining programs, and to alter community infrastructure to attract new inhabitants. To accomplish these goals, Federal policy will have to be complemented by regional and local policies.

Those rural communities that benefit from changes in agricultural technology and structure may do so in several ways. For example, as agriculture becomes more concentrated, some communities will emerge as areawide centers for the provision of new, high-value technical services and products. Likewise, some communities will emerge as centers for high-volume food packaging, processing, and distribution. In both cases, the economic base of these communities is likely to expand. However, unless total demand for agricultural commodities increases substantially, centralization of services, marketing, and processing will be like a zero-sum game in many areas; the market centers will benefit at the expense of other communities. Many of the communities that are bypassed will decline as a result of the process of centralization.

Communities may also benefit in those parts of the country in which the number of small and part-time farms is increasing. This phenomenon results in an increase in population in many rural areas and in an increase in total income and spending in some of these rural areas. The increase in small farms may sustain more retail establishments than would otherwise be the case, since purchases by small farmers may tend to be more local than those by larger farmers. The operators of these farms in many cases subsidize their own production from off-farm income.

This chapter assesses the impacts that emerging technologies and structural change have had on rural communities in the 1970s in five regions of the United States, outlines several areas of potentially adverse impacts, and provides a policy framework for options that may help mitigate the adverse impacts.

[1]Rural communities are defined as places with less than 20,000 inhabitants in a nonmetropolitan county.

STRUCTURAL CONCENTRATION AND INDUSTRIAL AGRICULTURE

A landmark study that addressed the relationships between increased concentration in agricultural production and community welfare was done by Walter Goldschmidt in 1944. Goldschmidt found a series of negative social effects associated with large-scale agriculture in the central valley of California. His research was based on a matched-pair comparison of a community of relatively large farms (Arvin) and a community of relatively small farms (Dinuba). He found higher median family income, lower poverty, better schools, more retail trade, stronger institutions—including churches, more recreational opportunity, and more newspaper readership—associated with the set of small farms surrounding Dinuba. Although numerous methodological and theoretical criticisms have been made about this study, the thesis of this study continues to frame the discussion of structure and community relationships. Goldschmidt and his supporters argue that when farming is practiced on a scale that exceeds a family's ability to provide the main source of labor and management, it tends to acquire industrial relations of production in which ownership and management are separated from labor. As a result, this industrialized form of agriculture tends to become disarticulated from the surrounding communities, increasingly relying on these communities as a source of low-cost hired labor. This, in turn, is thought to result in social inequality, poverty, and a range of associated pathologies. The general hypothesis is that increases in structural concentration of production, and especially industrial relations of production, are associated with decreases in social welfare.

There has been a long-run secular trend toward an increased proportion of hired labor and a decreased proportion of family labor in agriculture. Large and very large farms tend to rely more on hired labor than do moderate farms. As total agricultural production becomes more concentrated in the large and very large sales classes, the proportion of hired labor in U.S. agriculture is likely to continue to increase. The available data on regional changes in proportion of hired labor is limited. Data is available on changes in number of hired workers by the four regions of the United States defined by the 1970 and 1980 Census of Agriculture. Hired labor increased by 21 percent in the North Central region during this period. In the West hired labor stayed about the same. Hired labor decreased by about 18 and 5 percent in the Northeast and South, respectively (Pollack, et al., 1983).

Increasing structural concentration in U.S. agriculture is not necessarily synonymous with agricultural industrialization. In many areas of the United States large and very large farms are owned and operated primarily with family labor. This is generally most true where the farming system is land-extensive and not labor-intensive, such as cash grain production in the Midwest and ranching in the Great Plains and the West. Increasing concentration is expected to continue to take place in these regions without large increases in hired labor. However, adverse effects may also occur in regions where continued concentration is likely to result in the loss of a substantial proportion of the moderate farms. Adverse impacts may result simply from the loss of a substantial proportion of the local population if many farm families relocate to other parts of the country and they are not replaced by inmigration. This in turn will result in a reduction in the population base that supports civic activities and patronizes the small businesses and services in the local rural communities.

Retail establishments in rural communities that provide goods and services to these moderate farms may decline when the consolidated farms choose to purchase goods and services at greater distances. The operators of the large and very large farms that emerge from the process of structural change are considered more likely to purchase goods and services and to market their products over greater distances than their more moderate predecessors were. The argument is that they are able to receive volume discounts and premiums from more centralized purchases and sales.

REGIONAL DIFFERENCES

The structure of agriculture and the characteristics of rural communities vary greatly across the United States, owing to major differences in soils, climate, population density, pattern of land use, economic and social history, availabiliy of irrigation water, topography, availability of low-cost labor, and the level of education of the population. Following from this it can be expected that changes in agricultural structure will vary in different parts of the country and that the impacts of structural change on rural communities may vary in different regions. To avoid overgeneralizing in its assessment of impacts of structural change on rural communities, OTA analyzed five regions of the United States: the Northeast; South; Midwest;

the Great Plains and the West; and the CATF (those counties with the most industrialized agriculture in four Southern and Western States—California, Arizona, Texas, and Florida). This division of the United States differs in several respects from the regional division used in the U.S. Census of Agriculture. In particular, the Great Plains and the West in this chapter include most of the Western region of the Census of Agriculture and also have parts of the North Central and Southern regions from the census. The CATF region in this chapter has no close counterpart in the Census of Agriculture.

Although the intended focus was on rural communities, information on the welfare of individual communities and on linkages between individual communities and surrounding farms was not directly available on a regional basis for most States. In general, it was necessary to do the analysis in terms of rural agricultural counties instead of individual communities. This was a distinct disadvantage, since county-level data tend to obscure the details of linkages and impacts at the community level.[2] With the exception of the CATF region, the set of counties defined as rural and agricultural was drawn from the set of nonmetropolitan counties. Metropolitan counties as defined by the Office of Management and Budget are also known as Standard Metropolitan Statistical Areas, with the exception of the New England region. All other counties may be considered to be rural counties. Rural counties with a significant proportion of total income from agricultural sources were considered as candidates for inclusion in the set of rural agricultural counties. The minimum proportion of agricultural income that was considered significant varied by region and ranged from 5 to 20 percent.

[2]However, if statistically significant relationships can be shown between structural change in agriculture and changes in social welfare when the unit of analysis is the county, then it is likely that the associations are even stronger for a proportion of the communities in the county. This is generally true because not all communities are likely to be equally affected within a county. The county level of analysis aggregates the different impacts on the individual communities and tends to level out strong relationships with respect to particular communities. Therefore, statistically significant associations at the county level are a strong test of a hypothesis concerning associations at the community level.

The CATF Region

The CATF region includes all of Florida and the industrial-agricultural counties of California, Arizona, and Texas.[3] The counties studied are the 98 counties that were either in the top 100 counties nationwide in sales or that had $2,000 or more per year in per capita income from agriculture. Twenty-six Texas cattle and grain counties that met these criteria were grouped with the region covered by the Great Plains and the West, since they fit the land-use intensity and farming system types of this region better than the CATF region. The CATF regional set of counties[4] differs considerably from the nonmetropolitan counties in the rest of the United States in that agricultural development is relatively recent in these counties. Many CATF counties have not gone through a period in which moderate farms dominated agricultural structure. These counties are also unique in the extent to which they depend on subsidized irrigation water from State and Federal water projects.

CATF counties are of particular interest because agriculture in these counties has already evolved into a highly concentrated structure. The great majority of agricultural sales comes from large-scale, industrial-type farms. In 1982, farms in the very large sales class had 66 percent of regional sales as compared with 25 percent for the rest of the United States. Moreover, the share of sales from very large farms increased from 58 to 66 percent between 1978 and 1982. The number of farms in all acreage categories in CATF counties has recently declined, with the exception of the very large sales class. This reduction in the number of farms with fewer than 2,000 acres increased the concentration of production by 17 percent at the same time that total agricultural sales in the region increased only 6.5 percent (see tables 11-1 and 11-2). The counties selected as the data set for

[3]"CATF counties," "CATF regional set of counties," and "CATF region" are used synonymously.

[4]The findings in this section are contained in Dean MacCannell and Edward Dolber-Smith, "Report on the Structure of Agriculture and Impacts of New Technologies on Rural Communities in Arizona, California, Florida, and Texas," prepared for the Office of Technology Assessment, Washington, DC, 1985.

Table 11-1.—Comparison of Selected Farm Characteristics, All Counties in California, Arizona, Texas, and Florida, for 1969 and 1978

Attribute	1969		1978		Change from 1969 to 1978
Number of farms:	***332,878***	***(856)[a]***	***290,977***	***(746)***	*– 9.88*
Sales:					
Sales categories:					
>$100,000			31,983	(82)	
$40-90,999			29,693	(76)	
>$40k	34,028	(88)	61,676	(158)	81.25
$20-39,999	27,672	(71)	28,536	(73)	3.12
$10-19,999	34,973	(90)	34,856	(90)	– 0.33
$ 5-9,999	43,733	(112)	42,960	(110)	– 1.77
$2,500-4,999	46,947	(121)	45,679	(117)	– 2.70
<$2,500	145,476	(374)	76,903	(197)	–47.14
Total sales per county	$20,509,402	(1,000)	$21,838,408	(1,000)	6.48
Average sales per farm (1980 dollars)	$66,180		$77,334		16.85
Acreage:					
Acreage categories:					
≥2,000 acres	14,826	(38)	14,869	(38)	0.29
1,000-1,999 acres	16,969	(44)	16,468	(42)	– 2.95
500-999 acres	32,419	(83)	27,772	(71)	–14.33
180-499 acres	70,513	(181)	56,773	(146)	–19.49
50-179 acres	96,974	(249)	81,939	(210)	–15.50
10-49 acres	69,878	(180)	62,002	(159)	–11.27
<10 acres	31,319	(81)	31,154	(80)	– 0.53
Average acreage per farm	1,939		1,957		0.93
Type of ownership:					
Family ownership	166,911	(429)	176,448	(452)	5.71
Partnership	25,283	(65)	26,443	(68)	4.59
Corporate ownership	5,209	(13)	9,529	(24)	82.93
Other type of ownership	1,866	(5)	1,287	(3)	–31.03

[a]First value is the sum for all counties; the mean value per county is in parentheses.

SOURCE: Office of Technology Assessment.

this analysis are shown in detail in a paper by MacCannell and Dolber-Smith (1985).

Background on the CATF Region

Current agricultural patterns were not fully established in the CATF area until after World War II. Arizona was not even a State until 1912, and many of the new agricultural regions of all four States were not settled until this century. Historically and nationally, the original pattern of Spanish land grants and dependence on irrigation systems are the two main underlying causes of present-day farming systems. Prior to this century, agricultural land holdings were enormous, and the main products were range animals and nonirrigated grains. The first crude irrigation systems were built at the beginning of this century by land developers and speculators. The Imperial Valley of California went through several successive cycles of irrigation, land sales based on the promise of cheap and plentiful water, irrigation system failure, farmer bankruptcy, land repurchase or repossession by the original speculators, irrigation system overhaul, land resale, and so forth. These and similar abuses eventually resulted in Federal intervention and public involvement in the construction and management of irrigation systems throughout CATF. A number of these systems are enormous and of very recent construction. The San Luis Unit of the California Central Valley Project, completed in 1969, is one such system. This farm and rural community system in the most productive area of California (West Fresno County) has only a 15-year history.

Since World War II, the large agricultural operators of the CATF region have exploited their natural, historical, and political advantage by combining new agricultural technologies, modern irrigation techniques, Government sup-

Table 11-2.—Comparison of Selected Farm Characteristics, Agricultural Counties in California, Arizona, Texas, and Florida, for 1969 and 1978

Attribute	1969		1978		Change from 1969 to 1978
Number of farms:	***97,557***	***(995)***	***84,951***	***(867)***	***−12.92***
Sales:					
Sales categories:					
>$100,000			16,184	(165)	
$40-90,999			13,703	(140)	
>$40k	16,298	(166)	29,887	(305)	83.38
$20-39,999	13,082	(133)	11,048	(113)	−15.55
$10-19,999	14,473	(148)	10,522	(107)	−27.30
$ 5-9,999	14,240	(145)	9,942	(101)	−30.18
$2,500-4,999	11,696	(119)	8,766	(89)	−25.05
<$2,500	27,768	(283)	14,705	(150)	−47.04
Total sales per county	$10,307,860	(1,000)	$11,855,632	(1,000)	15.02
Average sales per farm (1980 dollars)	$113,068		$153,750		27.14
Acreage:					
Acreage categories:					
≥2,000 acres	5,418	(55)	5,744	(59)	6.02
1,000-1,999 acres	6,420	(66)	6,647	(68)	3.54
500-999 acres	11,792	(120)	9,839	(100)	−16.56
180-499 acres	18,968	(194)	14,926	(152)	−21.31
50-179 acres	21,507	(219)	18,303	(187)	−14.90
10-49 acres	24,757	(253)	21,047	(215)	−14.99
<10 acres	8,702	(89)	8,445	(86)	− 2.95
Average acreage per farm	2,074		2,192		5.69
Type of ownership:					
Family ownership	60,503	(617)	55,639	(568)	− 8.04
Partnership	10,015	(102)	10,019	(102)	0.04
Corporate ownership	2,096	(21)	4,048	(41)	93.13
Other type of ownership	648	(7)	459	(5)	−29.17

[a]First value is the sum for all counties; the mean value per county is in parentheses.

SOURCE: Office of Technology Assessment.

port programs, and an abundant supply of cheap, foreign labor. At the present time, CATF counties occupy a preeminent position in the national agricultural economy and international trade. Half of the top 100 agricultural counties nationwide are found in these four States. Agricultural products are the principal "industry" and export of California, Arizona, and Texas. In Florida, agriculture ranks behind tourism and manufacture, but it still employs 77,000 workers and has an annual sales of $1.3 billion. In Texas, the value added in agriculture is 1.3 times that of all manufacturing. The economic position of agriculture within the CATF region is all the more remarkable when the high level of industrial development of these same States is taken into consideration.

The agricultural commodity mix of the CATF counties is extremely diverse. With the exception of Florida citrus, there is no statewide monoculture or clear dominance of entire regions by a single crop or commodity. Leading commodities of the four States are cotton, sorghum, beef, wheat, citrus, row-crop vegetables, rice, sugarcane, sugar beets, grapes, melons, avocados, strawberries, nuts, peanuts, and corn. Over 9 million acres of cotton are grown in the CATF region, amounting to 70 percent of the U.S. total and about 30 percent of the world trade in cotton. One hundred percent of U.S. citrus and 55 percent of all noncitrus fruits are grown in the CATF region.

The CATF counties have four dominant forms of agricultural operations: 1) large-scale, family-owned corporations; 2) large-scale, corporate farms and partnerships; 3) highly sophisticated "part-time" operations owned by investors (usually urban-based professionals), which have high gross sales from small acreages; and 4) small-scale, unsophisticated, part-time farming operations with low sales. There has never been a widespread pattern of moderate-scale, family-owned farms in the CATF region.

Findings for the CATF Region

Examining the impact of structural change that has already occurred on the rural communities among these counties may give insight into the impacts of continuing concentration in other regions of the country. The analysis of these counties has benefited from the availability of data on poverty, unemployment, and standard of living at the community level (census tract data) in contrast to the analyses of other regions, which are based almost entirely on county-level data.

The primary finding is that there is a strong correlation between increased concentration and substandard social and community welfare in this regional set of counties. However, this relationship is not strictly linear. As agricultural scale increases from very small to moderate farms, the quality of community life improves. Then, as scale continues to increase beyond a size that can be worked and managed by a family, the quality of community life begins to deteriorate. Increasing concentration in this region results in increasing poverty, substandard living and working conditions, and a breakdown of social linkages between the rural communities that provide labor and the farm operators. In other words, the relationship of community welfare to agricultural structure resembles an inverted U curve (MacCannell and Dolber-Smith, 1985). This finding is a modification of the basic Goldschmidt hypothesis that any increase in concentration is associated with a decrease in community well-being.

The most extreme poverty in CATF counties is found in those counties with the most concentrated and productive agriculture. Up to 70 percent of the population of the most highly concentrated counties live in poverty. Up to 40 percent of the population live in houses without plumbing in the same counties.[5]

It was found that the types of rural communities in the CATF counties could be usefully placed in one of three categories:

1. Communities in which the population living on farms is wealthy and the associated rural communities are impoverished. This pattern is found in the central valley of California and in parts of Texas.
2. Communities that are internally segregated; the wealthy and the poor live in segregated neighborhoods in the same community. This pattern was found in and near Lubbock and Brownsville, Texas.
3. Communities that are externally segregated. In this pattern, entire communities are dominated by a single social class or ethnicity. The result of this pattern is a regional set of counties within which some of the towns are lower working class, farm worker, and transient communities, while other towns nearby exhibit the classic pattern of the rural trade center, and are the communities of choice for middle-class inmigrants and nonagricultural, industrial relocation.

In general, CATF communities that are surrounded by farms that are larger than can be operated by a family unit have a bimodal income distribution, with a few wealthy elites, a large majority of poor laborers, and virtually no middle class. The absence of a middle class at the community level has a serious negative effect on both the quality and quantity of social and commercial services, public education, and education. Rothman and others (1977) find that hired agricultural laborers are always located on the bottom of community status hierarchies, are always transient to some degree (even if not technically migratory), and are never treated as full-fledged members of the rural community. On the other hand, the large-scale farm owner-operators tend to bypass local public and commercial services and establishments, preferring

[5]The measure of agricultural concentration that is the best predictor of change in median family income between the 1970 and 1980 census years is the proportion of farms in each county with greater than $40,000 in annual gross sales of farm products. This variable is inversely related to the variation in median family income and accounts for 31 percent of the change in income. That is, counties that had a high proportion of farms with sales in excess of $40,000 in 1970 had a strong tendency toward low growth in family income in the decade of the 1970s. This finding lends strong support to the Goldschmidt hypothesis.

to shop in distant cities. These same large-scale farm owner-operators purchase education, police protection, recreation, and other public sector amenities from the private sector for their own exclusive use. As a result, their needs and desires are not translated into community well-being. The recent public involvements of the largest farmers in CATF are not based in the local community, but in lobbying and selling at the State, Federal, and international levels.

In sum, CATF rural communities in the most productive agricultural areas do not share in economic or social benefits from increased production and sales. Instead, the rural community stagnates or declines in the context of increasing agricultural productivity. Under these structural conditions, CATF agriculture is increasing its dependence on foreign labor. Continued importation of labor, operating within a different value system than the rest of the United States, is the only possible support for an agricultural economy that has become disarticulated from the local community. Increasingly, the rural communities in CATF agribusiness areas are not local in the sense of participating in U.S. social and cultural traditions. Instead, they resemble Honduran plantation communities more than their rural counterparts in other areas of the United States. In effect, social and economic relations from the developing world have been adopted to maintain the world market position of CATF agriculture.

A major cause for concern in CATF counties stems from the potential for substantial additional displacement of labor in the production of certain fruit and vegetable crops and dairy products. Historically, fruit and vegetable production in CATF has been a large, steady source of employment, although low paying and often seasonal in nature. However, one of the anticipated impacts of emerging technologies is a reduction in the labor required to prepare fields and seedbeds, plant, cultivate, treat, harvest, sort, and process fruits and vegetables. This situation could result in substantial increases in unemployment in CATF counties among farm laborers who have few employment alternatives. Since the rural-labor communities in this region are already impoverished and alienated from mainstream U.S. society, substantial increases in unemployment are likely to result in increased unrest and discontent among the farmworker population. This problem will be offset to some extent if CATF counties succeed in capturing additional production shares from other regions of the country.

The Great Plains and the West Region

Background on the Great Plains and the West Region

The region of the Great Plains and the West encompasses the 17 States—North Dakota, South Dakota, Nebraska, Kansas, Oklahoma, Texas, New Mexico, Colorado, Montana, Wyoming, Idaho, Utah, Arizona, Nevada, California, Oregon, and Washington—that potentially had agricultural counties in which farming systems of grains, or livestock (excluding dairy), or combined grains and livestock predominated.[6] In these 17 States, 351 counties were classified as agricultural. The basic units of analysis are the 234 counties with at least 20 percent of total proprietor and labor income from agriculture and counties with at least 25 percent of their economically active population in agriculture in 1982. Each county in the region is classified by its predominant type of agriculture, according to data from the 1974 and 1982 Census of Agriculture. The four classes of agriculture used are wheat farming (33 counties), livestock (84 counties), wheat and livestock (72 counties), and mixed grains (corn, sorghum, and wheat) and livestock (45 counties). Counties with 25 percent or more of production in other grains and field crops, hogs, or dairy production are eliminated from the analysis. Those counties of California, Arizona, and Texas that are characterized by industrial agriculture are also excluded. The industrial-agricultural counties of these States are considered separately as part of the CATF region. The counties used to rep-

[6]The findings reported in this section are contained in: Jan L. Flora and Cornelia Butler Flora, "Emerging Agricultural Technologies, Farm Size, Public Policy, and Rural Communities: The Great Plains and the West," prepared for the Office of Technology Assessment, Washington, DC, 1985.

resent the Great Plains and the West are shown in detail in the paper of Flora and Flora (1985).

Counties are further subdivided into those with a predominance of moderate farms and those with a predominance of large farms. Farms in this region are generally much larger per dollar of production than farms in other regions of the United States. Consequently, the definition of large and moderate size is different than that in other regions. Farms with acreages in the range of 500 to 999 acres represent moderate farms, whereas farms larger than 2,000 acres represent large farms.

The analysis was based on agricultural counties in the Great Plains and the West. The Great Plains includes most of both the northern and southern plains. Northeast New Mexico and the eastern half of Colorado are included in the Great Plains. The western half of South Dakota and northwestern Nebraska, areas in which grazing on federally owned land is common, are considered part of the West. Annual precipitation in the Great Plains ranges from more than 40 inches, in eastern Texas, to less than 20 inches, along the western border of the region. In the area of the West considered agricultural, annual precipitation ranges from 25 inches or less in most river valleys to 6 inches or less in the intermountain basins and southern desert areas. Much of the land outside the river valleys is owned by Federal or State governments that regulate access to the land, which is used primarily for grazing cattle and sheep. The frost-free period in this region ranges from less than 100 days in North Dakota to over 300 days in southern Texas.

Wheat production (spring wheat in the northern plains and winter wheat in the southern plains) dominates the Great Plains. Corn, an increasing proportion of which is irrigated from west to east, is an important crop in southeastern South Dakota and eastern Nebraska and Kansas. In those areas where integrated grain-livestock operations rotate corn and soybeans, and are the dominant farming system, agriculture is very similar to that in the Corn Belt. Grain sorghum is an important crop farther west in Nebraska, Kansas, and Texas, where precipitation levels are less. Cotton dominates agriculture in parts of Oklahoma and Texas. However, much of the land, especially in the western half of the region, is in native range, whose productivity is limited by low precipitation.

Findings for the Great Plains and the West Region

Growth in the number of large farms in each type of county was associated with a decline in the number of moderate farms. In general, there was a kind of homogenization; more counties had a similar structure in 1980 than in 1970. A number of significant findings parallel those found in CATF counties, although they are not as dramatic. Support for the Goldschmidt hypothesis was provided by two specific findings:

1. The decline in the number of retail services and retail sales was greatest in those counties with the largest increase in very large farms. Conversely, more retail services and sales were retained in counties in which the number of moderate farms increased in the 1970s.
2. Counties with a predominance of moderate farms in 1970 experienced a greater increase in hired labor than large farm counties during the 1970s, where farm consolidation occurred, and an even greater increase in mechanization investment.

There is evidence that the growth in the number of very large farms was associated with moderate declines in rural, nonfarm, and total population. This suggests that concentration is occurring at the same time that the total number of agricultural workers is declining. That is, expansion of farm size in the Great Plains and the West is taking place without a substantial increase in hired labor as has been the case in industrial agriculture of the CATF counties.

Counties with growth in the number of moderate farms tended to retain rural nonfarm and total population. Unlike CATF counties, there was no significant association between change in size of farms and change in measures of income.

The counties with a predominance of moderate farms in 1969 had a much greater decline in the numbers of commercial establishments between 1969 and 1982 than did those counties initially dominated by large farms. This was largely because moderate farms in the 1970s declined in numbers during the process of land concentration.

Median family income was generally positively associated with the proportion of land in large farms and negatively associated with land in moderate farms. This suggests that population has declined in large farm counties and that the reduced number of large farms that remain generally have higher net incomes than the farms in counties where moderate farms still predominate. During the decade of the 1970s, the gap in median family incomes between large and moderate farm counties generally tended to close, especially for wheat and wheat/livestock counties.

A number of differences emerged between the four different types of counties:

1. Livestock counties were the only counties that gained population as a group. The livestock counties had the greatest increase in rural nonfarm population and did not lose as much of their farm population as the other three types of counties. However, the livestock counties had the lowest initial population base.
2. Wheat counties that were dominated by large farms in 1969 had a greater increase in hired labor than wheat counties dominated by medium farms in 1969. This is the opposite of what occurred in the other types of counties.
3. The strongest correlation between increases in median family income and increased farm size occurred in wheat counties. This was consistent with the relationship of poverty and farm size in wheat counties. Wheat counties with a higher percentage of large farms had a lower percent of poverty. Interestingly, wheat/livestock counties had the opposite relationship between poverty and farms size. Poverty was positively correlated with percent of large farms in the wheat/livestock counties. Moreover, the correlation between poverty and percent of large farms became more strongly positive during the 1970s. The dominant size class of farms was unrelated to poverty for the livestock and mixed crop/livestock counties.

The basic conclusion about the Great Plains and the West is that there is some support for the modified Goldschmidt hypothesis, but that the outcome there is likely to be much different than that in CATF counties. Incomes improved as concentration increased, but there were declines in population and number of retail establishments. There is a strong potential for the development of a high concentration of agricultural production in the Great Plains and the West—especially in terms of farm size, if not gross sales per farm. The most likely adverse impact will be the loss of population and small retail firms in the region. At the same time, there is likely to be lower availability of alternate employment options in manufacturing and service industries as compared with other regions of the country. As a result, many small rural communities are expected to become substantially less viable. As in the CATF region, the trend is toward increasing sales and net income per farm as farm concentration and consolidation continue.

The region of the Great Plains and the West is unlikely to develop the highly intensive, diversified agriculture with high dependence on low-cost, hired labor that characterizes the CATF counties. The people that remain in the agricultural counties of the West are likely to have higher median incomes as a result of concentration of production in this region. The type of farming systems in the counties included in the Great Plains and the West will simply not have the labor requirements of the intensive fruit, vegetable, and livestock production of CATF industrial counties.

The Northeast Region

Background of the Northeast Region

The Northeast region comprises six New England States (Connecticut, Maine, Massachusetts, New Hampshire, Rhode Island, and Vermont) plus three Middle Atlantic States (New Jersey, New York, and Pennsylvania).[7] This region is characterized by a relatively uniform distribution of farm types and sizes and a relative absence of large-scale or industrialized agriculture. Agriculture in this region is dominated by the production of dairy products, followed by fruit, vegetable, and poultry production for nearby urban markets. There has been a longstanding decline in the amount of land in agriculture in this region; however, this decline was attenuated in the early 1970s. The farm population as a percentage of the nonmetropolitan population has been lower than that of the other regions of the United States since the turn of the century.

Urban economic and social influences have a relatively dominant role over the well-being of rural communities in this region. There are only 105 nonmetropolitan counties in the whole region, and only 30 counties in which 5 percent or more of labor-proprietor income was derived from agriculture. The most agriculturally productive counties in this region are not the most rural, but are instead closely linked with major urban centers. In consequence, it is reasonable to expect that structural change in agriculture in the Northeast is much less likely to be associated with adverse effects on rural communities as it is in other regions of the United States, since opportunities for off-farm employment are likely to continue to be better for more rural residents in this region than in many other parts of the country.

The Northeast is also quite diverse. One source of diversity is agroecological in nature. The six New England States generally have low-quality soils and short growing seasons, with a few exceptions, such as the Connecticut River Valley. The Middle Atlantic States generally have more favorable agricultural conditions. The second source of diversity is socioeconomic in nature and relates to the dramatic variations in urban-metropolitan influence in the region. The contrasts are striking between the Boston to Washington, DC, megalopolis, with its densely settled 35 million inhabitants, on the one hand, and the State of Vermont, which is composed entirely of nonmetropolitan counties, on the other.

The farm structure of the Northeast showed increased strength in the small and part-time farm component during the 1970s and early 1980s. The number of farm operators whose principal occupation was not farming, or who worked any days off-farm, or who worked 100 or more days off-farm increased more rapidly in the Northeast than in the remainder of the United States. The Northeast also exhibited a larger increase in the number of individual or family farms than in the rest of the United States, primarily in the small and part-time classes of farms. Table 11-3 shows a comparison of the Northeast with the rest of the United States for selected characteristics for 1974 and 1982.

Findings for the Northeast Region

The results of the analysis of relationships between measures of structural change and measures of community well-being support the expectation that structural change in agriculture is not likely to have great impact on rural communities. This finding is in stark contrast with the findings from CATF counties and also differs considerably from the findings for the Great Plains and the West region. The analysis does provide some useful insights of a more detailed nature:

1. There was no strong pattern of social or economic decline in rural counties between 1970 and 1980.
2. The rural population of the Northeast region has relatively high income levels and access to services.
3. The rate of technical change as measured by expenditures on machinery and chemicals was relatively low during the 1970s.
4. There was no significant relationship between: a) change in technology and farm structure, and b) rural community welfare changes during the 1970s.

[7] The findings reported for the Northeast region are contained in: Frederick H. Buttel and Mark Lancelle, "Emerging Agricultural Technologies, Farm Structural Change, Public Policy, and Rural Communities in the Northeast," prepared for the Office of Technology Assessment, Washington, DC, 1985.

Table 11-3.—Comparison of Northeast Regional Farms With Total U.S. Farms by Selected Characteristics, 1974 and 1982

Farm structure characteristics	Northeast region		Percent change, 1974-82	Total United States		Percent change, 1974-82
	1974	1982		1974	1982	
Number of farms	127,531	131,991	3.5	2,314,013	2,241,124	−0.3
Land in farms (acres)	23,359,889	23,061,163	−1.3	1,017,030,357	984,755,115	−0.3
Average size of farm (acres)	183	175	−4.4	440	439	−0.1
Value of land and buildings:						
Average per farm (dollars)	121,227	214,623	77.0	147,838	347,974	135.2
Average per acre (dollars)	662	1,236	86.7	336	791	135.4
Farms by size:						
<10 acres	7,689	10,599	37.8	128,254	187,699	46.3
10-49 acres	19,416	26,421	36.1	379,543	449,301	18.3
50-179 acres	54,901	51,866	−5.5	827,884	711,701	−14.0
180-499 acres	37,864	34,533	−8.8	616,098	526,566	−14.5
500-999 acres	6,421	7,070	10.1	207,297	203,936	−1.6
1,000-1,999 acres	1,046	1,282	22.5	92,712	97,396	5.1
≥2,000 acres	194	220	13.4	62,225	64,525	3.7
Land use:						
Total cropland (acres)	13,851,473	13,972,802	0.8	440,039,087	445,527,557	1.2
Woodland (acres)	5,809,958	5,899,750	1.5	92,527,627	87,133,026	−5.8
Agricultural products sold:						
Market value ($1,000)	4,291,380	7,179,543	67.3	81,526,124	131,810,903	61.6
Average per farm	33,650	54,394	61.6	35,231	58,815	66.9
Crops	1,440,397	2,181,303	51.4	41,790,360	62,274,394	49.0
Livestock	2,216,436	4,998,240	125.5	33,301,560	69,536,509	108.8
Poultry	616,094	844,395	37.1	6,202,291	9,732,222	56.9
Farms by type of organization:						
Individual or family	82,142[a]	115,713	40.9	1,517,573[a]	1,945,724	28.2
Corporation	2,615[a]	4,098	56.7	28,656[a]	59,788	108.6
Tenure of operator:						
Full owner	83,389	82,043	−1.6	1,423,953	1,325,931	−6.9
Part owner	36,112	40,005	10.8	628,224	656,219	4.5
Tenant	8,030	9,943	23.8	261,836	258,974	−1.1
Principal occupation:						
Farming	78,144	75,111	−3.8	1,427,368	1,234,858	−13.4
Nonfarming	46,390	56,442	21.5	851,902	1,006,266	18.1
Operators reporting: days of work off farm						
Any days	56,670	67,751	19.6	1,011,476	1,187,490	17.4
≥100 days	46,691	56,048	20.0	814,555	963,728	18.3
Selected production expenses ($1,000):						
Commercial fertilizer	207,433	309,769	49.3	5,137,361	7,689,577	49.7
Other agricultural chemicals	74,225	140,301	89.0	1,757,776	4,282,795	143.6
Hired labor	401,846	712,383	77.3	4,652,074	8,434,399	81.3
Workers working ≥150 days: farms	21,775[a]	29,242	34.3	223,093[a]	312,621	40.1
Numbers of workers[b]	66,149	88,547	33.9	712,715[a]	950,112	33.3
Machinery and equipment:						
Estimated value ($1,000)	2,879,414	5,337,081	85.4	48,402,626	93,686,308	93.6
Average per farm	23,470	40,435	72.3	22,303	41,930	88.0

[a]Among farms with sales ≥$2,500.
[b]Computed from the preliminary reports for the nine Northeastern States.

SOURCES: Data for 1974: *1978 Census of Agriculture Preliminary Report,* Northeast region and United States (Washington, DC: U.S. Department of Commerce, Bureau of the Census, 1980); Data for 1982: *1982 Census of Agriculture: Preliminary Report,* nine Northeastern States and United States (Washington, DC: U.S. Department of Commerce, Bureau of the Census, 1983).

5. Farm population change, the proportion of full-owner farms, and the proportion of part-time farmers had a positive effect on community well-being in rural counties, as measured by the poverty rate and median family income. That is, counties in which farm population declined, the proportion of full-owner farms declined, or the proportion of part-time farmers declined had a moderate increase in the poverty rate. The proportion of farms that were fully owned had the strongest relationship to the poverty rate.
6. Counties in which the percent of farms owned by corporations increased also had increases in poverty rates and declines in median family incomes.

The primary structural change that is likely to occur in the Northeast is due to technological changes in the dairy industry. As discussed in chapters 2 and 8, new dairy technologies, primarily bovine growth hormone and computerized feeding technologies, are expected to increase production greatly and lower production costs substantially for those dairies that are able to adopt them. The result will be greatly increased production at the currently administered milk price levels. This will in turn trigger price support reductions and increased failure rate among dairy farms. Over the next 10 years, over half of the small-to-moderate dairies in the Northeast may be forced to leave agriculture. The production of milk will become concentrated in the larger dairies in the region, and more milk will be shipped into the region. Unlike the CATF region, where the bulk of milk production is expected to be concentrated in very large-scale dairies, with thousands of cows and with industrial relations of production, dairy production in the Northeast is more likely to remain concentrated in dairies with herds in the 100- to 500-cow range and with relatively few hired workers per dairy.

The Midwest Region

Background of the Midwest Study Area

The Midwestern region is composed of Illinois, Indiana, Iowa, Michigan, Wisconsin, Minnesota, Missouri, and Ohio.[8] The 565 nonmetropolitan counties served as the data set for the analysis of this region. These counties were segregated into three groups; those in which 10, 20, and 30 percent of total labor and proprietor income came from agriculture. Table 11-4 shows the distribution of the nonmetropolitan counties according to the level of dependence on agriculture by State. As can be seen, the counties most dependent on agriculture are located in the western part of the region. The range of dependence on agriculture among the seven States varies considerably. For example, 29 percent of the counties with at least 30 percent of income from agriculture are in Iowa alone, whereas Ohio and Michigan have no counties with 30 percent income from agriculture and only one and two counties, respectively, with at least 20 percent income.

With the exception of CATF counties, this region differs from the other regions of the United States in the extent to which agriculture dominates its landscape. Nonetheless, this region also has a large industrial base. Table 11-5 provides aggregate statistics for the Midwest region as well as for the nonmetropolitan counties and agricultural counties of this region. The nonmetropolitan counties account for 30 percent of the Midwest's population and for over 75 percent of the farm population, farming acreage, and farm sales. When the proportion of county income derived from agriculture is taken into account, it becomes apparent that the agricultural counties have only a modest percentage of the regional population. Around 12 percent of the region's population live in counties with at least 10 percent of income derived directly from agriculture. The counties that depend most on agriculture account for less than 2.5 percent of the region's total population and less than 10 percent of the nonmetropolitan population (van Es, Chicoine, and Flotow, 1985).

Because change in agricultural technologies will not have uniform impacts on agriculture—

[8]The findings reported for the Midwest region are contained in: J.C. van Es, David L. Chicoine, and Mark A. Flotow, "Agricultural Technologies, Farm Structure, and Rural Communities in the Midwest: Policy Choices and Implications for 2000," prepared for the Office of Technology Assessment, Washington, DC, 1985.

Table 11-4.—Distribution of Rural Counties, by Varying Levels of Dependence on Agriculture, for States in the Midwest

	Rural counties		Dependence on agriculture					
			At least 10%		At least 20%		At least 30%	
	Number	Percent	Number	Percent	Number	Percent	Number	Percent
Ohio	49	8.67	14	4.19	1	0.52	—	—
Indiana	62	10.97	30	8.98	15	7.81	5	5.00
Michigan	56	9.91	9	2.69	2	1.04	—	—
Wisconsin	57	10.09	35	10.48	16	8.33	3	3.00
Illinois	80	14.16	50	14.97	30	15.63	17	17.00
Minnesota	71	12.57	58	17.37	35	18.23	26	26.00
Iowa	91	16.11	76	22.70	56	29.17	29	29.00
Missouri	99	17.52	62	18.56	37	19.27	20	20.00
Total	565	100.00	334	100.00	192	100.00	100	100.00

[a]Dependence on agriculture is measured in terms of the proportion of labor and proprietory income derived from agriculture. The figures are based on the years 1975-79.

SOURCE: Office of Technology Assessment.

for example, technological changes in cash grain production will have little direct impact on dairy farms and vice versa—a more accurate analysis of technological impacts entailed the subdivision of the nonmetropolitan agricultural counties in the Midwest into grain counties, dairy counties, and mixed agriculture counties. The dominant type of agriculture in each county (defined as 50 percent or more of sales) determined a county's classification.

Dairy production in the Midwest accounts for about 13 percent of the regional agriculture and about 41 percent of dairy production nationwide. Dairy farms account for about 9 percent of the farm population but only about 2 percent of the regional population. There are more dairy farms (about 60,000) in the part-time sales class in this region than in any other sales class in the other regions. Almost all of the dairy counties are concentrated in Wisconsin. In 1978 there were 28 counties in the Midwest in which 50 percent or more of total agricultural sales were from dairy products. Only 3 dairy counties derived 30 percent or more of county income from agriculture, all in Wisconsin.

Grain counties realize more than 50 percent of their agricultural sales from the sale of grain. In the Midwest, grain sales include corn, wheat, soybeans, and minor specialty grains. Within the Midwest, grain farming is predominantly carried out by family-based enterprises, with a considerable input of outside capital. These families rent the land but contribute much of the labor and management. The grain counties account for about 20 percent of the agricultural agreage and sales and about 16 percent of the farm population and number of farms of the agricultural counties.

Once the counties characterized by the grain and dairy industry have been separated, the remaining 181 counties are characterized as being mixed livestock, dairy, and grain counties. The exact nature of these counties varies considerably. Most of the counties are in the western part of the Midwest and are characterized by integrated grain and livestock farm enterprises. In a few cases, the counties contain different types of enterprises that are primarily specialized, none of which account for more than 50 percent of agricultural sales. The mixed agricultural counties account for 6.5 percent of the region's total population but about 30 percent of the region's farm population and sales. The 53 mixed agricultural counties that are most dependent on agriculture (30 percent or more income from agriculture) have less regional significance than might be expected. These counties have only 1 percent of the regional population and less than 10 percent of its sales from agriculture.

Findings for the Midwest Region

The impact of farm structure changes at the community level are difficult to isolate from other societal changes that may occur simultaneously. For example, agricultural counties

Table 11-5.—Characteristics of the Midwest Region, Rural Counties in the Region, and the Rural Agricultural Counties

		Rural counties		Dependence of agriculture[a]								
				10%			20%			30%		
	Region	Value	Percent of region	Value	Percent of region	Percent of rural	Value	Percent of region	Percent of rural	Value	Percent of region	Percent of rural
Population 1980	53,101,523	15,446,750	29.09	6,561,206	12.35	42.48	3,000,400	5.65	19.42	1,277,287	2.41	8.27
Farm population, 1980	2,308,613	1,760,442	76.26	1,192,596	51.66	67.74	670,299	29.03	38.08	331,551	14.36	18.83
Total farms, 1978	757,484	576,713	76.13	378,485	49.97	65.63	211,411	27.91	36.66	106,459	14.05	18.46
Acreage in agriculture (in thousands) 1978	182,579	146,967	80.49	103,445	56.66	70.39	63,236	34.63	43.03	33,984	18.61	23.12
Total sales (in millions), 1978	33,468	26,245	78.42	19,395	57.95	73.90	12,084	36.11	46.04	6,399	19.12	24.38
Number of countries	732	565	77.19	334	45.63	59.12	191	26.09	33.81	100	13.66	17.70

[a]Dependence on agriculture is measured in terms of the proportion of labor and proprietary income derived from agriculture. The figures are based on the years 1975-79.

SOURCE: Office of Technology Assessment.

(those receiving at least 10 percent of their income from farming) in the rural Midwest experienced differential population change and a sharp increase in unemployment during the 1970s; neither change can be attributed to structural changes in agriculture.

Comparisons within homogeneous groups (grain, dairy, and mixed farming) of agricultural counties show that those counties with larger farms tend to be somewhat better off than those counties where the process of concentration in agriculture has not progressed so far.

The data available for this analysis does not provide evidence of negative county-level consequences associated with the historical direction of change in farm structure in the Midwest during the 1970s. Although the process of structural change continues in the Midwest, the major changes in population, agricultural structure, and impacts on rural communities appear to have taken place before the 1970s.

The concentration of sales was used as the principal indicator of structure and structural change in the Midwest. Concentration of sales is measured with reference to the mean percent of farms with sales over $100,000 in 1980 for each set of counties deriving 10, 20, and 30 percent of their income from agriculture. That is, in each set, those counties whose percentage of farms with sales over $100,000 was greater than the median percentage were considered to have greater concentration than those counties whose percentage was less. The degree of concentration is measured by the difference between the median percent and the individual percentage for each county. In terms of the sales classes used elsewhere in this report, this measure of concentration groups counties into those with small and part-time farms on the one hand and those with more moderate to large and very large farms on the other hand.

According to this basic dichotomy, dairy counties with higher concentrations of moderate-to-large dairy farms consistently have higher median family incomes and lower median percentages of poverty than dairy counties with large percentages of small and part-time farms. These findings are shown in table 11-6. This measure does not provide any definite understanding of the association between increased sales from large and very large farms and measures of welfare such as median family income or poverty.[9] It is surprising that in the dairy counties, where the labor demands of the agricultural enterprise are high, there is a very large amount of off-farm employment.

Table 11-7 indicates that in grain counties with a higher concentration of sales of the moderate, large, and very large farms, it is more likely that farms are fully rented and that more labor will be hired. The farm operators are less likely to work at least 100 days off the farm. These factors are what would be expected in an agricultural setting characterized by moderate, large, and very large farms. Table 11-7 also shows that the grain counties characterized by higher concentration tend to have larger populations and to be more urbanized. Median family incomes are higher and the occurrence of poverty is less. In the counties with higher concentration, employment in manufacturing is higher, and unemployment is lower. Overall, it can be argued that grain counties with a higher concentration of sales have a higher level of economic well-being. It should be noted again that this definition of higher concentration includes farms in the moderate, large, and very large sales classes.

As shown in table 11-8, mixed agricultural counties with a heavier dependence on agriculture are somewhat different from counties that depend less on agriculture. Farms in the most dependent counties are slightly more likely to be rented fully and to hire labor. In these counties, the farm operator is less likely to work more than 100 days off the farm. The most dependent counties tend to contain fewer people, have a higher percentage of farm population, have lower median incomes, and have higher levels of poverty. Their retail sales and manufacturing employment are lower. These kinds of differ-

[9]Because of the small number of dairy counties, a comparison of the different levels of agricultural dependence among the dairy counties is not meaningful. Similarly, data on counties with increased concentration in sales among large and very large sales class farms would probably not be statistically significant, since the Midwest does not have many very large dairies.

Table 11-6.—Comparison of Midwestern Counties With Greater and Lesser Concentration of Sales on Farms With Sales of More Than $100,000 for Counties Dominated by Dairy Production

	Counties with sales concentration					
	10%		20%		30%	
Mean county values (ca 1980)	Below the regional mean	Above the regional mean	Below the regional mean	Above the regional mean	Below the regional mean	Above the regional mean
Agriculture:						
Percent of renters of farms	4.09	5.63	6.54	4.50	8.69	6.45
Hired labor, 150+ days per farm	0.29	0.43	0.32	0.46	0.21	0.54
Percent of farms hiring some labor	43.53	47.27	45.15	48.17	55.56	48.91
Percent of operators working 100+ days off farm	35.20	30.39	27.46	30.71	31.25	24.76
Demographic:						
Total population	20,576.00	31,950.00	20,280.00	21,929.00	25,642.00	14,309.00
Percent of farm population	18.93	16.34	26.19	19.45	31.55	27.33
Percent of urban population	19.07	27.27	15.60	22.83	14.00	18.00
Median family income ($)	15,790.00	16,996.00	15,703.00	16,262.00	15,703.00	16,996.00
Percent at poverty level	10.60	8.60	11.00	9.17	10.60	8.60
Business and employment:						
Retail sales per capita ($)	2,391.00	2,687.00	2,379.00	2,154.00	1,807.00	1,893.00
Percent employment: manufacturing	22.54	23.40	20.00	21.17	16.00	16.00
Percent employment: services	18.15	18.53	17.20	18.50	19.00	18.00
Percent of unemployed	8.62	7.47	7.40	8.00	6.00	7.00
Number of counties	13	15	5	6	1	1

SOURCE: Office of Technology Assessment.

Table 11-7.—Comparison of Midwestern Counties by Concentration of Sales on Farms With Sales of More Than $100,000 for Counties Dominated by Grain Production

	Counties with sales concentration					
	10%		20%		30%	
Mean county values (ca 1980)	Below the regional mean	Above the regional mean	Below the regional mean	Above the regional mean	Below the regional mean	Above the regional mean
Agriculture:						
Percent of renters of farms	15.35	25.57	17.84	26.44	18.70	27.23
Hired labor 150+ days per farm	0.19	0.32	0.21	0.30	0.22	0.26
Percent of farms hiring some labor	38.18	46.85	40.63	47.67	42.80	48.00
Percent of operators working 100+ days off farm	32.57	26.14	27.53	24.47	24.26	22.59
Demographic:						
Total population	18,074.00	21,022.00	14,258.00	18,507.00	13,269.00	14,829.00
Percent of farm population	20.03	16.80	23.49	18.89	26.44	22.75
Percent of urban population	24.95	32.19	18.35	28.78	17.00	21.95
Median family income ($)	16,706.00	18,665.00	16,637.00	18,623.00	15,844.00	18,221.00
Percent at poverty level	9.46	8.05	10.00	8.07	11.60	8.30
Business and employment:						
Retail sales per capita ($)	2,521.00	2,711.00	2,370.00	2,627.00	2,519.00	2,446.00
Percent employment: manufacturing	20.93	22.04	18.15	21.80	14.78	19.36
Percent employment: services	18.72	18.38	19.10	17.53	18.95	18.00
Percent of unemployed	7.94	6.96	7.43	6.64	7.43	6.14
Number of counties	65	57	40	36	23	22

SOURCE: Office of Technology Assessment.

Table 11-8.—Comparison of Midwestern Counties by Concentration of Sales on Farms With Sales of More Than $100,000 for Counties Dominated by Mixed Agriculture

	Counties with sales concentration					
	10%		20%		30%	
Mean county values (ca 1980)	Below the regional mean	Above the regional mean	Below the regional mean	Above the regional mean	Below the regional mean	Above the regional mean
Agriculture:						
Percent of renters of farms	9.90	19.43	11.76	22.11	13.30	22.85
Hired labor 150+ days per farm	0.18	0.26	0.19	0.26	0.17	0.25
Percent of farms hiring some labor	36.85	43.87	37.67	45.87	38.35	47.03
Percent of operators working 100+ days off farm	36.34	27.65	32.54	22.76	29.15	21.85
Demographic:						
Total population	16,776.00	21,874.00	13,174.00	16,794.00	10,213.00	13,290.00
Percent of farm population	22.93	21.62	26.95	25.15	39.54	27.52
Percent of urban population	18.82	30.01	14.20	27.81	8.97	21.29
Median family income ($)	15,363.00	18,184.00	15,363.00	17,768.00	14,729.00	17,070.00
Percent at poverty level	11.63	8.68	11.92	9.36	13.50	10.33
Business and employment:						
Retail sales per capita ($)	2,300.00	2,699.00	2,162.00	2,616.00	2,197.00	2,398.00
Percent employment: manufacturing	20.16	19.72	17.18	16.95	13.97	15.81
Percent employment: services	17.95	18.40	17.87	18.63	18.21	18.10
Percent of unemployed	7.18	5.53	6.23	4.98	6.03	4.76
Number of counties	109	72	61	43	32	21

SOURCE: Office of Technology Assessment.

ences were not expected in advance of the analysis. Counties that are most heavily dependent on agriculture in general turn out to be less prosperous.

The general finding that measures of social welfare improve as farm structure moves away from a predominance of small to part-time farms is consistent with the findings in other regions of the country. However, negative associations between farm scale and social welfare might emerge if data were available to distinguish between counties with a predominance of moderate farms and those with a predominance of large to very large farms. Also, the decade of the 1970s was generally a very prosperous period for the Midwest. It is not easy to find adverse associations between social and economic factors during periods of relatively little economic adversity.

An analysis of the relationships between factors as factors changed during the 1970s yields results similar to those of the static, cross-sectional analysis described above. As average farm size increased in the direction of moderate to large farms, median levels of income increased in dairy and mixed agricultural counties. Associations between the change in poverty rates and other factors did not yield consistent or significant results. There was a negative association between population change and change in the share of full ownership and part ownership of farmland. As the share of part ownership increased, county populations tended to decrease. The percent of manufacturing and service employment in 1970 was positively associated with population change during the 1970s.

The biotechnologies for animal agriculture will be of less significance to technological change in the Midwest because monoculture cash grain farming characterizes much of the agriculture in the region. Since the biotechnologies for plant agriculture are expected to bring changes relatively more slowly to this subsector, past trends in the Midwest will characterize much of the farm structural change and related community effects of the rest of the century (OTA, 1985).

The combined impacts of biotechnologies on the mixed crop and animal counties of the Midwest is more difficult to discern. This is particularly true in much of Iowa and parts of Illinois and Missouri, where pork production is the primary type of animal agriculture. As discussed in chapter 3, the rate of productivity increase in pork production is not significantly different from past trends, even under the most optimistic scenario of technological change. The hog industry is already in the process of restructuring in the direction of more concentration, vertical integration, and specialization of production technology. The impact of new technologies in the mixed crop and animal counties may be simply to accelerate the changes that are already taking place.

In general, the Midwest lies between the Northeast and the Great Plains and the West with respect to expected impacts on rural communities. At the regional level, the Midwest is similar to the Northeast in that certain areas have a concentration of dairy production that is likely to undergo considerable change. Also, like the Northeast, employment opportunities are likely for displaced farmers over large parts of the region. Unlike the Northeast, the Midwest does have some areas—primarily the western counties of Iowa and Missouri—in which agriculture dominates the economic base of the rural communities. These areas are likely to be more similar to the Great Plains and the West in that a decline in population and number of retail establishments will be associated with a continuing concentration of agricultural resources. Like the Great Plains and the West, the counties with at least 30 percent income from agriculture are likely to have fewer opportunities for off-farm employment.

The South Region

In this study, the Southern region comprises Alabama, Arkansas, Delaware, Georgia, Kentucky, Louisiana, Maryland, Mississippi, North Carolina, South Carolina, Tennessee, Virginia, and West Virginia.[10] Florida is excluded from this region and treated separately as part of the CATF counties. The South is more difficult than the other regions to characterize in terms of farm structure, owing to the relative diversity and complexity of areas within the South. For this reason, the structure of agriculture for this region will be presented in more detail than it was for the other regions.

Farm Structure in the South Region

The South has a high degree of diversity in commodities and in structures of production. It has the largest concentration of small, low-income farms in the United States, particularly in the Appalachian and tobacco-producing States. About 35 percent of the Nation's farms with sales less than $10,000 are located in the South. In contrast, 21 percent of all farms in the United States with sales in excess of $250,000 per year are also located in the South. In 1982, farms with sales of less than $40,000 made up 82 percent of the farms in the South and 68 percent of the farms in the rest of the United States. Commodities such as cotton and poultry are highly concentrated, whereas other commodities such as cattle and tobacco have very low concentrations. Hog and pig production is in the middle of a transition period from widely dispersed, small-scale production to concentrated, large-scale confinement operations.

Agriculture is more diversified in the South than in the other major regions of the United States. Cattle sales in the South account for the largest percentage of total sales in the South. However, at 27 percent in 1982, this share was less that that of the top commodity in each of the other four regions.

Different commodity groups have different structures of production in the South. In the deep South, where the plantation system was strongest and both cotton and sugarcane are still raised, there is a greater preponderance of industrial-type farms and larger-than-family farms. However, cotton and sugarcane have never been raised by family farms. On the other hand, both Kentucky and North Carolina are centers for tobacco production and are characterized by a proportionately larger number of small family farms. The South also has an especially large number of farms that raise cattle—363,994 out of a national total of 618,270

[10] Findings reported for the South are contained in: Jerry R. Skees and Louis E. Swanson, "Examining Policy and Emerging Technologies Affecting Farm Structure in the South and the Interaction Between Farm Structure and Well-Being of Rural Communities," prepared for the Office of Technology Assessment Washington, DC, 1985.

cattle farms in 1982. Fifty-three percent of all U.S. farms that had sales or inventory of cattle in 1982 were in the small farms class in the South. In contrast, poultry and egg production is relatively concentrated in the South in large and very large farms, with 26 and 39 percent of regional sales in 1982, respectively.

The South is also undergoing change at a faster rate than the rest of the United States. A comparison of basic farm structure statistics between nonmetropolitan counties in the South and in the rest of the United States is shown in table 11-9. The number of farms declined faster in the South than in the rest of the United States between 1969 and 1978. Average farm sales increased by 87 percent in the same period in the South, yet the increase was only 58.1 percent in the rest of the United States. Average acres per farm increased 37 percent in the South and actually decreased in the rest of the United States. One of the most dramatic changes in the structure of agriculture in the South involves the percent of full owners: the rate of decline was twice as great in the South as in the rest of the United States. The South went from 71 percent full ownership in 1969 to 39 percent in 1978, a 33-percent decline. The rest of the United States went from 56 to 39 percent, a 17-percent decline.

Many rural areas of the South still have low standards of living compared with other regions of the United States. Table 11-10 shows several quality-of-life variables in nonmetropolitan counties of the South and the rest of the United States. Poverty in the South was greater in both 1970 and 1980. In 1980 the average rural county poverty level was approximately 6 percent higher in the South than in the rest of the country. Median family income was also significantly lower than that in the rest of the United States. Because of the diversity of agriculture and the wide dispersion of production scales, there is a strong a priori expectation that definitive conclusions may be difficult to achieve for the region as a whole.

Table 11-9.—Comparison of Basic Farm Structure Statistics in Rural Counties of the South With Those of Total United States

Variable	Rest of U.S. county mean	South county mean
Farm numbers	860[a]	849
	741	624
	(−119)	(−225)
Farm size (sales)	$48,788	$24,675
	77,140	46,112
	(28,352)	(21,437)
Farm size (acres)	1,232	205
	1,220	232
	(−12)	(27)
Part-time farming	36.1%	44.3
	38.2	47.4
	(2.1)	(3.1)
Chemical and fertilizer use per farm	$2,697	$2,646
	5,713	5,517
	(3,016)	(2,871)
Machinery value per farm	$27,584	$17,023
	54,710	36,043
	(27,126)	(19,020)
Hired labor per farm	$4,781	$3,049
	4,638	3,936
	(−143)	(887)
Farms below 180 acres	43.4%	71.0%
	43.5	67.2
	(0.1)	(−3.8)
Percent full owners	55.8%	70.6%
	38.9	38.0
	(−16.9)	(−32.6)
Percent tenant operators	13.9%	9.9%
	11.9	8.0
	(−2.0)	(−1.9)
Percent grain sales	18.4%	11.0%
	25.9	20.0
	(7.5)	(9.0)
Percent livestock sales	38.0%	17.7%
	36.9	20.6
	(−1.1)	(2.9)
Percent unemployment	4.5%	4.9%
	6.5	7.8
	(2.0)	(2.9)

[a]1967-70 values are listed first.
1977-80 values are listed second.
NOTE: The change between the two time periods appears in parentheses.
SOURCE: Office of Technology Assessment.

Findings for the South Region

The analysis of associations between structural change and rural community welfare is based on the complete set of 706 rural counties in the region. Nonmetropolitan counties with low proportions of income from agriculture were not excluded from the set of counties.

The principal findings are as follows:

1. The Goldschmidt hypothesis is not confirmed at the regional level for the majority

Table 11-10.—Quality of Life Variables in Rural Counties

Variable	Rest of U.S. county mean	South county mean
County population	21,738 25,064 (3,326)	23,036 26,723 (3,687)
Percent families below poverty	15.0% 11.8 (−3.2)	26.6% 17.6 (−9.0)
Total year housing units	7,550 9,840 (2,290)	7,500 9,907 (2,407)
Property taxes per capita	$346 396 (50)	$103 135 (32)
Retail establishments	247 248 (1)	226 237 (11)
Median family income	$17,547 20,860 (3,313)	$14,055 18,112 (4,057)
Percent unemployment	4.5% 6.5 (2.0)	4.9% 7.8 (2.9)
Farm/rural population	27.1% 18.8 (−8.3)	17.5% 9.1 (−8.4)
Percent employed in manufacturing	16.0% 15.8 (−0.2)	30.4% 29.3 (−1.1)
Percent employed in services	7.0% 18.9 (11.9)	7.7% 16.8 (9.1)

[a]1967-70 values are listed first.
1977-80 values are listed second.
NOTE: The change between the two time periods appears in parentheses.
SOURCE: Office of Technology Assessment.

of indicators for community welfare, although there is support for a modification of the Goldschmidt hypothesis with respect to levels of unemployment. The nonmetropolitan counties are more dependent on manufacturing and service sector employment than on employment in agriculture. The structure of the manufacturing and service sector has a greater impact on social welfare than does agriculture in these counties. Manufacturing industries are associated with low levels of unemployment, but also with lower median family incomes. Service industries in the South are associated with both low levels of median family incomes and high rates of poverty.

2. There is a strong association between average farm size and unemployment in southern agricultural counties in both 1970 and 1980. However, this association is not strictly linear as is predicted by the Goldschmidt hypothesis. A pattern similar to an inverted U emerges when the agricultural counties are compared as a cross-section in 1970 and 1980. In each year, unemployment decreases sharply over the range from small farms to moderate farms. However, unemployment is also strongly associated with increasing average farm size over the range from moderate to large-scale farms. Other basic measures of social welfare, such as percent of poverty and median family income, do not appear to follow the same pattern. The basic conclusion is that the lowest rate of unemployment is associated with a farm structure dominated by moderate farms in the South. Unemployment tends to be substantially higher when the average farm size in a county is especially small or large.
3. Counties in which the average farm size increased the most during the years between 1970 and 1980 were likely to have declining levels of unemployment but greater increases in poverty. This analysis of changes over time provides some weak evidence in support of the Goldschmidt hypothesis.
4. Counties that had a substantial decrease in farm population have increased unemployment, poverty, and decreased median family incomes.
5. Levels of part-time farming are associated with county well-being. Those counties with high levels of part-time farming in 1969 and 1978 were more likely to have lower levels of poverty and higher levels of median family income. Furthermore, counties with the most rapid increase in the proportion of operators working 100 or more days off of the farm were more likely to have had a faster rate of decline in unemployment and poverty, along with a faster rate of increase in median family income. It is likely that the part-time farms have a welfare func-

tion: they provide their operators with supplemental income and some security in the context of employment variability.

Potential for Adverse Impacts on Rural Communities in the South

The potential for structural change varies considerably in different parts of the South. Large sections of the South are hilly and mountainous, terrain more similar to that of the Northeastern region of the United States. Like the Northeast in general, there is relatively low potential for the concentration of production in large and very large farms in these areas. The geography of these areas prevents the creation of large contiguous parcels of land on which large machinery can operate effectively. However, there is cause for concern with respect to the potential for developing highly concentrated industrial-scale agriculture in the coastal areas of the South. The topography and climate of this area lends itself to the establishment of agricultural structure with a pattern similar to that found in the other Sunbelt States included in the CATF region. This coastal area also has a labor force with the same kind of characteristics as those found in the CATF region—that is, relatively poorly skilled, segmented, and impoverished. Therefore, agriculture in the coastal plains has the potential to develop a similar structure and a similar set of adverse impacts on rural communities in this area as has already occurred in CATF counties. This in turn may result in substantial worsening of living standards and community welfare in this area. Detailed research on this area would be necessary to assess the potential for adverse structural change and the extent to which adverse impacts have already occurred in rural communities in this area.

In summary, the South is more similar to the CATF region than to the other regions of the United States. Unlike the CATF, the South has a relatively high percentage of small farms and rural poverty in areas that are not dominated by industrial agriculture. The availability of services, levels of education, and income levels is substantially lower in the rural counties of the South compared with those in the Northeast and Midwest. Unlike the Northeast, Midwest, and CATF regions, specific technologies are not seen as having a clearly identifiable impact on rural communities in the South, since production of particular commodities does not predominate regionally or within a particular State in the region. One moderate exception to this lies in the soybean/rice rotations in Louisiana, which accounted for almost all of the cash grains produced in this State in 1982. However, cash grain production is already highly concentrated in Louisiana, and relative to other parts of the South, the structure of agriculture is not likely to change greatly in Louisiana. Public policies that pertain to tobacco production and cattle raising may have a detrimental impact on small-scale farms that depend on these commodities.

The economic fortunes of the rural South are tied to its position in the national and international economy. Given the relatively poor position of this region in the national economy it is not reasonable to expect these areas to improve their social and economic conditions with their own resources. Rural sociologists and agricultural economists have argued for a comprehensive rural development program for the South. They argue that the rural areas are already experiencing extreme social and economic problems. The dire social consequences of these depressed conditions and the persistence of social inequalities include "intolerably high rates of infant mortality and homicide . . . inadequate jobs and income, inadequate services" and a decline in effective grassroots, self-help initiatives (Wilkinson, 1984). Those communities that remain primarily dependent on their farming hinterlands are thought to be the most likely to experience a decline in their populations, quality of services, and retail establishments (Whiting, 1974).

THE RURAL DEVELOPMENT POLICY CONTEXT

Concern about the impact of changing agricultural technologies and structural change in agriculture on rural communities should be placed in the context of changes that have occurred and are likely to continue to occur in the general economic structure of rural areas. For purposes of this discussion, all nonmetropolitan counties and areas of the United States are aggregated together into the category of rural areas. Two basic trends have clearly been operating in rural areas for several decades. First, since about the time of World War I, there has been a long-run displacement of labor from agriculture, primarily due to mechanization and consolidation of agricultural production. Second, since about 1940 there has been a steady growth in the number of manufacturing and service sector jobs in rural areas. The rate of industrialization of rural America increased greatly in the 1970s, especially in the South. Reduction in the population of farm operators slowed in the 1970s and reversed in some areas. Overall, rural areas in the United States are much less dependent on agriculture in 1980 than they were in 1940. In 1983 the natural resource sector, which includes all of agriculture as well as forestry, fisheries, and mining, accounted for only 11 percent of the wage and salary income of nonmetropolitan areas (U.S. Department of Commerce, 1984). Government accounted for 19.4 percent of income, and the service and manufacturing sector accounted for 70 percent. The economy of rural areas has therefore become more diversified and less sensitive overall to economic cycles in agricultural prices.

Along with the changes in population and economic structure in the 1970s, there were several trends of improvement in the welfare of rural areas and communities. The gap between incomes of rural and urban workers narrowed, to within 12 percentage points of each other, availability of services has improved, and poverty rates in rural areas declined from 17 percent (in 1970) to 14 percent (in 1980). Even at the historical low point in rural poverty, nearly all of the poorest counties in the United States were nonmetropolitan. There were only 8 metropolitan counties with poverty rates of 25 percent or more in 1980, whereas there were 339 nonmetropolitan counties with this rate of poverty.

Improvement in the welfare of rural areas is due to investment by both private and public sectors. The private sector has invested in new housing, manufacturing, and service facilities, while the public sector invested primarily in the basic infrastructure of rural areas: roads, water and sewer systems, health and educational facilities, and so forth. Much of this development has been interrelated; water projects such as dams and reservoirs tend to attract investment in retirement housing, which in turn provides these rural areas with a relatively stable increase in service sector jobs. In turn, increases in industrialization and retirement housing as well as increases in land values in general tend to increase the tax base, which provides for improvements in many rural services such as education, health facilities, water treatment plants, and so forth.

The improvements in rural welfare have not been evenly distributed. Some regions and subregions have not improved nearly as much as other areas. Figure 11-1 shows the incidence of nonmetropolitan counties with a poverty rate of 25 percent or higher as determined by the 1980 Census of Population. Nonmetropolitan counties with this high percentage of poverty appear to occur in five groups. The largest group is the southern "black belt" counties that run across the States of North Carolina, South Carolina, Georgia, Alabama, Louisiana, Mississippi, and Arkansas. Another group occurs in the cluster of Appalachian counties in eastern Kentucky and northeastern Tennessee. Poverty is also prevalent in a more diffuse pattern in rural counties along the Rio Grande Valley in Texas, and among many of the counties dominated by Indian reservations in Arizona, New Mexico, Utah, North Dakota, and South Dakota.

Moreover, the incidence of poverty falls disproportionately on minorities and women in rural areas. In 1982, only 15 percent of non-

Figure 11-1.—Nonmetropolitan Counties With a Poverty Rate of 25 Percent or Higher

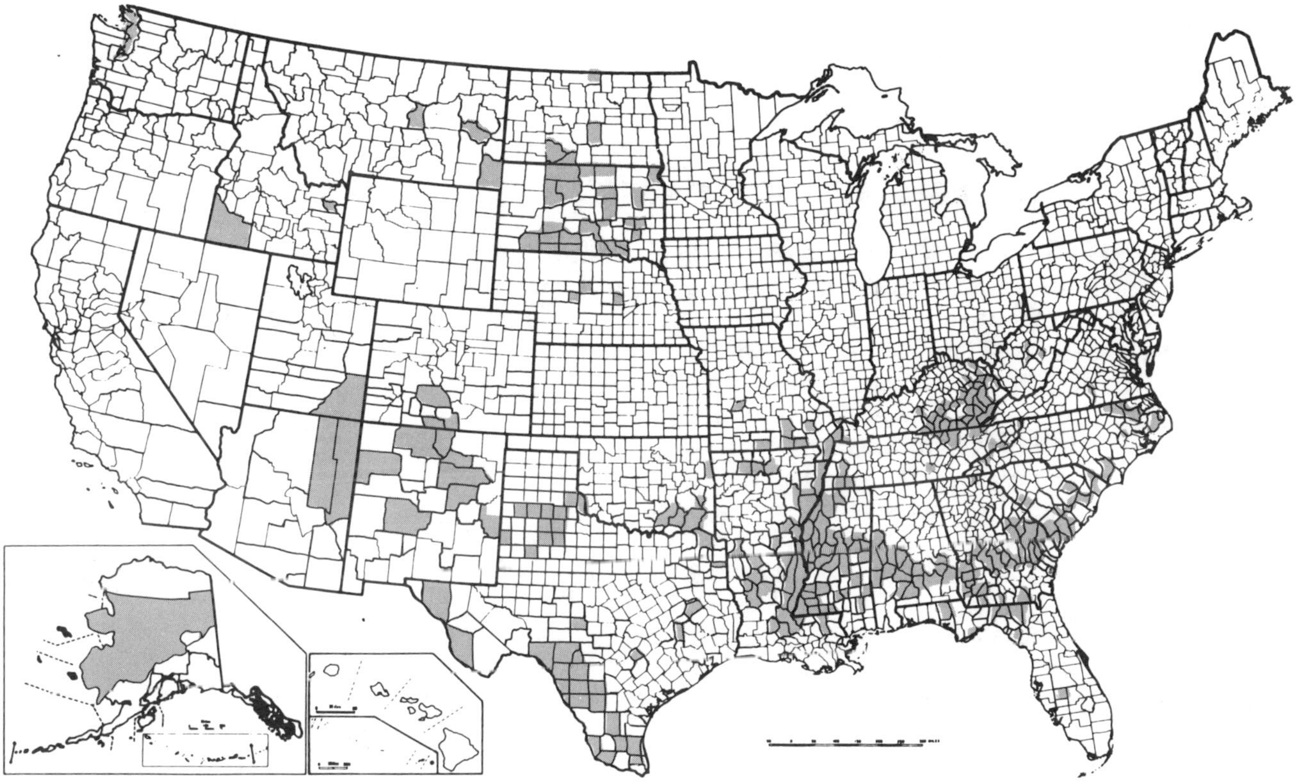

SOURCE: U.S. Department of Commerce, Bureau of the Census, *1980 Census of Population.*

metroplitan whites were poor, compared with 42 percent of nonmetropolitan blacks and 31 percent of nonmetropolitan Hispanics. In 1979, 34 percent of nonmetropolitan Indians had poverty-level incomes. The incidence of female headed households in nonmetropolitan areas increased by 25 percent in the 1970s. In 1980, 35 percent of these households were classified as being at or below the poverty line (Skees and Swanson, 1985).

Minorities have had some relative gains in welfare in some areas. In the South, the poverty rates among nonmetropolitan blacks decreased by 10 percentage points from 1970 to 1982, while the poverty rates among nonmetropolitan whites increased somewhat, from 13 to 15 percent over the same period.

Trends toward improvements in rural welfare have taken a turn for the worse in the 1980s. Rural poverty increased again in the recession of 1980-81 to 18 percent in 1982. Many rural service and manufacturing industries that relocated in rural areas in the 1960s and 1970s began to move overseas in search of still lower labor rates. The tax base that supports many rural services such as schools and hospitals was eroded as cropland values fell across the country. There is also evidence that the movement of population from urban to rural areas in the 1970s has reversed in many areas. Throughout the 1970s, population grew more rapidly in rural areas than in urban areas. The rate of population growth in rural areas fell rapidly in 1980 through 1982 and is now significantly lower than that of urban areas.

In summary, rural poverty is still very prevalent at high levels compared with metropolitan areas. Changes in the economic structure of rural areas have increased the economic base of many communities and counties but have a long way to go before the welfare of rural areas is equivalent to that of urban areas. Moreover, the economic and financial base of many rural communities is significantly less strong in 1985 than in 1980. Overall, agriculture plays a much smaller role in the rural economy than it did in previous decades, but the economic situation of many rural counties is sufficiently precarious that substantial changes in agriculture will undoubtedly have an impact on the welfare of these areas.

SUMMARY AND POLICY IMPLICATIONS

A wide range of diversity is evident in the character, agricultural structure, patterns of change, and patterns of impact on rural communities in five different regions of the United States. A clear picture of adverse relationships between agricultural structure and the welfare of rural communities is evident in the CATF counties. Large-scale and very large-scale industrialized agriculture in these counties is strongly associated with high rates of poverty, substandard housing, and exploitive labor practices in the rural communities that provide hired labor for these farms. Very large-scale agriculture has been a strong source of employment in the CATF region for many years, although at very low wage rates. Emerging technologies may reduce the labor requirements throughout much of the CATF region between now and year 2000. Increased unemployment will greatly increase the strain on these communities. There is potential for CATF to increase its share of national agricultural production, which would mitigate the trend toward increasing unemployment. However, increased agricultural production in this region will tend to be constrained by the cost of irrigation water and the need to control environmental impacts.

There is a substantial potential for a pattern similar to that of the CATF region to occur in the coastal zone of the South. The topography and climate favor large-scale, labor-intensive production of fruits, vegetables, and dairy products. The South also has a segmented, relatively unskilled labor force that could provide a source of low-cost labor similar to that of the CATF region. It is difficult to draw generalizations about the rest of the South, owing to the diversity of agricultural structure and production. There is evidence of a relatively strong association between rates of unemployment and agricultural structure. Unemployment rates tend to be lowest in counties with a predominance of moderate farms. Unemployment rates are higher in counties with a predominance of small or large farms.

Dairy products are the single most important agricultural commodity group of the Northeast. Dairy farms are likely to experience widespread failure because of the combination of technological change and public policies. The structure of agriculture in the Northeast is therefore likely to change substantially during the next 10 to 15 years. However, rural communities in the Northeast have a low overall dependence on income from agriculture. Almost all of the most productive agricultural counties in the Northeast are in metropolitan areas where employment opportunities and services are relatively available. The most rural counties are not the most agricultural. Therefore, rural communities in the Northeast generally are not likely to experience adverse consequences from structural change, with the exception of a few localities with especially high dependence on dairy production.

There is no clear-cut evidence that rural communities in the Midwest were adversely affected by structural change during the 1970s. In general, alternative sources of employment in the manufacturing and service sectors have been relatively prevalent and are expected to continue to be relatively good in the Midwest. In general, indicators of social welfare tended to improve as farm structure moved from small and part-time farms toward moderate to large farms during the 1970s. However, there was a tendency for the population to decline in counties where the share of part-ownership of farms increased. As with the Northeast region, there is a reasonable expectation that technological change in the dairy industry will result in a mass exodus of small to moderate dairy farms during the next 5 to 15 years. Rural communities in dairy counties may not be adversely affected, since off-farm employment is quite high in these counties. Mixed agricultural counties on the western edge of the Midwest that are relatively dependent on agriculture are the most likely to suffer adverse consequences from structural change. If the percent of part-ownership increases as agriculture becomes more concentrated, population, median income, and retail sales may decline in these counties.

There is a strong potential for the development of a high concentration of agricultural production in the Great Plains and the West, especially in terms of farm size, if not gross sales per farm. In turn, the number and percent of hired managers in this region is likely to increase. Unlike the South, there is low potential for development of industrialized agriculture with large numbers of hired field workers. The most likely adverse impact will be the loss of population and small retail firms in the region. In general, fewer alternate employment options will be likely in manufacturing and service industries in this region than in the other regions of the country.

One of the most important findings is that it is very difficult to generalize across regions of the United States about the impacts of changing agricultural technology and structure on rural communities. As a consequence, policies designed to prevent or ameliorate adverse impacts and promote beneficial impacts will run the risk of being inappropriate unless they are crafted with consideration for regional differences.

Regional Rural Development Policy

Rural development policies are carried out at the national, State, and local levels. Over the years, rural development policies have received high priority and at other times, including the early 1980s, they have received relatively little attention in terms of leadership and resources. This policy of "benign neglect" is based on the view that rural communities have strong, cohesive social institutions and can help themselves better than the Federal Government and States can. There is strong evidence that many, if not most, rural communities have suffered a decline in their strength, cohesion, and capabilities and are now much less able to help themselves. Urbanization has reduced the traditional bases of cohesion in many rural communities (Wilkinson, Hobbs, and Christenson, 1983). Many of the gains in social welfare that were achieved in rural communities in the 1960s and 1970s are in danger of being lost. Moreover, the examination of indicators of social welfare of rural communities in this study has shown that poverty, substandard housing, unemployment, and lack of access to basic services continue to be widespread problems in major regions of the United States. There is strong potential for further declines in the welfare of rural communities in some areas. It follows that policymakers who are concerned about the quality of life in rural areas should give renewed consideration to regional development policies. Regional development policies that address the quality of life in rural areas will benefit from higher political priority and a new focus on the issues. It will take the cooperation and coordination of policymakers at all levels of Government to achieve this increase in priority and this improvement in focus.

The national role is critical in providing leadership and in setting national standards for the improvement of conditions in rural areas and regions that cut across State lines. This responsibility is shared by the Federal Government and national organizations. State governments and State organizations have the responsibility for selecting particular areas for assistance and for assisting State and local organizations in the delivery and use of services. Local governments are responsible for direction and implementation of programs to meet local needs and in the use of local capacities and resources (Bradshaw and Blakely, 1983). In this section, the general roles and capabilities of the three levels of Government are outlined, followed by a discussion of policy considerations for each region.

Roles and Capabilities of Government

The National Role in Regional Rural Development

The Federal Government can promote regional rural development in a number of ways:

1. Develop a strong Federal rural policy that would help coordinate the various activities of the several Federal agencies active in rural development and set a clear direction for Government involvement.[11]
2. Develop rural human capital by targeting resources for training and skills development to minorities and the poor. The Federal Government is in a better position to do this than many State governments.
3. Integrate programs of economic development. The Federal Government can provide incentives for establishing in rural areas new industries that are integrated with the need for human resource development in those areas.
4. Directly provide resources for the most needy rural areas (especially the South). The Federal Government has the special ability to reallocate resources from areas and regions of affluence to areas in which poverty and depressed conditions prevail. The largest proportion of rural poverty in the United States is in the South. As a consequence, actions by individual State governments in this region are not likely to be as effective as national policies targeted at this region. At the national level many of the programs that have been most successful in achieving improvements in social welfare in rural areas have not operated under the label of rural development per se. Examples are the Interstate Highway System, the Social Security System, Environmental Protection Agency grants for pollution abatement, Corps of Engineers' waterway development and flood control projects, the Rural Electrification Administration, the Farm Credit System, and the Farmer's Home Administration.
5. Create a context for improved assessment and analysis of rural development problems and policies. The U.S. Department of Agriculture is in a better position than single State institutions to promote improved scholarship on issues and policy options for rural development. At the same time there is a great need to sustain and improve the Federal collection and dissemination of information on rural communities. Only the Federal Government can establish the uniform national database and analytical standards required for an adequate definition of the problems of rural communities and rural development. Public policies are best established on the basis of well-defined problems. Public policies toward rural development in the past have been poorly formulated, in part due to the lack of consistent definitions and data about rural communities.
6. Provide certain governmental services with indirect but potentially substantial impact on regional rural development. For example, the welfare of many poor rural communities in the South and the CATF region is affected by the rate of influx of immigrant farmworkers. It will be very difficult to improve the incomes and housing standards of hired labor in these two regions in the face of uncontrolled competition from or nonregulation of immigrant labor.

[11]Federal agricultural commodity policies, other income support policies, and Federal research and extension policies also have an impact on regional rural development. These policies are discussed in other chapters.

The State Role in Regional Rural Development

Each State can play a pivotal role in many respects in the process of regional rural development. While the Federal Government can provide leadership and funding, regional development policies will be carried out to a large extent through State agencies and programs. To the extent that the States increase their level of responsibility and activity in rural development, they will also have to increase their organizational capabilities. States also have the opportunity to organize themselves into regional federations to coordinate programs and share resources for regionwide development programs. The States have roles that the Federal Government cannot perform:

1. Only individual States or regional groups can adequately coordinate the different interest groups and opportunities within their boundaries.
2. Strategies that are politically feasible can only originate with the States; they cannot be successfully imposed by the Federal Government.
3. The States are uniquely capable of improving the organizational capacity of rural development groups in those places where the need for development is greatest.
4. States can exercise leadership in creating multijurisdictional organizations.
5. State responsibility for land use assessment and zoning can be an effective way to minimize some of the disadvantages of growth in rural areas.
6. Legislative and administrative actions by State governments within the broad policy guidelines of the Federal Government are necessary to ensure that benefits from development programs reach the most needy rural residents.

The Local Role in Regional Rural Development

The basic economic development activities that work to improve the quality of rural life are conducted by jurisdictions that lie below the level of the State government. These local efforts must work within national and State guidelines and priorities, but they must have a great deal of flexibility to create programs appropriate to local conditions and resources. Local organizations working at the local level ultimately have a great deal of responsibility to make sure that the needs of disadvantaged rural residents are met (Bradshaw and Blakely, 1983). Local organizations have some strengths relative to State and Federal agencies. Local governments and agencies are capable of developing more diversity in sources of funding and types of services that are delivered. Localities are better able to identify and use particular local resources in the process of development.

Regional Policy Considerations

The CATF Region

The social welfare of many rural communities in CATF counties is already very poor. Public policies aimed at rectifying the existing problems are needed in addition to policies to mitigate adverse impacts from continued concentration and technical change.

There are several essential elements of any program directed toward correcting existing problems in the CATF region:

1. Community development, cooperative extension, and poverty programs might be focused on the specific needs of the small communities and of displaced individuals.
2. Building codes could be enforced on rental properties, and grants might be provided to owner-occupants to bring their dwellings up to code.
3. Safe and sanitary public housing could be provided to migrant agricultural labor.
4. More rigorous monitoring and enforcement of water and air quality is needed in rural communities. Specific controls could be enacted on environmental problem areas —burning of crop stubble, disposal of pesticide containers, and drainage of irrigation water.
5. The general issue of below-minimum wages should be addressed:
   a. barriers to unionization of agricultural labor could be removed;
   b. benefit packages could be adapted for use by migratory labor;
   c. job costs (charges for transport to the fields, lodging, and food) could be disallowed if they depress wages below the minimum levels; and
   d. professional standards and licensing could be established for labor contractors.

The Great Plains and the West

The analysis of the Great Plains and the West indicates that public policy rather than technology per se accounts for most of the recent shifts in agricultural structure and will have the greatest impact in the foreseeable future. The principal impact of technology was the adoption of larger machinery on moderate farms, which resulted in predominantly medium-sized farm counties becoming more like large farm counties during the 1970s. Much of this change can be attributed to Federal incentives for the substitution of capital for labor. While this has been true since the 1940s, the process accelerated during the inflationary 1970s and received further impetus through increases in investment tax credits in 1981. The more recent reversal of monetary policy has resulted in a great deceleration of capital investment, but has also greatly decreased net farm income. Interest payment write-offs have provided a major subsidy for growth, especially in irrigated mixed crop and livestock counties.

The Northeast

The structural change in dairy production will have a substantial impact on the overall structure of agriculture in the Northeast because the dairy farm is the predominant type of agriculture in this region. One possible way to mitigate this impact will be to convert dairy farms to the production of fruits, vegetables, and poultry. These commodities are produced in large quantities in the Northeast, the markets are well developed, and demand is likely to be more elastic than the demand for dairy products.

The Midwest

As in the Northeast, there is a need to consider public policies for the Midwest that will address the major structural changes expected in the dairy industry. Public policies that assist dairy farmers to shift resources into alternate types of production will benefit communities in areas that are relatively dependent on income from dairy production. Programs that enable dairy farmers to retrain for employment in new occupations and to leave agriculture may be of more benefit to farmers than to the communities in which they reside if these programs result in outmigration to other parts of the country.

The South

There seems to be a consensus among the specialists in rural affairs about the character of a national or regional rural development program. Such a program would require a "twofold attack, one that combines Federal initiatives with local initiatives—the former to increase resources, the latter to build a sense of community" (Wilkinson, 1984). Four general criteria are:

1. there must be a program aimed at the creation of jobs that generate a livable income;
2. basic rural services such as health care, education, water, sewer, and power must be provided or upgraded;
3. labor and civil rights laws must be strengthened and enforced; and
4. local participation must be included in any rural development program.

CHAPTER 11 REFERENCES

Bradshaw, Ted K., and Blakely, Edward J., "National, State and Local Roles in Rural Policy Development," *The Rural Sociologist,* vol. 3, No. 4, July 1983.

Flora, Jan L., and Flora, Cornelia Butler, "Emerging Agricultural Technologies, Farm Size, Public Policy, and Rural Communities: The Great Plains and the West," prepared for the Office of Technology Assessment, Washington, DC, 1985.

Goldschmidt, Walter, *As You Sow: Three Studies in the Social Consequences of Agribusiness* (Montclair, NJ: Allanheld, Osmun & Co., 1944).

MacCannell, Dean, and Dolber-Smith, Edward, "Report on the Structure of Agriculture and Impacts of New Technologies on Rural Communities in Arizona, California, Florida, and Texas," prepared for the Office of Technology Assessment, Washington, DC, May 1985.

OTA, Office of Technology Assessment, *Technology, Public Policy, and the Changing Structure of American Agriculture: A Special Report for the 1985 Farm Bill,* OTA-F-272 (Washington, DC: U.S. Government Printing Office, March 1985).

Pollack, Susan L., and Jackson, William R., Jr., *The*

Hired Farm Working Force of 1981, U.S. Department of Agriculture, Economic Research Service, Economic Development Division, Agricultural Economic Report No. 507, 1983.

Rothman, Robert A., et al., "The Undulating Community: A Typology of Recurrent Migrations," *Rural Sociology* 42(1):93-100, 1977.

Skees, Jerry R., and Swanson, Louis E., "Examining Policy and Emerging Technologies Affecting Farm Structure in the South and the Interaction Between Farm Structure and Well-Being of Rural Communities," prepared for the Office of Technology Assessment, Washington, DC, 1985.

U.S. Department of Commerce, Bureau of the Census, *Current Population Survey,* 1984.

van Es, J. C., Chicoine, David L., and Flotow, Mark A., "Agricultural Technologies, Farm Structure, and Rural Communities in the Midwest: Policy Choices and Implications for 2000," paper prepared for the Office of Technology Assessment, Washington, DC, 1985.

Whiting, Edward (ed.), *Communities Left Behind: Alternatives for Development* (Ames, IA: Iowa State University Press, 1974).

Wilkinson, Kenneth P., "Implementing a National Strategy of Rural Development," *The Rural Sociologist* 4(5):348-351, September 1984.

Wilkinson, Kenneth P., Hobbs, Darryl J., and Christenson, James A., "An Analysis of the National Rural Development Strategy," *The Rural Sociologist* 3(6):384-391, November 1983.

Chapter 12

Impacts on Agricultural Research and Extension

Contents

Tables

Figure

Chapter 12

Impacts on Agricultural Research and Extension

Much of the success of American agriculture is attributable to the creation of the Nation's agricultural research and extension system (Ruttan, 1982; Cochrane, 1958). For well over a century this system has contributed to a plentiful, low-cost supply of food and fiber, and to the positive U.S. balance of agricultural trade, through the system's research on new agricultural technologies and practices and through its transfer of technology to farmers and other members of the agricultural community. The technological innovations brought about by agricultural research and extension increased agricultural output in 1945 through 1979 by 85 percent, with no change in the level of agricultural inputs (USDA, 1980).

The public has invested substantial sums of money (currently about $3 billion annually) in agricultural research and extension at Federal and State levels. This investment has been no accident. Several important events have helped make the agricultural research and extension system an integral and long-standing part of U.S. agricultural policy—the first Federal appropriations to agricultural research in 1856, the establishment of the U.S. Department of Agriculture (USDA) and land-grant university system in 1862, the funding of a State agricultural experiment station system in 1887, and the creation of the Federal-State-local extension partnership in 1914.

However, agriculture's entrance into the era of biotechnology and information technology raises several questions about the impact of technical advances on the performance of the research and extension system and about how that performance will ultimately affect the structure of agriculture. For example, in the past, public research was the driving force for agricultural production. Now, with the private sector becoming more involved, the public sector is emphasizing more basic research while the private sector is focusing on certain areas of applied research and development.

This situation leaves open the question of who will do other aspects of applied research in the public sector. Although the public sector has allocated resources to research in biotechnology and information technology, extension has done little to make information about these technologies available to farmers. Extension must thus decide what its mission will be, for extension policy will determine how effective moderate farmers will be in gaining access to new technology. Without such access moderate farms will disappear even faster.

The role of extension raises additional questions:

- Who gains and who loses from the process of technological change in agriculture?
- Is agricultural research and extension structurally neutral, or does it favor the growth of large industrialized farms?
- What are the roles of the various components of the agricultural research and extension system as they relate to technological change in the biotechnology and information technology era?
- What are the implications of increased private sector involvement in agricultural research?
- What are the implications of patents being conferred on biotechnology and information technology discoveries, that is, for the social contract under which the agricultural research system was created?
- How is a proper balance to be struck between public and private sector components of the agricultural research and extension system?

These major questions will be addressed in this chapter. The answers to the questions are based on previous OTA studies, on an extensive body of literature on the impact of technology on agriculture, and on papers commissioned by OTA regarding the status of the agricultural research and extension system as it relates to developments in biotechnology and information technology.[1]

[1]The OTA papers were prepared by George Hyatt, Roy Lovvorn, Ronald Knutson, and Fred White. The findings from these papers were integrated into this chapter by Ronald Knutson.

THE FUNCTIONS AND CHALLENGES OF RESEARCH AND EXTENSION

Increasing demands are being placed on the agricultural research and extension system. These demands result largely from pressures to increase food and fiber production in the face of an ever-expanding world population, the goal of eliminating hunger and malnutrition, higher levels of consumer income, agriculture's impact on the environment and worker safety, policies designed to expand exports, the desire for a safer food supply, and reduced availability of water for irrigation.

Technological change is necessary for solving each of the problems implied by these public concerns. The process of achieving technological change in agriculture involves three basic steps, each a function of the research and extension system:

1. basic research—discovery of new ideas, concepts, and relationships;
2. applied and developmental research:
   - development of ideas, concepts, and relationships into products (where a product is the output of technology);
   - adaptation of new technologies to as many agro-ecosystems as possible; and
   - maintaining newly achieved productivity from evolving pests, disease, decline in soil fertility, and other factors (sometimes referred to as maintenance research); and
3. adoption of products (transfer of technology).

Discovery is primarily the function of basic research. Most basic research has traditionally been done in the public sector. There appears to be a general assumption that the private sector will not support sufficient amounts of high-risk basic agricultural research because that research is unlikely to yield a near-term payoff. However, this assumption is now being challenged by large private sector investments in biotechnology and information technology.

Developmental and applied research is conducted by both the public and private sectors. The marked increase in the quantity of applied private sector research has resulted in suggestions that public sector support for agricultural research might logically be reduced. Such a suggestion, however, is overly simplistic. Research policy decisions like this require an understanding of the relative payoffs from various types of research, the interrelationships between basic and applied research, and the types of research undertaken by the public and private sectors (White, 1984). Most of the applied research conducted by the private sector is development of ideas, concepts, and relationships into products. Very little private sector applied research is allocated to the adoption of new technologies to a specific agro-ecosystem or to defense of newly achieved productivity from enemies of the agro-ecosystem (maintenance research). This responsibility falls to the public sector.

The function of encouraging technology adoption has traditionally been shared by the public and private sectors. In the public sector, extension educators at the Federal, State, and county level work directly with farmers to test and demonstrate the usefulness of new products flowing out of both sectors. Private firms tend to concentrate their adoption strategies on more conventional promotion and advertising strategies.

Over time, the effort and resources required to achieve a technological breakthrough, as a general rule, increase. This is true because the simpler problems naturally tend to be solved first. More difficult problems require more complex tools of analysis and thus a larger research commitment in time, effort, and resources. The entry of agriculture into the contemporary biotechnology era illustrates this increased complexity. For years, agriculture has depended on chemicals to control pests, diseases, and weeds. These chemicals have been applied without a full knowledge of either precisely how they work or how they affect the environment. This practice has increasingly been questioned as chemical residues have become more associated with environmental contamination and safety concerns. Moreover, biotechnology research has increased the understanding of the specific effects of chemicals, such as atrazine on weeds. As a result of such research, it is becoming possible to develop chemical control agents for specific needs. Potentially, all agricultural plants could, for example, be made resistant to "Roundup" herbicide. With all cultivated plants resistant to the herbicide and all undesirable grasses susceptible to it, the potential exists for nearly complete control of grassy weeds on a farm. Higher output and/or reduced inputs would result from improved weed control. In addition, fewer and safer chemicals, and chemicals in smaller quantities, could be used. The result could be a safer food supply and environment, less use of valuable resources, and a higher level of output.

To achieve these benefits, large investments must be made in basic research. Much of this research uses techniques not common to agriculture. New scientists having modern biotechnology research skills must be trained for agricultural research, and existing scientists must be retrained. Laboratories and related equipment will be more complex and expensive. The educational levels of the producer clientele will have to be improved to adopt and use effectively the more complex new technologies.

Such needs will not be accomplished overnight. Research and education are, of necessity, long-term processes. Interruptions in research and education create gaps in the flow of technology into agriculture that are of a considerably longer duration than the interruption itself. For example, if a line of research designed to pinpoint molecular defects in genes that make poultry and cattle vulnerable to leukosis (a form of cancer) were interrupted, it could increase the time required for discovery, development, and adoption of leukosis control methods by several years.

Agricultural research and extension educational programs compete with other demands for both public and private funds. In the private sector, support for research depends on overall firm profitability and the potential for near-term cost recovery and contribution to profits. When firm profits fall, research funds are traditionally among the first to be cut. This variability in private sector research investment increases the need for stability of funding by the public sector. It also increases the need for policymakers to evaluate the comparative payoff from various forms of Government expenditures—recognizing that all requests for Government assistance cannot be satisfied. Weighing the payoffs from the many alternative demands on the public treasury may be the most complex task facing policymakers (Knutson, 1984).

COMPONENTS OF THE AGRICULTURAL RESEARCH AND EXTENSION SYSTEM

The U.S. agricultural research and extension education system contains many research and education agencies, grouped in the following five categories:

- U.S. Department of Agriculture (USDA);
- other Federal agencies;
- land-grant universities;
- non-land-grant universities; and
- private firms, individuals, and foundations.

The agencies can be viewed both from the perspective of the sources of funds and from the perspective of the performers of research and educational activities (users of funds) (figure 12-1). Each of the components of the agricultural research and extension system has its unique role, although all components are interrelated and tied to the central objective of technology discovery and transfer for the benefit of farmers and of society as a whole.

USDA

The 1977 farm bill designated USDA as the lead Federal agency for research, extension education, and teaching in the food and agricultural sciences. This action confirmed by law what had been true since before the turn of the century. It did not, however, mean for USDA to provide a majority of the funds for these functions. In fact, the proportion of funds provided by USDA for agricultural research and extension has declined from about 54 percent in 1966 to 47 percent in 1982 (CSRS, 1984).

Research

USDA provides funds both to its own research agencies and to universities. Its own agencies include mainly the Agricultural Research Service (ARS) and the Economic Research Service (ERS); together they use about 75 percent of USDA's research funds. The remaining 25 percent goes almost entirely to universities. Most of the university funds go to land-grant universities, established by law in 1862 and 1890. USDA funds to non-land-grant universities, are limited to a relatively few competitive grants used to support high-priority research.

USDA's agricultural research is carried out at 148 locations across the United States. About two-thirds of USDA's agricultural research scientists are located in USDA laboratories, with the remainder being located in the land-grant universities' agricultural experiment stations. In contrast to its agricultural research, USDA's economic research tends to be heavily concentrated in Washington, DC. This concentration is increasing with the recent policy decision to eliminate the regular ERS field staff. In the future short-term detail to university sites will only be possible. It remains to be seen how compatible this notion is with the kind of long-term commitment much research requires.

Figure 12-1.—Agricultural Research and Extension Funding (in million dollars), 1982

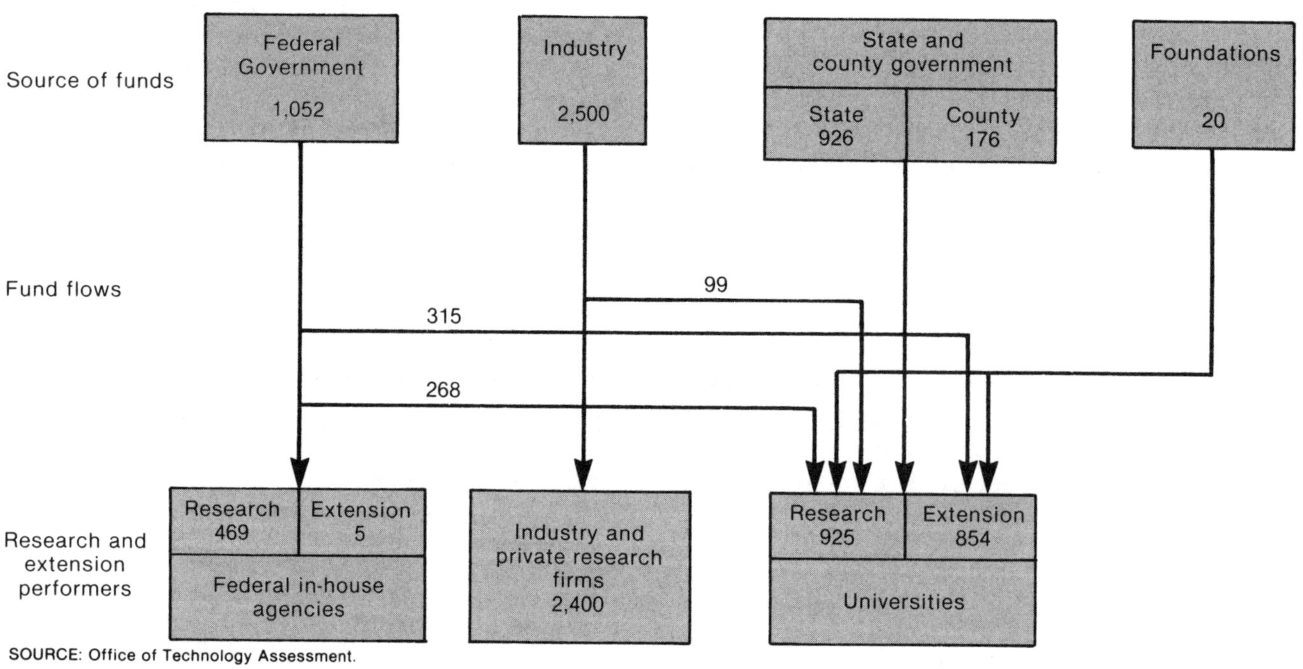

SOURCE: Office of Technology Assessment.

Over time there has been considerable debate regarding the role of USDA in both agricultural and economic research. During the 1950s through the 1970s, agricultural research had a tendency to become increasingly decentralized, given the proliferation of agricultural research facilities located throughout the United States. Administration was also decentralized, with substantial authority for program development being established at the regional level. As a result, questions arose about the role of USDA and about potential duplication of research functions between USDA and the land-grant universities (OTA, 1981). After a number of special studies and recommendations, the issue of the role of USDA in agricultural research came to a head in the debate on the 1977 farm bill, when Congress designated USDA as the lead agency of the Federal Government for agricultural research and directed the Secretary of Agriculture to coordinate all agricultural research, extension, and teaching activities conducted or financed by Federal funds.

The 1977 farm bill did not specifically address the functions of USDA versus those of the land-grant universities, although it established the Joint Council on Food and Agricultural Sciences and the National Agricultural Research and Extension Users Advisory Board to assist in planning the research and extension program agenda. The responsibilities of the Joint Council include a formidable list of tasks:

- evaluating research, extension, and teaching program impacts;
- identifying high-priority research;
- developing memoranda of understanding among the participants;
- establishing priorities;
- recommending responsibilities for research, extension, and teaching programs; and
- summarizing achievements.

In a recent comprehensive study, OTA concluded, "There is concern whether the functions assigned the Joint Council are attainable" (OTA, 1981).

The Users Advisory Board has a somewhat less formidable task of determining the needs and priorities for agricultural research and extension. Its major mandates include:

- reviewing USDA's policies, plans, and goals for research and extension;
- examining relationships between private and public sector research and extension programs;
- recommending policies, priorities, and strategies for research and extension; and
- assessing distribution of resources and allocation of funds for research and extension.

While it is generally agreed that the functions of the Users Advisory Board are more attainable than those of the Joint Council, the impact of the board in establishing research priorities is unclear (OTA, 1981).

In addition, OTA concludes that there is still no satisfctory long-term process for evaluating existing research activities and potential research opportunities and for the development of a new set of research priorities. At the same time, OTA recognizes the potential for too much planning and organization. Agricultural research is sufficiently complex that research administrators have difficulty evaluating the relative merits of particular projects. Therefore, specific decisions on what research is to be undertaken are generally made by the research scientists. The administrator's comparative advantage is in establishing policy, organizing to get the job done, obtaining and allocating funds, and coordinating to eliminate unnecessary duplication (OTA, 1981).

One of the most important contemporary issues that the Federal Government has to deal with is that of establishing broad-priority research and extension needs and the roles of the components of the research and extension system. The Joint Council and the Users Advisory Board, if given time and sufficient encouragement to perform, have the potential for effectively dealing with the priorities issue. Positive progress is indicated by the Joint Council's *Needs Assessment for Food and Agricultural Sciences.*

The primary question about the roles issue involves the line of demarcation between USDA and land-grant programs. This issue has been treated quite differently by research and extension. The OTA study concludes that USDA research should concentrate on those agricultural problems that are important to the Nation and that no one State or private group has the resources, facilities, or incentive to solve (OTA, 1981). Such a role can logically be assigned to both ARS and ERS. For ARS, however, a shift in the focus of agricultural research to research only on national and regional problems would represent a marked departure from that agency's increasing emphasis on research having a State or local focus.

Available evidence suggests that the progress of the agricultural research community in establishing priorities is more advanced than that of the extension community. The agricultural research community has been extensively studied and critically evaluated in a series of projects extending back to the mid-1960s. This series of internal and external analyses has led to adjustments in the distribution of the research system's resources in recognition of potential advances evolving in biotechnology and information technology (Knutson, 1984).

Extension

As the rate of technological change accelerates, access to information plays a more important role in agricultural productivity and farm survival. In the evolving biotechnology and information technology era the trend is to substitute information for time, capital, labor, land, and energy throughout agriculture (Warner and Christenson, 1984).

In the agricultural research system researchers have traditionally been the producers of new technology, whereas extension personnel have been the agents of technology transfer (through their roles as adopters, evaluators, disseminators, and trouble shooters). An accelerating rate of technological change thus places increased demands on performance by the Extension Service, making it more important that extension sort out its priorities.

The extension community has not made the same progress in sorting out its priorities that the research community has. Identified national extension objectives play little role in program development at the State and local level. (Most extension planning takes place at the local level through advisory committees and other forms of direct contact with clientele [Marshall, et al., 1985].) One major, congressionally mandated, extension evaluation project culminated in a series of reports that concentrated more on past benefits than on future needs, priorities, and required adjustments (Extension Service, no date). Moreover, there was relatively little reference to the functions or programs of extension in the reports of either the Joint Council or the Users Advisory Board.

Federal extension has dramatically deemphasized its direct educational role in the past 20 years (Hyatt, 1984). Although Federal extension specialists were once generally viewed as having a vast subject matter base in their own right and were frequently called on to engage in staff training and to conduct educational programs, they are now viewed more as program leaders, coordinators, and facilitators. The technology transfer and education function is thus left to the State specialists and agents. These changes were at least partially forced by reductions in personnel ceilings and by limited appropriations. Nevertheless, this change in strategy has not been beneficial to the overall national extension education program. In addition to the lack of progress in national planning and needs assessment, the quality of educational service to the States has deteriorated.

As in research, there are issues of national significance that the USDA Extension Service is better able to cope with educationally than are the States. While ultimately the States must take the leadership in extending information to farmers, USDA extension can play an important role in making the information and related educational materials available on a timely basis. (For another perspective see Hyatt, 1984, pp. 17-18.) This role is currently being played on, at best, a spotty basis. The need is particularly critical for facilitating technology transfer between USDA research agencies and the State extension services as well as facilitating technology transfer between States. Facilitating communication between the USDA and State specialists should be a key mission of the USDA Extension Service. Unless this function is adequately performed, Federal research agencies such as ARS will be encouraged to develop their own outreach programs. The need is for increased integration of the research and extension function, not greater fragmentation.

With these needs in mind, if it is decided that a portion of the USDA Extension Service staff will be state-of-the-art national program leaders, the following changes would be required:

- support for Federal extension would have to be substantially increased;
- the designated leaders would have to be recognized as national extension program coordinators by the States and be provided compensation consistent with that role; and
- the program leaders would have to have access to resources allowing them to coordinate with researchers and State specialists to develop state-of-the-art educational materials that could be used in all States.

Other Federal Agencies

Although other Federal agencies have become more important sources of funding for agricultural research in universities, they still provide less than 3 percent of the total agricultural research funds. The main sources of these funds are the National Institutes of Health, the Department of Defense, the Department of Energy, the National Aeronautics and Space Administration, and the National Science Foundation. The National Institutes of Health and the National Science Foundation support basic university research, largely in the biotechnology area. Their grants tend to go to leading scientists working on the frontiers of promising new areas of basic research.

Land-Grant Universities

Land-grant universities represent a joint Federal-State partnership in research, extension, and teaching. Land-grant universities (1862 and 1890) perform the majority of total public sector agricultural research. About 52 percent of their funds are from State-appropriated sources—a marked increase from the past. Fourteen percent were formula funds (explained later); 19 percent were other Federal funds; and 16 percent were funds from farm sales, private grants, and contracts.

Research

Land-grant university research is performed primarily in the academic departments (e.g., animal science, soil science, agronomy, agricultural economics, biochemistry) of the land-grant universities. Land-grant universities combine the training of future scientists (graduate and undergraduate) with their research programs. Having the research scientists teach in classrooms increases the relevance and timeliness of those universities' curricula.[2]

Research planning and priority setting is much more decentralized in the land-grant university system than it is in USDA. This decentralization results largely from the number of research institutions involved, the orientation toward problems of the State, the increased proportion of funding from individual States, and the higher level of academic freedom afforded university scientists compared with that of most Federal and private sector scientists.

Most land-grant universities now have or are developing long-range research plans. These plans are normally developed from the scientist up rather than from the administrator down. Because of the increased complexity of projects, experiment station directors and other high-level research administrators are frequently not in the best position to evaluate the relative merits of particular projects. The more removed the administrator's training and expertise is from that of the scientist, the more imperfect is his or her level of knowledge in dealing with specific research problems. Academic heads of departments are thus generally in a better position to judge the potential value of specific research

[2]The same reasoning can be applied to split appointments involving extension and research or to extension and teaching. In each instance, relevance and timeliness are fostered.

than are experiment station directors (Knutson, et al., 1980).[3] Administrators achieve their research priorities and goals through the funding, position description, and hiring processes.

On the other hand, some hold the view that scientists are becoming more isolated in basic research (Marshall, et al., 1985). At the same time, administrators are being held more accountable for the performance of the system in meeting public needs. They must develop a sense of the broad needs of the public and build the case for continued public support. A delicate balance must be struck between the needs perceived by research administrators and the needs of the scientists. In a system where communication is good, these needs should converge. In fact, communication and consensus development is the key to performance, particularly in a system where one unit depends on other units of the system for information and coordinated action.

In this setting, the potential for unnecessary duplication of research among universities and between the State and Federal levels is reduced by communication and by the reward system within the scientific community. There is little or no reward in the scientific community for research that simply duplicates what has already been discovered and confirmed. Failure to advance the frontiers of knowledge becomes the basis for outright rejection of proposed scientific publications used as criteria for promotion and tenure. Communication within professional societies provides an important information base on which future research decisions are based. However, this is not to be confused with the need for adaptive and maintenance research. Many technologies in agriculture need to be modified to be successful in various agroecosystems. Likewise, once established, maintenance research is needed to prevent yield declines as a result of the evolution of pests and pathogens, decline in soil fertility and structure, and other factors. These areas of needed research are at times viewed as unnecessary duplication or replication of research. In fact, the time may come when a relatively large share of the public agricultural research effort will have to be devoted to maintenance and adaptive applied research. More communication on the need for this research is warranted.

One avenue for research communication that has been substantially curtailed by restricted funding and the way funds are handled within the system is regional research. Regional research allows scientists who have mutual interests in a problem area that concerns more than a single State to work together. By bringing these scientists together, the critical mass of knowledge, research skill, and resources can be assembled to tackle a particular problem.

However, persistent problems have prevented the fulfillment of the potential payoff from regional research, because even research funds earmarked for regional research are generally handled by universities in the same manner as other funds. In most States, scientists or departments receive no additional support for engaging in regional research activities. As a result, scientists must conduct regional research, which is often more costly, with the same funding base. When regional research funds were relatively plentiful, regional research was frequently undertaken and completed because of scientist initiative and the perceived administrative obligation to support regional research. But as research budgets tightened, the interest of both scientists and land-grant universities in regional research declined.

Those who suffer the most from the declining interest in and commitment to regional research are the smaller, less well-financed land-grant universities. These universities frequently do not have the critical mass of research talent required to tackle larger research problems. They can, however, get involved on a regional basis. In contrast, the larger universities are more likely to have that critical mass. As a result, in the absence of regional research, the larger universities are in a position to compete for the grants involving priority research on the cutting edge of knowledge.

[3]The same reasoning can also be applied to the administration of extension programs—those closer to the work are better able to evaluate it.

The superior ability of larger universities to compete for grants, combined with the increasing complexity of agricultural research, has from time to time led to proposals for establishing universities that are regional "centers of excellence" in either specific or broad areas of agricultural research. The center concept was expanded by Marshall in recommending the establishment of Centers of Research and Extension Excellence to methodically analyze, synthesize, and disseminate research findings and to identify high-priority research needs (Kendrick, 1981).

It can be argued that the marketplace, combined with contemporary public and private research funding policies, is already leading to the development of such centers. Questions, however, exist about whether the marketplace will generate enough centers of excellence and whether the result will be the creation of a set of "have and have not" university research and extension programs. Since the land-grant universities are public institutions, it would appear appropriate that this be an overt public policy decision rather than one left to the marketplace. This does not mean that there would be no role for even the smallest, poorest funded land-grant university. It plays an important role in a national system designed to deal with thousands of agro-ecosystems and is vital to the existence of a decentralized system with nationwide capability.

Extension

Extension education of farmers is also an integral part of the land-grant universities' functions. Extension receives about 63 percent of its funds from State and county sources, with the remainder provided by USDA, largely under formula funds.

How to apply new research findings is seldom obvious. It cannot be assumed that once research findings are available, they will be quickly and effectively put to use.[4] The process of developing and using research is complex and requires a close working relationship between the research and education functions. Extension plays a critical role in alerting farmers to new discoveries and products, evaluating the discoveries and products, and determining how they can best be used in combination with existing products and techniques. This is particularly true for the vast majority of farmers (likely, at least 95 percent of them) who do not have direct access to research results and do require extension interpretation of them.

Because of these complexities, extension activities go beyond a public information role. At the State level, extension has technically trained applied scientists (generally referred to as specialists) who are headquartered primarily at a land-grant university. These scientists may also have research and/or teaching responsibilities. Their extension role is to develop educational programs, prepare applied publications, conduct meetings, and provide technical assistance at the request of county staff.

Extension is involved not only in educating farmers but also in providing important feedback to research scientists about farmers' problems and further needed research. The proximity of extension specialists to research scientists is deemed critical for developing a working knowledge of the scientific developments and for closing the "feedback loop" between extension and research.

Available evidence suggests, however, that the feedback loop concept is operating unsatisfactorily. Marshall (1985) and his colleagues found that extension's ability to influence what research was done in the agricultural research system was inadequate. His study projected that more research coordination problems could be anticipated with the expected increased orientation toward basic research. This finding appeared to be the main origin of Marshall's recommendation for the need for Centers of Research and Extension Excellence.

Because of their direct contact with agricultural producers and agribusiness clientele, extension programs tend to be more grassroots oriented than research programs. In most States, educational needs are determined predominant-

[4]This analysis is limited to the agricultural component of the extension program. Other functions include home economics, 4-H, and community development.

ly by producer advisory committees. Programs are then developed to address these needs using county agent and State specialist expertise. In addition, individuals from the private sector are often called on by extension to provide a working knowledge perspective on solving particular problems.

State extension specialists are normally highly skilled scientists trained at the doctoral level in specific agricultural disciplines such as agronomy, animal science, entomology, or agricultural economics. In addition, these scientists develop skills in educational methodologies, including the ability to use computer and other electronic technology as they become available, to deliver research findings in an educational context. With these interdisciplinary skills, specialists develop educational programs designed to fill the needs of extension's clientele. They may prepare educational materials (including the development or adaptation of computer software), bulletins, press releases, and radio or television tapes. Such educational materials may be used directly in farmer and rancher programs or in training county agents who in turn work with farmers.

Of equal importance to extension programs is extension's use of the result demonstration. The typical result demonstration involves the planting of different crop varieties, the application of different fertilizer levels, or the application of different pest control methods to relatively small plots of land on an actual farm. The result demonstration is open for inspection, and field tours are periodically conducted for interested farmers to observe the progress of the crop.

Result demonstrations are not limited to products developed in university laboratories. As private sector-branded products enter the market, they are also used in result demonstrations to compare their effectiveness with that of established products and practices. Extension thereby serves as a public sector evaluation of new products and practices. Without such evaluation individual farmers and ranchers would incur the costs of experimenting to determine the optimum input combinations to use in production. These costs would be converted into reduced farm numbers (for those who used the wrong input combinations), higher food costs, and reduced competitiveness in international commodity markets.

With renewed emphasis on basic agricultural research, substantial concern arises over the potential for the development of an applied research gap (Christenson and Warner, 1985; Marshall, et al., 1985; and Feller, et al., 1984). This gap could occur because applied scientists are attracted to higher rewarded basic research, leaving open the question of who will do the applied research. The potential for such a gap may be reduced by increased private sector interest in biotechnology research and development. However, as the private sector performs a larger share of the applied research in the development of new products, extension has the potential for becoming even more involved in the evaluation of technologies and products flowing out of the private sector.

Substantial challenge is involved in extension's adjusting to this new role. Although in some States extension is already deeply involved in the evaluation of new products, in other States product evaluation is primarily the function of experiment stations. In the future, experiment stations will likely be doing less of this work, and extension's responsibilities will correspondingly increase. This increased responsibility will entail a larger specialist staff with modern scientific training.

Some States may be inclined to forego the responsibility of getting involved in conflict-oriented product evaluation programs. Some probably already have.[5] To the extent that this occurs, the usefulness of extension to the farmer clientele will decline. Leadership at the Federal level will be required to assure that technology transfer is facilitated in the farmer's interest.

As agriculture becomes more complex, filling the gap between research and extension will entail a larger role for extension in applied re-

[5]The problem of foregoing conflict-oriented product evaluation is by no means limited to extension. For example, private firms supporting university research may place restrictions on the university's conducting and/or publishing evaluations of the impacts or the economic feasibility of particular discoveries.

search. This is already occurring. Marshall and his colleagues found that 56 percent of extension agricultural specialists with 100-percent extension appointments are involved in applied research. Despite the need for extension involvement in applied research, the Smith Lever Act provided no explicit authority for extension to conduct research. However, an amendment in the recently passed Food Security Act of 1985 (farm bill) gives extension explicit authority to conduct applied research. The intent of this amendment is to clarify extension's role in the process of technology transfer, not to duplicate the mission of the experiment stations.

Extension has a regional counterpart to research, whereby specialists meet to develop educational materials on a multi-State basis. As in research, the funds committed to such activities (frequently referred to as "special need" or "pilot project" funds) have been substantially curtailed. The decision to reduce these funds occurred during the late 1970s when the Science and Education Administration was in control and when Federal funding was being substantially squeezed. As a result, communication between extension specialists in different States is more limited, and the quantity of educational materials produced by regional committees has been substantially reduced. Once again, this occurrence has not had as much of an adverse effect on the educational programs of the larger, better funded universities as it has on the smaller universities.

At current funding levels, one of the most difficult issues facing extension in their agricultural program is that of limiting its role and coverage to those functions for which it has the greatest expertise (Feller, 1985). Without criteria for limiting the role of extension, there is danger that extension activities in agriculture will become so dispersed and out of focus that their effectiveness will be impaired. Danger exists that extension will be called on to solve any problem, whether related to agriculture or not. In the process, the agriculture program of extension could become more of a social program than an instrument of technology transfer.

This is not a new issue but a continuing and progressively more complex one. It is made more treacherous by the politics of funding and the reality that once a new program is established it develops its own constituency and is difficult to cut back (Feller, et al., 1984). It is not possible for extension to be everything to everybody, particularly in times of limited resources. Yet, additional functions are frequently dictated by political realities at the Federal, State, and local levels. (For further discussion of the difficulty in delimiting the clientele and roles of extension see Hyatt, 1984, pp. 14-19 and 33.)

The Joint Council has not given sufficient attention to the role of extension. As a starting point for defining extension's role, it must be remembered that the root of extension is research. Similarly, extension is a primary outlet for research, after an appropriate level of product development. Extension is, therefore, delimited by the scientific endeavors of the research components of the agricultural research system, including both the public and private sector components. This delimiting role is illustrated in figure 12-2.

Figure 12-2.—Research and Extension Roles in the Technology Discovery and Transfer Process

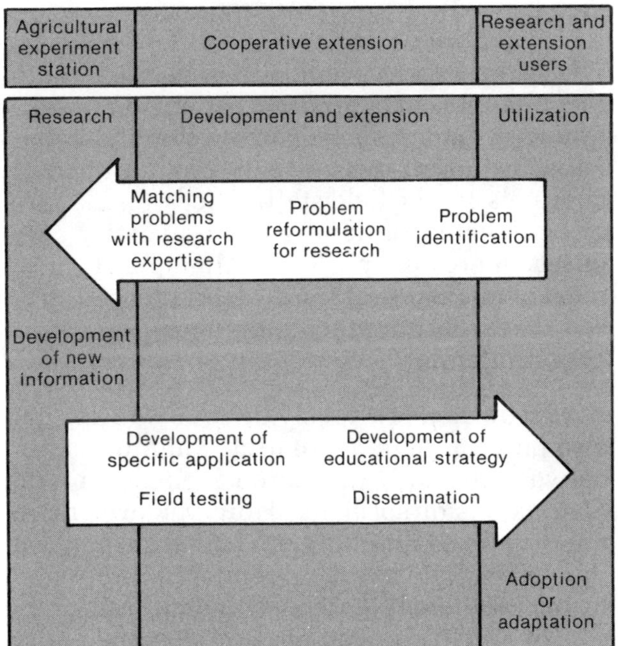

SOURCE: University of California, *Cooperative Extension Long Range Planning Statement*, Berkeley, CA, August 1982, p. viii.

Feller (1985) defines the mission of extension:

> The core mission of extension is, therefore, one of developing, extending, and bringing about the use of research-based knowledge. The core source of that knowledge is the agricultural experiment stations. Viewing extension in a broader context than this runs serious risk of reducing its overall effectiveness.

This is particularly the case when it is recognized that extension is likely to play an increasing role in filling a portion of the gap between research and extension. Another dimension of this role problem involves the tendency for experiment stations to become more involved in extension-type education programs as a means of gaining public recognition and support. Considerable care must be taken not to foster such duplication of efforts.

Research done in the land-grant universities is in direct proximity to extension specialists and can therefore be directly channeled into the State extension program. USDA research is often done at locations distant from State extension programs, which sometimes creates an incentive for USDA research agencies to reach out and develop their own educational channels. Such initiatives generally amount to an unnecessary duplication of effort.

USDA research agencies that do not have direct channels of communication and cooperation with extension need to develop them. Perhaps the most important such communication channels are the field staff, offices, and laboratories located on land-grant university campuses. Interestingly, ERS has attempted to move most of its field staff into Washington—a strategy that runs counter to the need to improve communication.

As indicated previously USDA extension can also play a role in facilitating communication between the USDA research agencies and the State extension specialist. However, even maintenance (to say nothing of needed strengthening) of this role has been rendered impossible by the previously discussed deemphasis of the role of USDA extension's staff in subject matter education.

Non-Land-Grant Universities

Non-land-grant universities include a broad range of higher education institutions, ranging from strictly private and autonomous State universities having little or no direct relationship to agriculture, to State universities having agriculture, forestry, and food-related programs but not having land-grant status (1862 or 1890). Some of these institutions have had significant applied research programs in agriculture since their founding. The major expertise of most, however, lies in teaching and research in the biological, physical, and social sciences. When agriculture entered the biotechnology era, some non-land-grant universities such as Stanford were ahead of the land-grant universities in numbers of discipline scientists (such as molecular biologists and biochemists) who were involved in basic biological research having potential application to agriculture.

The non-land-grant universities support their research programs through State appropriations, Government grants, endowments, foundations, corporate grants, and contracts. Outstanding scientists in the non-land-grant universities often received biological research support from Government institutions such as the National Science Foundation and the National Institutes of Health. The 1977 farm bill opened up USDA competitive grant research to proposals from the non-land-grant universities.

Non-land-grant universities do relatively little in terms of extension-type adult education programs. Involvement in such programs is largely limited to "public service" conferences and adult outreach programs held near these universities or community colleges. Such services may be provided free as a public service, on a cost basis, or under consulting arrangements with individual faculty members.

Private Firms, Individuals, and Foundations

Private Sector Research

The land-grant university system was established largely because it was concluded that in a decentralized competitive structure, the private sector would not have the economic incentive to provide the level of funding needed to maintain an efficient, viable agriculture. Despite many changes in the structure of agriculture since the founding of the land-grant system, this premise went largely unchallenged until the 1970s. The presumption was that agricultural firms would not undertake sufficient basic research and applied research to keep American agriculture efficient, productive, and competitive.

Until recently, private sector research, therefore, has been limited largely to providing a small number of grants for university research and private sector developmental research associated with the introduction of new products. As a result, private sector grants for agricultural research have historically come primarily from foundations such as Ford and Rockefeller. With the advent of biotechnology, private firm interest in agricultural research increased sharply. While much of this interest appears to be a spinoff from biomedical human research, substantially expanded resources have also been committed to plant and animal reproduction designed to produce new varieties or to expand the rate of genetic improvement. In addition, increased interest is being shown in developing disease- and insect-resistant plants as well as organic methods of pest control.

One of the major reasons for this expanded private sector interest in agricultural research has been the extension of patent rights to plant varieties and other biological discoveries. The potential for capturing the benefits of the resulting patented discoveries has spurred private sector support of university research. Because such arrangements hold the potential for substantially changing the basic public service nature of the land-grant system, a separate section of this report is devoted to the implications of increased private sector involvement in biotechnology research. These implications are by no means limited to research: they affect the overall thrust of extension education and the availability of new research knowledge to extension.

The current magnitude of private sector commitment to agricultural research is largely unknown, although studies suggest that it may approach $3 billion, particularly with the recent increases in private funding (National Agricultural Research and Extension User's Advisory Board, 1983; and Agriculture Research Institute, 1985). That makes the private sector research commitment approximately equal to or potentially larger than the public sector commitment and represents a major shift toward private sector dominance of agricultural research.

Approximately half of the private sector research budget is spent on production agriculture and half on food production or postharvest technology research. Private sector research resources are obviously devoted to those areas having the highest short-run profit potential. Also, despite recent large increases in private sector agricultural research, questions remain about the long-term willingness of private sector firms to invest large sums of money in agricultural research and about the breadth and stability of investment in such research. As noted previously, private firms tend to cut back on research first in times of adversity.

Private Sector Promotional and Educational Programs

The private sector is playing a more important role in education (Christenson and Warner, 1985). For most agribusiness firms, this role is pursued in conjunction with their efforts to promote the products and services that they market. The educational value of these promotional activities is more in terms of alerting farmers to the availability of new products than in objectively evaluating the performance of those products.

The burden of new product evaluation then falls either on the farmer (through trial and error) or on the extension service (through result demonstration).[6] While extension involvement is more efficient, there is potential for increased antagonism between private sector firms and

[6]A considerable amount of new product testing is also done by the university research community under contracts, grants, or consulting arrangements. While product testing at one time was an important component of experiment station research, it is considerably less important today.

extension. Extension testing will not be appreciated by firms found to produce products having relatively lower levels of performance.

With a few important exceptions, such as integrated pest management (IPM) checkoff programs, private sector direct financial support for agricultural extension programs has been limited but appears to be growing. It might be argued that limitations on private sector funding are essential for keeping extension education programs objective. There may be greater dangers in increased private sector funding of extension than of research. In both cases it is critical that the objectivity and availability of information flows be maintained.

Two of the most important private sector supporters of extension programs are the Farm Foundation and the Kellogg Foundation.[7] Both of these institutions are maintained largely by endowment grants. Each foundation has played particularly important and unique roles during the recent period of reduced funding for extension programs.

The Farm Foundation has played a particularly critical role in filling the void created by the reduction in funds available for communication and program development on a regional basis. The foundation's support of regional extension committees in the areas of farm management, marketing, policy education, and community development has frequently been the only support for contact and coordination among specialists in neighboring States. The pressure on Farm Foundation funding has become increasingly intense as Federal extension decisions not to fund meritorious projects become more prevalent.

The Kellogg Foundation has periodically attempted to fill a portion of the void left by the reduction of USDA Extension Service pilot project funds. While Kellogg continues to support what it perceives to be the most innovative proposals for educational program development, an increasing backlog of proposals has developed with little hope of their being funded on a timely basis.

Increased pressure on funding from public and unbiased private sector sources discourages new program development by extension specialists. Potential and existing extension employees are increasingly being attracted by research positions and/or the private sector. A large infusion of new private support, without a vested interest, to institutions such as that provided by the Kellogg and Farm Foundations appears unlikely. The IPM checkoff concept may hold promise for increased, direct producer funding of specific educational programs. The only remaining option then becomes the establishment of a new thrust for public support of extension education. Such a thrust is needed particularly at the specialist and program development level, which is a logical level for increased Federal support and leadership (Knutson, 1984).

[7]General farm organizations and commodity groups have been important supporters of both research and extension programs at the Federal and State levels. This support has, however, been largely one of influencing Federal, State, and county government appropriations. This important private sector role is frequently not recognized.

TRENDS IN LEVEL OF SUPPORT AND RELATED ISSUES

In the 10 years from 1966 through 1975, the level of support for agricultural research and extension programs increased 215 percent in current dollars and by 30 percent in terms of constant dollars (table 12-1). During this period, research and extension resources increased at nearly the same rate. From 1975 through 1982, total expenditures on research and extension increased 87 percent in current dollars and 9 percent in constant dollars.

Research-Extension Balance

From 1966 through 1975, Federal support for extension increased considerably more than Federal support for research. However, since

Table 12-1.—Trends in Agricultural Research and Extension Funding by Source and User, Selected Years

	Research						Extension			Total research and extension
	State agricultural experiment stations						State agricultural extension service			
Year	Federal	State	Private	Total	USDA	Total research	Federal	State and county appropriations	Total extension	
Millions of current dollars:										
1966	79	118	9	206	153	359	75	126	201	560
1975	135	331	23	489	266	755	179	269	448	1,203
1982	268	563	94	925	469	1,394	315	539	854	2,248
Millions of constant dollars:										
1966	80	119	9	208	155	363	76	128	204	567
1975	83	203	14	300	163	463	110	164	274	737
1982	95	201	34	330	167	497	112	192	304	801

SOURCE: Cooperative State Research Service, *Inventory of Agricultural Research FY 1982*, vol. II (Washington, DC U.S. Department of Agriculture, 1982); U.S. Congress, Office of Technology Assessment, *An Assessment of the United States Food and Agricultural Research System*, OTA-F-155 (Washington, DC: U.S. Government Printing Office, December 1981).

1975, research expenditures have increased somewhat more than extension expenditures. Increased support for agricultural research relative to extension has been particularly unbalanced when the surge of private sector investment in agricultural research is considered. This increased emphasis on research likely reflects the following:

- a higher level of sensitivity to the needs for high-priority biotechnology research;
- the potential for major breakthroughs in productivity;
- a reaction to concerns about the availability of an ample supply of food;
- a desire to maintain competitiveness in international trade; and
- the higher costs associated with conducting biotechnology research, which has been used to justify higher appropriations.

In a time of tight budget constraints, policymakers (particularly at the Federal level) have apparently made a decision that research has a higher priority than extension. Longer run questions, however, exist regarding the need to maintain a balance between research and extension activities.[8]

Research and extension are part of a complex agricultural system designed to discover, adopt, evaluate, and (where favorable) facilitate technology transfer to farmers and ranchers. All parts of the system are equally important for accomplishing this mission.

The biotechnology and information technology era presents at least as many, and probably more, challenges for extension as it does for research. Many of the technologies that are on the horizon are exceedingly complex and foreign to many extension staff. In the foreseeable future, embryo transplant technology may be as important to the dairy industry as artificial insemination has been over the past three decades, growth regulators will increasingly be applied in minute quantities to plants to increase productivity, and new strains of genetically engineered plants and animals will be entering commercial production channels. Extensive staff training and development will be required at both the specialist and county levels for extension to play an effective role in technology transfer of biotechnology and information technology. Without such training, extension will play an increasingly less important role in production agriculture. Technology transfer will occur less efficiently and with more structural impacts—larger farms will benefit at the expense of smaller farms.

Another important effect of the research-extension imbalance in emphasis is to attract the best scientists into research rather than extension. While the public sector agricultural research community is experiencing increased difficulty competing with private sector firms for the services of qualified scientists, extension is having even more difficulty competing with both interests.[9] At the specialist level, extension draws on the same pool of doctoral-level scientists as does research. Because it is receiving increased emphasis, research is able to compete more effectively for the services of the top scientists.[10] Over time, unless corrected, the result will be a lower quality of extension staff. The same principle applies at the county level, where extension must likewise compete for its professional staff with both public and private sector employment alternatives. With relatively less extension support, the best county and area extension staff will be attracted to the private sector or to other better endowed agencies in the public sector. These effects are already occurring, at a time when extension is being called on to transfer a larger quantity of increasingly complex technology.

[8]It is interesting to note that the relative increase in emphasis on research began during the Carter-Bergland Administration. Previously, the Nixon-Ford-Butz Administration had put relatively greater emphasis on extension programs, while the Kennedy-Johnson-Freeman Administration had favored research. The impacts of these shifts in emphasis in terms of productivity have not been adequately studied.

[9]There is a concurrent concern that the best research and extension scientists are being attracted into private sector managerial jobs.

[10]One method by which extension might adjust to this competition is to reduce the number of staff and concentrate more resources around a smaller number of highly qualified staff. Extension has not, as a general rule, employed this strategy.

Federal-State Balance

The States have been picking up a larger share of the cost of the agricultural research and extension system. From 1966 through 1975, Federal support for research and extension declined as a proportion of the total, from 55 to 48 percent. In 1982 the Federal share was 47 percent. The historic commitment to a national system of developmental institutions in agriculture is fading.

This trend is consistent with the philosophy of a reduced overall Federal role. However, it is inconsistent with the role of U.S. agriculture nationally in terms of maintaining stable prices, contributing to a favorable balance of trade, and meeting world food needs. These are important national goals that require a higher level of Federal involvement and support.

During the period 1975 through 1982, most of the relative reduction in Federal support has been in extension appropriations. While Federal support for research increased by 7 percent (constant dollars), extension support increased by less than 2 percent (table 12-1). State and county support for extension, on the other hand, increased by 17 percent. The Federal share of extension support, thereby, fell from 40 percent, where it had been since the early 1950s, to 37 percent. Appropriations for extension in 1984-86 suggest a further drop in extension's share of Federal support.

The rationale for reduced Federal support for extension relative to research is unclear. Although education has traditionally been viewed as a State and local community function, extension was formed on the principle of a Federal-State-county partnership. The ability and willingness of State and county governments to support extension adequately in the face of reduced Federal support is questionable.[11] Clearly, if the biotechnology and information technology era justifies higher levels of support for agricultural research, it also justifies higher levels of support for agricultural extension—particularly because of the increased private sector commitment to agricultural research.

Research and Extension Professional Staff

Despite increases in real appropriations for agricultural research, the number of professional research staff has declined. This decline results from the continuously increasing cost of supporting a research scientist with research equipment and materials. Greater cost increases can be anticipated in the future as agricultural research progresses into the biotechnology era and as the demand of the private sector for newly trained scientists continues to accelerate.

Extension experienced an 11-percent increase in the numbers of professional staff from 1966 to 1975, and a subsequent 6-percent decline through 1984 (table 12-2). Nearly all of this decline was in the specialist staff, which experienced a 15-percent decline in numbers. This reduction in number of specialists is particularly alarming since the specialist staff has the highest level of training and is the best equipped to educate both county agents and farmers on evolving agricultural technologies.

The disproportionate reduction in the number of specialist staff is probably best explained by budget considerations and the lack of direct State control over county staff. As budgets tighten, considerably more funds are made available to the State director when a specialist position is eliminated or not filled. In addition, competition for specialist staff has become increasingly keen. For extension program administrators, the avenue of least resistance compared with the option of reducing the number of county

Table 12-2.—Trends in Numbers of Extension Professional Staff, Selected Years

Year	Specialist	County	Total
1966	3,641	10,451	14,092
1975	4,224	11,357	15,581
1984	3,581	11,140	14,721

SOURCE: Extension Service, U.S. Department of Agriculture.

[11] Interestingly, the 1977 farm bill contained authorization for USDA to be the lead agency in university education programs related to agriculture. While there was a transfer of staff and offices from the Department of Education to USDA, this initiative has received very limited USDA support and is essentially dead.

staff is reduction in force at the specialist level—precisely what has occurred. Therefore, while the need is for an increased emphasis on specialist staff, just the opposite is occurring. In fact, one can forcefully argue that in the biotechnology and information technology era, without substantially increased emphasis on county agent development, the specialist will have an ever-increasing and comparative advantage in educating farmers. The model for cooperative education could shift from county agent-to-farmer education to specialist-to-farmer education. This is probably already happening and has the potential for substantially changing the structure and role of extension. Without substantially increased State or Federal support for extension, counties will have to pick up a larger proportion of extension's costs, or the counties' impact and effectiveness in education will gradually erode.

An alternative strategy for extension would involve an intense, continuing program of staff development at the county level designed to provide county agents with state-of-the-art research findings and related information. A decision to emphasize this strategy is based on the premise that the strength of extension lies in the county agent. Historically, the county agent has been one of the best educated persons in the county. Questions increasingly arise as to whether this era is gone.

Christenson and Warner (1985) put the issue in the following very cogent terms:

> If county staff are not providing relevant and timely information, if they do not have access to innovative ideas, if they are not seen as outstanding educators in the county, they will not have the trust and respect of the people. County staff who are seen as just another information disseminator who hands out pamphlets, gives advice on fertilizing lawns and gardens, and holds meetings for "expert" speakers from the State university, may not survive in an information society.

This is not to contend that all county agents are out-of-date. Many continue to carry out state-of-the-art programs. However, as the rate of technological change accelerates, research results will become more complex and difficult to comprehend. County agents will find it increasingly difficult to keep up.

Such observations are not limited to county agents. Many researchers will also find their knowledge level bypassed (antiquated) by biotechnology; specialists will thus also need to update their knowledge. However, the cost of retraining specialists will be less than that of retraining county agents because specialists are fewer and have closer day-to-day contact with research.

Fund Allocation

The funds that land-grant universities receive from the Federal Government can be allocated either on the basis of competitive grants or formulas. Historically, about two-thirds of the funds have been allocated to the States by formulas. While there are formula differences between research and extension, the principal factor in both formulas is rural population and farm numbers. As a result, States having a larger rural population and greater farm numbers receive more formula funds.

The specifics of the formula have been the subject of considerable debate. Large rural populations and farm numbers in the Southeast do not correspond with the quantity or value of production. Midwestern and Western States feel that formula funds ought to be allocated on the basis of the value of commodities produced. Senator Lugar (Indiana) has become a champion of debate to change the formula (GAO, 1983).

Since agricultural research deals more with the products of agriculture, not population, Midwest advocates suggest that inequities result from research funds being allocated on the basis of population. Current formula funding procedures have tended to promote regional crops such as cotton, tobacco, and peanuts as opposed to wheat, corn, soybeans, milk, beef, and hogs. Yet, those States that produce the majority of the wheat, corn, soybeans, milk, beef, hogs, fruits, and vegetables have been more competitive in achieving competitive research grants. In some instances, strong State research support has compensated for less Federal support.

A change to a product value-based formula would accelerate the trend toward increased centralization of research in the major agricultural States such as California, Iowa, and Indiana. It can also be argued that the effect would be to shift the allocation of research resources in the direction of moderate-size and larger scale farms. However, questions exist regarding whether the size distribution of the clientele being served by research and extension is any different in those Southern States that receive a larger proportion of formula funds just because of a larger number of smaller farmers. Maybe the needs of small farmers are not being served in the South any better than in the rest of the country.

The case for a population-based formula appears to be stronger for extension than for research. Education deals more with people than with the value of products. However, even here the urgency of education can be argued to be product value-based. That is, education is more urgent where more products are produced. In addition, as in research, questions arise as to whether extension is effectively serving the educational needs of farmers having smaller scale operations.

Competitive grants are a much discussed method of allocating USDA agricultural research funds. Prior to 1970, Federal contracts and grants generally represented about 10 percent of the USDA funds going to the State agricultural experiment stations and about 2 percent of total experiment station funds. However, the world food crisis and advances in biotechnology created greatly increased interest in grants. By 1982, contracts and grants had increased to 16 percent of experiment station funding.

In 1977 Congress authorized a special competitive research grants program primarily to support basic research in food and agricultural science. The competitive grants program was available to any research institution, land grant or not. In 1982 experiment stations received only 38 percent of competitive grant appropriations, accounting for less than 1 percent of experiment station funds. The land-grant universities accepted the grants concept only on the condition that grants not displace formula fund appropriations. At least partially because of land-grant resistance to formula fund reductions, the competitive grants program has received a low level of appropriations.

Although competitive grants are made on the basis of a peer review system, basic research scientists complain that the grants are generally so small that they cannot sustain even a middle-size biotechnology research project. In 1982 the average size of a grant was approximately $70,000 (CSRS, 1984). The program is frequently referred to by researchers as the "small grants program." The underlying reason for the small size of these grants probably lies in political pressure on USDA research administrators to distribute the grants geographically among the States.

For many years, extension has used savings from Federal administration funds, plus approximately $500,000 in so-called special needs funds, for allocation to the States in the form of competitive grants to support, among other priorities, the development and testing of innovative concepts of extension education. This important, highly successful (albeit, informal) counterpart of the competitive research grants program has been severely restricted since 1978—ironically starting about the same time as the research grants program was initiated. The reasons for this restriction lie in the interaction of such factors as reductions in Federal administration funds, the subversion of extension funds to support a vast experiment to coordinate research and extension at the Federal level, a congressionally mandated evaluation project, and the subsequent emphasis on increased ongoing evaluation, which had to be absorbed out of existing funds.

One of the unique features of the extension special grants program was that the projects supported by it were frequently regional or national in scope, thus facilitating the production of educational programs that could be replicated and applied on a multistate basis. Since the restriction of this program, innovative extension program development has been severely curtailed, particularly for programs having a regional or national focus. Individual States have not, and probably cannot, fill this void. Several recent attempts to provide funds for these purposes as a designated item in the Federal extension budget have been unsuccessful.

Interaction Between Non-Land-Grant Universities and Land-Grant Universities

The world food crisis and the biotechnology era have fostered increased non-land-grant interest in agricultural research. This interest was further heightened by the establishment of the competitive grants program in the 1977 farm bill. Experience indicates that the non-land-grant universities are fully competitive with the land-grants in receiving these funds. However, competitive grants have not been expanded sufficiently to augment significantly most non-land-grant agricultural research programs.

Increased funding for human research in the biotechnology area holds the potential for rapid technology transfer of medical discoveries to agriculture at a relatively low cost. Potential also exists for fortifying existing non-land-grant basic research in photosynthesis, plant embryology, genetics, and animal physiology. This will, however, require significant increases in funding beyond current levels as well as a movement away from the "small grants" philosophy discussed previously.

One of the factors hindering the contribution of the non-land-grant universities to discoveries in agriculture is the traditional competition within States between land-grant and non-land-grant universities. Because of increasingly limited funding, competition for the allocation of appropriated funds and the establishment of new educational programs has become increasingly intense. Over time, substantial conflicts have developed over the favored position of land-grant universities in having access to formula funds. Such conflicts even exist within land-grant universities between experiment station-related agricultural departments and academic departments having no ties to the experiment stations, such as biology departments.

Such conflicts are difficult to overcome. Danger exists that in attempting to "force" cooperation, policymakers could destroy productive elements of the existing system that have served agriculture well. Yet constant pressure to obtain a higher level of cooperation would appear to be warranted. Perhaps the most effective means of applying such pressure would involve the development of programs that provide financial rewards for cooperative land-grant/non-land-grant research programs. However, if funding levels remain low little progress is likely.

PROPERTY RIGHTS, EXCLUSIVITY, AND THE LAND-GRANT UNIVERSITIES' SOCIAL CONTRACT

Land-grant universities were created to serve the public. The agricultural component of land-grant universities has unique responsibilities to conduct research and extend the results of agricultural research for the public benefit. Traditionally, those research results have been readily and freely available to the public, inasmuch as the results have no private property or exclusivity rights attached to them. Policy changes that have occurred over the past 15 years, however, hold the potential for substantially changing this traditionally ready-and-free-access concept of land-grant university research. Some changes have already occurred; others could occur very rapidly. In other words, changes in the rules may have also changed the very concept of the land-grant system (Knutson, 1984).

Questions of how the land-grant universities might adjust to the new concept of research property rights and the related opportunities for increased private sector funding have been the subject of extensive study. However, the impact of this concept on the unique nature or "social contract" of the land-grant system has received little attention. A discussion of both dimensions follows. This discussion is important because it has a potentially profound effect on the land-grant system and its relationship to the public.

The Development of Discovery Property Rights

Policy changes regarding property rights in agricultural research had their origin in the enactment of the Plant Variety Protection Act of 1970. Previously, patent protection in plants was limited to asexually reproduced material—mainly orchard fruits and ornamental flowers. The Plant Variety Protection Act provided that a breeder of a new, stable, and uniform variety of sexually reproduced plants could restrain other seedsmen from reproducing and selling that variety for 17 years.

Of possibly greater significance was the 1980 landmark U.S. Supreme Court decision, *Diamond* v. *Chakrabarty,* which held that the inventor of a new micro-organism, whose invention otherwise met the legal requirements for obtaining a patent, could not be denied a patent solely because the invention was alive. This decision opened the door for patenting potentially all new products of the biotechnology era.

Since the passage of the Plant Variety Protection Act and the *Chakrabarty* decision, private sector interest in agricultural research has mushroomed. OTA, for example, found that in 1983 there were 61 companies pursuing applications of biotechnology in animal agriculture and 52 companies applying biotechnology to plants. The companies involved ranged from established agricultural chemical suppliers such as Monsanto, DuPont, Dow, Eli Lilly, and American Cyanamid to new biotechnology firms such as Genentech, Biotechnica International, MGI, and Genex (OTA, 1984).

Most of these firms have developed their own in-house research capability by employing molecular biologists, biochemists, geneticists, plant breeders, and veterinarians. Whereas past emphasis in plant and animal science was on selecting and breeding for specific, desired traits, the emphasis has changed to understanding the factors that control the genetic traits and overtly changing them. Progress is already being made with growth hormones, vaccines, and herbicide-resistant varieties of plants. Several genetically engineered products are very close to being marketed commercially.

Relationships are also developing between many of these firms and universities. For example, Monsanto has a 5-year, $23.5 million contract with Washington University under which individual research projects are conducted. At Stanford University, five corporate sponsors (General Foods, Koopers Co., Inc., Bendix Corp., Mead Corp., and McLoren Power & Paper Co.) contributed $2.5 million to form the for-profit Engenics and the not-for-profit Center for Biotechnology Research.

Such relationships are not limited to private universities. Michigan State University (a land-grant university) created Neogen to seek venture capital for limited partnerships to develop and market innovations arising from research. The formation of Neogen points up a significant problem being encountered by the universities. Neogen was formed, in part, to retain faculty members who were getting offers from biotechnology companies. In Neogen, faculty members are allowed to develop their entrepreneurial talent and reap the associated financial rewards while remaining at the university.

The establishment of biotechnology property rights has substantially heightened scientists' interest in private sector employment opportunities. In the process, questions have arisen over who maintains the property right—the university, the private firm, or the scientists? In the Washington University-Monsanto case, the University retains the patent rights while Monsanto has exclusive licensing rights. In Engenics, Stanford likewise gets the patent rights while the center and the six corporate sponsors receive the licenses and pay royalties. Neogen will buy patent rights from Michigan State University, while the inventor will get a 15-percent royalty or a stock option in Neogen.

Land-Grant University Adjustments

The potentially profound implications of such developments on the land-grant university system seem clear. Such private sector arrangements integrate business into the university fabric, raising questions about the control of the university research agenda, the allegiance of scientists to their university employer, the willingness of scientists to discuss research discoveries that have a potentially patentable products associated with them, and potential favoritism shown particular companies by the university because of their research ties.

This controversy has caused the land-grant Agricultural Experiment Station Committee on Policy (ESCOP) to express its concerns publicly and to develop guidelines to deal with these biotechnology issues. The statement of ESCOP concerns includes the following:

- As publicly supported institutions, the SAES (State Agricultural Experiment Stations) will need to assure that industrial relationships generate an end result in the interest of the general public. This end result should reward the industrial investor but avoid placing such an investor in an unwarranted position of financial advantage through privileged use of information or technology partly derived from research using public funds; neither should a curtailment of new information to the public occur.
- The SAES are greatly concerned about the curtailment of communication on early research results and about the constraints on sharing of germplasm emerging due to concerns on the part of scientists and institutions for protecting potentially patentable research results. Industry sponsorship of this kind of research tends to exaggerate this problem.
- There is general concern in the academic community about the drain of scientific manpower from the universities to industry. The ability to continue to conduct basic research in an academic environment and the concurrent interdependent ability to continue educating scientists are key issues.
- There is concern that individual scientists may place themselves in positions of compromise or conflict of interest as they establish personal relationships with industry as contractors, consultants, or institutional officers.
- There is concern on the part of both scientists and the SAES that through industrial sponsorship of research, there may be introduced an undesirable level of direction of effort by industry (ESCOP, 1981).

Out of these concerns ESCOP developed the following interim policy guidelines:

- *Maintain SAES management control of research:* Consensus: SAES should retain the ability to manage research programs, and control the direction of new investigations, regardless of the source of support, including situations in which one or several firms may sponsor research at several institutions.
- *Strong basic research and graduate education capability*: Consensus: SAES should maintain and expand the basic research capability in genetic engineering and related areas within the domain of publicly supported institutions.
- *Faculty-industry relationships*: Consensus: Scientists should maintain close communication with institutional administrators in the development of relationships and commitments with the commercial sector. Institutional guidelines should be developed that assist the scientists in avoiding institutional or personal conflicts of interest.
- *Publication and communication*: Consensus: The ability to publish and exchange information is essential and must be secured in agreements. In some instances, publications or information exchange may need to be temporarily delayed to allow time for an institution or sponsor to assure adequate patent protection. The final decision to defer or modify a publication should reside with the public institution.
- *Trade secrets and confidential information*: Consensus: Protection of "trade secrets" or "confidential information" for more than a very limited period should be avoided by public institutions. Advance review by a private sponsor, to avoid premature release of information, may be advisable but should not become a mechanism to "shelve" use-

ful information or unpatentable technology.

- *Patent rights and premature disclosure*: Consensus: SAES should retain the right to participate in the decisions related to the disposition of intellectual and real property and patent rights resulting from research. Retained ownership of patents by the SAES is preferred. In any agreement, the SAES should retain the right to use discoveries and inventions from SAES research to extend and enhance public research and education. The need of private sponsors to obtain a return on investment must be recognized, and agreements may provide for special licenses for patents originating from sponsored research.
- *Biosafety of recombinant DNA*: Consensus: SAES must retain responsibility for review and decisions in the release or distribution of laboratory research products, although some research may be supported by outside sponsors.
- *Grants and income earnings*: Consensus: Extending knowledge and developing new technology while serving the public interest should be the prime motivations in agreements between SAES and the private sector. Royalty income from discoveries originating under such agreements should be recognized as a secondary consideration.
- *Licensing responsibilities and performance expectations*: Consensus: SAES should assure that "due diligence" clauses are included in contracts to assure that new technology is not shelved and the public interest is served while private investment in commercialization is respected. Assignments, rights, or licensing of patents for commercial use should be considered separately from contractual definition of research to be conducted. Initial or developmental processes and pervasive technology ultimately leading to improved biological materials generally should not be assigned for sole use by a sponsoring firm.
- *Tax code implications*: Consensus: When sponsored research is motivated by certain interpretations of Tax Code Section 1235, exclusive licensing or co-ownership of patent rights is a preferred alternative for the institution, since the institution maintains a vested interest and some ownership of patent rights involving the scientist, the institution, and the firm may require unique documentation. Careful attention to these rights and relinquishments is suggested (ESCOP, 1981).

Impact on the Land-Grant Social Contract

Potential basic changes in the relationship between land-grant universities and the public are implied by the preceding adjustments, although not explicitly discussed. The land-grant university system was established on a public service basis different from that of other universities. Its tradition has implied a social contract that makes its discoveries freely available to the public.

The advent of patent rights, exclusive licensing, and private sector investment in public sector research has the potential for changing the distribution of benefits from land-grant research discoveries.[12] These changes warrant direct public discussion and consideration by policymakers. They occur for at least five reasons:

1. By exclusive licensing or transfer of patent rights to private firms, the right to use discoveries is no longer freely available—even if information on the discovery itself is freely available.
2. Certain individuals or firms are conferred the benefits of specific land-grant research to the potential detriment of others. Prior to the transfer of discovery rights, the benefits were available to anyone who adapted a land-grant discovery to commercial usage.
3. The costs of the resulting discoveries are internalized in the price of the resulting product. The price the public pays for the product also includes any monopoly rents associated with the conferral of the rights.

[12]Similar implications may also exist for ARS research to the extent that patent rights and exclusive licensing arrangements are created by ARS.

Society thus pays twice: once for the cost of the research and then again for its benefits. Without the conferral of property rights, rents are minimized by competition.

4. Private sector-public sector inequities are virtually assured in any granting of research property rights to an individual firm. This occurs because with a relatively small private sector investment there is access to a much broader range of current and prior research.
5. The existence of patent rights, trade secrets, and confidential information has as many potentially adverse implications for extension in terms of the increased burden for product testing, the potential lags in information, and the absence of research information that would have previously been readily available.

The argument does not, however, flow exclusively against the conferral of private sector property rights by the land-grant universities. There are three main counterbalancing arguments:

1. With the conferral of private property rights and the associated private sector investment, the quantity of research discoveries may increase. Evenson (1983), for example, found a sharp acceleration in private plant breeding programs after the 1970 Plant Variety Protection Act was enacted into law. Over 1,088 patent-like certificates were granted by February 1, 1983.
2. Without land-grant university involvement in private sector-funded research, the universities may not be able to retain the top-quality scientists needed to conduct agricultural research that is on the frontier of knowledge. In the process, agricultural research, extension, and teaching programs would suffer.
3. Patent monopoly rights may be necessary to attract the capital investment needed to translate land-grant university scientific advances into commercial reality. Without such proprietary protection, new discoveries may not be able to compete for resources to develop marketable products or technologies. The public availability of such products could thereby be affected.

If policymakers do not want land-grant universities to confer property rights, policymakers must provide the level of funding necessary for competing with other non-land-grant universities that confer such rights. This decision is a basic public policy decision—maybe the most basic decision since the land-grant system was created. Once the land-grant system begins actively competing for private sector grants and conferring licensing rights, there will be no turning back.

PRICING INFORMATION SERVICES

Although seldom recognized as such, one of the most critical aspects of U.S. agricultural policy is that of information policy. Much of what USDA does is provide information. Until the 1970s, most agricultural information available to farmers had its origin in USDA. The department gathered the information, interpreted it, and published it. Extension Service personnel at State and Federal levels and private sector media made the information freely available to the public. The information covered a very broad range—technology developments, public policy changes, statistical data, economic trends, and price forecasts. USDA was respected for having the best information system in the world. In many respects, USDA had a monopoly on information that was freely available to anyone—small farmer or large agribusiness firm.

The information policy of USDA began to change in the 1970s. Tight Federal budgets resulted in cutbacks in the quantity and quality of information at a time when, because of greater instability, more information and information of better quality were needed. New methods of communication made timely transfer of information to the producer possible. Such communication could be accomplished in closed, often computerized, systems where the benefits could be captured by the supplier. Larger farm units required information of a more specific nature, tailored to their operations. Information had captured a value, yielding private sector profit opportunities.

Without substantially increased appropriations, neither USDA nor the land-grant system could adequately respond to these new demands. Perhaps more significantly, private sector firms, seeing increased profit opportunities, did not encourage increased funding for information. In the process, they indirectly (some might argue, directly) discouraged increased funding. Their philosophy was basically one of "give us (the private firms) the raw data and we will interpret it."

At a time when policymakers sought opportunities to transfer functions from the public to the private sector, it seemed quite logical to cut back on public sources of information. Since the information that was being collected by USDA had acquired greater value, it also seemed logical to begin charging for all (or nearly all) USDA publications.

Increasing quantities of information are now available only to those farmers and agribusiness firms who can afford to pay for it. Those who can afford to pay for it are the larger farm operations and agribusiness firms. Those who cannot afford to pay for it are the moderate-size and small farms as well as the moderate-size and small agribusiness firms. Since information is a lifeline for success in today's agriculture, its absence accelerates the trend toward a more highly concentrated agriculture.

For many moderate and smaller farmers the Extension Service was the only continuing reliable and consistent source of information. But even that source was curtailed by a USDA policy requiring State extension staff to pay for USDA publications they had the responsibility for distributing. Many States did not have the funds to obtain reports that were vital to timely producer decisions. Such policy changes are difficult to justify or excuse.

This problem is by no means limited to the Federal Government. Many States have also been forced by budget constraints to charge for publications as well as for many of their educational programs. Such policies aggravate the comparative disadvantage of moderate farms competing in agriculture.

DISTRIBUTION OF BENEFITS AND STRUCTURAL IMPACTS

Technology is one of the driving forces behind structural change in agriculture. This point has perhaps been most clearly argued by Willard Cochrane (1958), who points out that the first adopters of new technology are the immediate beneficiaries in that their costs per unit of production are lowered and their profits thus rise. The profits of those firms supplying the products of new technology also rise. In addition to reducing costs per unit of production, technology generally expands output. Also, higher profits encourage the adopting farmers to expand output—even to the extent of increasing the scale of their farm operations. But as output expands, prices decline. Later technology adopters thus realize less profit. In fact, those farmers who are the last to adopt new technologies may actually be forced either to adopt or to get out of agriculture.

Two important lessons arise from this description of the process of technological change:

1. Those farmers who are most aggressive in effectively adopting and applying new technologies are the most likely to survive. Their size or scale of operation thereby influences the structure of agriculture. Likewise, to the extent that research discoveries or extension programs favor certain size farmers, structure is affected. White (1984) finds that this impact is less than has sometimes been asserted (Hightower, 1973). These findings do not, however, negate the concern about the neutrality issue. The importance of technology in fostering structural change makes constant awareness and consideration of technology's potential impacts important in designing research and extension programs.

2. The ultimate beneficiary of agricultural research and extension has and will likely continue to be the consumer. Larger supplies, lower food prices, and better quality have almost invariably been the main end result of research. This does not mean that research operates contrary to the interest of farmers. Research directly benefits the more progressive farmers. Research is critical to expanding markets for farm products and to maintaining the competitiveness of U.S. agriculture internationally. Research overcomes the constant threat of new disease and other vagaries of nature that threaten the increased productivity created by science and its application.

These lessons present a difficult problem for policymakers and land-grant university administrators. While the returns on investment in agricultural research and extension programs are high, their benefits are by no means uniformly distributed. Although farmers and agribusiness firms are frequently described as the main clientele of the agricultural research and extension system, they are not the long-term beneficiaries. The benefits enjoyed by farmers and agribusiness firms are not uniformly distributed. The adverse effects of technology on farmers who fail to adopt, agribusiness firms that fail to obtain the property rights, or on farm laborers who are displaced may be dismissed as one of the costs of progress. They are, however, accentuated by policies that:

- fail to provide sufficient resources and incentives to serve the research and extension needs of the full range of farmer and agribusiness clientele regardless of their ability to pay for those services;
- fail to provide alternative retraining and employment opportunities for those who are displaced by the effects of technological change; and
- fail to take into consideration the unique nature of the social contract under which the land-grant university system was formed in designing a system of property rights for its discoveries.

In other words, the trend toward industrialization may continue—but the scales should not be tilted by public policy to speed up the process or assure the final conclusion. Indeed, public policy should work to keep options open for conscious public decision.

SUMMARY AND CONCLUSIONS

To an important extent, U.S. agriculture has been very successful because of technological advance. Yet, consideration of specific changes in research and extension policy may be justified. The following areas have been identified as meriting consideration for policy changes:

- The social contract on which the agricultural research and extension system was created needs to be reevaluated. This issue should not be left for resolution by the courts. Specific guidelines must be developed that, while allowing the system to compete, protect the public interest and investment in the agricultural research and extension functions. Both Congress and USDA should have an input in this type of policy development.
- It is sometimes suggested that increased private sector support for agricultural research signals less need for public support. While private sector support complements public support, basic biotechnology and information technology research is very costly. A reduced role for public research and extension would provide a slower rate of technological progress and a lower level of protection for the public health and welfare. In addition, there is a strong public interest in maintaining an agricultural research component in each State to serve the problem-solving needs of State agriculture.
- Many agricultural problems are local or regional in scope. The applied nature of the system, having an agricultural experiment

station and extension service in each State, has provided a unique capacity to identify and solve local or regional problems. Reality suggests that only certain universities have sufficient resources to compete for private sector support in biotechnology and information technology. The result is a confluence of forces that is creating a dichotomy of "have and have not" universities. In the process, traditional extension-research interaction and feedback mechanisms could break down, particularly in States that are not in a position to command a major biotechnology component.

- The role of extension is even more important than it has been in the past. New, more complex products require evaluation and explanation. In States where experiment stations have attracted substantial private sector support, the product testing function can most objectively be performed by extension. The recently passed 1985 farm bill gives extension explicit authority to engage in applied research functions such as product testing and evaluation.
- While agricultural research is not inherently biased toward large-scale farms, lags in adoption by smaller and moderate farms have the same effect. Unless special attention is given to technology generation and transfer to moderate farms, major structural changes could result—leading to the eventual demise of a decentralized, moderate farm structure. To the extent that preservation of these farms is a policy objective, special funding for and emphasis on the problems of technology generation and transfer of technology to moderate farms is warranted.
- While the agricultural research system has received the benefits of increased funding from both private and public sources, extension funding has not materially increased. As a result, extension staff at the county and specialist levels are being caught up in a whirlwind of technological change. The result is a need for the injection of substantial staff development funding into the extension system.
- Basic organizational issues must be addressed by the Extension Service. The premise on which extension was developed was that of research scientists conveying the knowledge of discoveries to the extension specialist who, in turn, supplied information to the county agent who taught the farmer. Over time, this concept has gradually but persistently broken down as agricultural technology has become more complex, and insufficient resources have been devoted to staff development. As a result, more emphasis has been placed on direct specialist-to-farmer education. More specialists have been placed in the field to be closer to their clientele, but at the cost of less contact with research scientists. As these changes have occurred, the role of the county agent has become increasingly uncertain. Appreciation for and use of county agents as educators and technology transfer agents has declined. As a result of these changes, a basic structural reevaluation of the organization of the extension function of the agricultural research system is needed. Such a reevaluation will inevitably have to tackle the politically sensitive issues of the role of the county, State, and Federal components of the Extension Service.

CHAPTER 12 REFERENCES

Agriculture Research Institute, *A Survey of U.S. Agricultural Research by Private Industry III*, Bethesda, MD, July 1985.

Christenson, James A., and Warner, Paul D., "Extension Client Impact in Knowledge Transfer Processes" (Los Angeles, CA: American Association for Advancement of Science, May 28, 1985).

Cochrane, Willard W., *Farm Prices: Myth and Reality* (Minneapolis: University of Minnesota Press, 1958).

CSRS, Cooperative State Research Service, *Inventory of Agricultural Research FY 1982*, vol. II (Washington, DC: U.S. Department of Agriculture, April 1984).

Evenson, Robert E., "Intellectual Property Rights and Agribusiness Research and Development: Implications for the Public Agricultural Research System," *Amer. J. Agr. Econ.* 65:967-975, 1983.

ESCOP, Experiment Station Committee on Policy, *Genetic Engineering Policy for the State Agricultural Experiment Stations,* Washington, DC, Nov. 11, 1981.

Extension Service, *Extension in the '80s* (Washington, DC: U.S. Department of Agriculture, no date).

Feller, Irwin, "Organizational Innovation in Innovating Organization: The Case of Agricultural Extension" (Los Angeles, CA: American Association for the Advancement of Science, May 28, 1985).

Feller, Irwin, et al., *Agricultural Technology Delivery System: Executive Summary,* Institute for Policy Research and Evaluation (University Park, PA: Pennsylvania State University, December 1984).

GAO, General Accounting Office, *Federal Agricultural Research Funding: Issues and Concerns,* GAO/RCED-84-20 (Washington, DC: General Accounting Office, Oct. 20, 1983).

Hightower, Jim, *Hard Tomatoes, Hard Times* (Cambridge: Schenkman Publishing Co., 1973).

Hyatt, George, "Agricultural Extension Policy: Its Evolution and Current Issues," paper prepared for the Office of Technology Assessment, Washington, DC, 1984.

Joint Council on Food and Agricultural Sciences, *Needs Assessment for the Food and Agricultural Sciences* (Washington, DC: U.S Department of Agriculture, January 1984).

Kendrick, James B., Jr., "Revitalize U.S. Agricultural Food Research" (editorial), *California Agriculture,* Berkeley, CA, January 1981.

Knutson, Ronald D., Paarlberg, Don, and McCalla, Alex, "Forces Affecting Food and Agricultural Research Decisions," paper prepared for the Office of Technology Assessment, Washington, DC, 1980.

Knutson, Ronald D., "Agricultural Research and Extension Policy: Contemporary Issues," paper prepared for the Office of Technology Assessment, Washington, DC, 1984.

Lovvorn, Roy L., "Agricultural Research Policy: Its Evolution and Current Issues" paper prepared for the Office of Technology Assessment, Washington, DC, 1984.

Marshall, H. Peter, et al., *Strengthening the Research Base for Extension: Final Report* (Washington, DC: U.S. Department of Agriculture Extension Service, March 1985).

National Agricultural Research and Extension User's Advisory Board, *Agricultural Research, Extension, and Education Recommendations* (Washington, DC: U.S. Department of Agriculture, July 1983).

OTA, Office of Technology Assessment, *An Assessment of the United States Food and Agricultural Research System,* OTA-F-272 (Washington, DC: U.S. Government Printing Office, December 1981).

OTA, Office of Technology Assessment, *Commercial Biotechnology: An International Analysis,* OTA-BA-218 (Washington, DC: U.S. Government Printing Office, January 1984).

Ruttan, Vernon, *Agricultural Research Policy* (Minneapolis: University of Minnesota Press, 1982).

University of California, *Cooperative Extension Long Range Planning Statement,* Berkeley, CA, August 1982.

U.S. Department of Agriculture, *Developments in Marketing Spreads for Food Production in 1979, AER 449* (Washington, DC: U.S. Department of Agriculture, 1980).

Warner, Paul D., and Christenson, James A., *The Cooperative Extension Service: A National Assessment* (Boulder, CO: Westview Press, 1984).

White, Fred C., "Economic Impact of Agricultural Research and Extension," paper prepared for the Office of Technology Assessment, Washington, DC, 1984.

Part IV

Implications and Policy Options for Agriculture

Chapter 13

Implications and Policy Options for Agriculture

Contents

Chapter 13

Implications and Policy Options for Agriculture

The biotechnology and information technology revolution has been fostered by substantially expanded private sector investment in agricultural research, an investment complemented by increased public sector emphasis on basic research. The output of this revolution is in its infancy today but can be expected to blossom over at least the next 30 years.

The potential payoffs of this era include increased food production for domestic and export demand, a lower cost and more nutritious food supply, more agricultural exports, improved food quality, and reduced adverse environmental impacts. However, if current agricultural policies continue, this era also holds the potential for marked changes in the structure of agriculture and rural communities, including the demise of many small and moderate-size farms, increased centralization and integration of farm production, and the degradation of many rural communities.

THREE FARM CLASSES

Major structural change in agriculture has already begun. Based on a continuation of current policies, past trends, and future technological expectations, the likely net result of this structural change would be the development of a farm structure composed of three predominant agricultural classes:

1. The *large-scale farm segment* will be composed of a relatively small number of farms that will produce the bulk of the production. By year 2000 there could be as few as 50,000 large-scale farms producing as much as three-fourths of the agricultural production. Some of these large-scale farms will be owned by agribusiness corporations and some will not. This large-scale farm segment will be highly efficient in production, marketing, financial, and business management functions. The farms will be run by full-time, highly educated business managers. Most of their land may be rented. These managers will probably know their chances of making a profit even before planting or breeding.

2. The *struggling moderate-size farm segment* will strive to find a niche in the market and to survive in an industrialized agricultural setting. The difficulty for the moderate-size farm to find that niche is rapidly becoming the center of the farm policy debate. Traditionally highly productive, efficient, moderate-size, full-time farms have been referred to as the "backbone" of American agriculture. It is still true that a moderate, technologically up-to-date, and well-managed farm with good yields is highly resilient. One key to their success clearly lies in the management factor. But more often than not, management has to be willing to accept a relatively low return on invested capital, time, and effort. With ever-increasing educational requirements in farming, there will likely be less willingness by successful managers of moderate-size farms to accept a lower return for their services and for invested capital. Another key to their survival lies in access to state-of-the-art technologies at competitive prices. Cooperatives have traditionally performed that role. But today, cooperatives are generally not conducting or funding basic or applied research in biotechnology and information technology. Like their predominantly moderate-size farmer members, cooperatives, too, have encountered financial difficulty.

3. The *small, predominantly part-time farm segment* tends to obtain most of its net income from off-farm sources. However, this segment is highly diverse.[1] It includes wealthy urban investors and professionals who use agriculture primarily as a tax shelter and/or country home. It includes the would-be moderate farmers who are attempting to use off-farm income as a means of entering agriculture on a full-time basis. This modern version of the old farmhand-to-tenant-to-owner agricultural ladder is also fading. Finally, this segment also includes a number of poor, essentially subsistence, farmers, vestiges of the "war on poverty" from the 1960s. These farmers remain a significant social concern that needs to be dealt with from a policy perspective, although traditional farm price and income policy hold no hope for solving their problems.

[1]No analysis exists that accurately measures the diversity of this farm segment.

CONSEQUENCES OF CONTINUING CURRENT POLICIES AND PROGRAMS VERSUS NO PROGRAMS

Today's farm structure is partly the product of past policies and programs and partly the product of technology. Since the 1930s, farm program benefits have been allocated on the basis of cost of production. In the late 1960s the conversion of farm programs from supporting farm prices to supporting farm income resulted in the imposition of limits on the amount of payments a person involved in farming could receive. These payment limits proved largely ineffective at stemming the flow of benefits to large farms. Likewise, large farms have benefited disproportionately from other programs such as economic emergency credit and soil conservation. Large farms have been in the best position to take advantage of new technologies derived from the public sector agricultural research and extension system. If current farm policies and programs continue, the number of large farms will continue to grow and reap the majority of program benefits.

Without substantial changes in the nature and objectives of farm policy, the three classes of farms will soon become two—the moderate-size farm largely will be eliminated as a viable force in American agriculture. In addition, the problems of the small subsistence farm will continue to fester as an unaddressed social concern.

As this structural change occurs, the face of rural America will change. Large farms naturally tend to concentrate their activities in larger communities. Moderate-size and small rural communities inevitably die as the business conducted by farm implement, fertilizer, and chemical dealers as well as agricultural bankers declines. As a consequence, the rural-community tax base is eroded as business activity, employment, and property values decline. Children are bused longer distances to schools. The economic and social fabric of rural America erodes.

It is still unclear as to what the consequences of this change may be, because the vast majority of Americans have little or nothing to do with agriculture other than consume its products. Clearly, an increase in rural unemployment results in an increase in costly Government social programs. The uncertainty, however, arises over food production efficiencies and costs. This study shows that large farms can indeed produce at lower cost than smaller farms. The question is whether the only way this lower cost can be achieved is through scale of operation. Can a moderate farm with adequate educational advice and assistance from existing institutions achieve the same low production cost without creating the adverse rural community economic and social consequences that are a result of current farm policies and trends? The answer is not clear.

In the much longer run an agriculture dominated by a few corporate giants may not be desirable from a general public, taxpayer, or consumer perspective. A progressive, decentralized, competitive structure would be preferable. The task facing policymakers is to foster such a structure.

The results of this study clearly demonstrate that these adjustment problems would not be solved by a quick transition to no Government involvement in agriculture. In fact, the adjustment problems at the farm and rural-community level would be aggravated further by additional farm and rural agribusiness financial failures. While reduced Government involvement in agriculture may be a desirable long-term goal, longer term transition policies and programs are clearly required. Indeed, every industrialized nation manages their agricultural sector to some degree—none are free of Government intervention. In fact, the U.S. agriculture economy is managed less than most other industrialized nations.

The remainder of this chapter sets forth the policy changes that would be required if Congress and the related body politic decided that overt steps should be taken to foster a diverse, decentralized structure of farming where all sizes of farms had an opportunity to compete and to survive in a time of rapidly changing technology. It should be noted that the objective of giving every farm the opportunity to compete and survive does not imply an unchanging and stagnant farm structure. It does imply a political and social sensitivity to both the impact of current farm programs on farm structure and to the different needs of large, moderate, and small farms for Government assistance. It can be expected that regardless of what Government does, fewer commercial farms will exist in year 2000 than today. But at a minimum, Government can do much to ease the pain of adjustment.[2]

[2]The policy options presented are not all inclusive; e.g., international trade dimensions of agricultural policy are not covered in detail.

REQUIRED POLICY ADJUSTMENTS

Previous attempts to deal with the agricultural structure issue have been limited to actions such as limiting direct income support payments to some fixed amount per farmer, like the $50,000 cap in present programs. Such marginal policy changes, though thought to be beneficial, are not discrete enough to separate or distinguish between the different farm segments effectively.

More substantive changes in policy direction are required for addressing the structure issue. Specifically, separate policies and programs need to be pursued with respect to each of the three farm segments—large farms, moderate farms, and small farms. The choice of any one set of policies would imply that Congress desired to selectively enhance the status of one farm segment.

Policy for all farmers implies two basic policy goals:

1. All farmers need to operate in a relatively stable economic environment where they have an opportunity to sell what they produce. Restrictive trade policies or misguided macroeconomic policies impede this basic goal.
2. All farmers need a base of public research and extension support whereby they can maintain their competitiveness in the markets in which they deal. A loss of U.S. comparative advantage in the world agricultural product market would be a serious blow to the American economy. Similarly, a loss of consumer confidence in the ability of the food system to produce a safe and nutritious food supply efficiently would undermine public support for all of agriculture.

Policy for large farms need address only these two goals. Policies for moderate and small farms must address these same goals plus additional problems now facing these farm segments.

Policy for Large Commercial Farms

A basic conclusion of this study is that large-scale farmers do not need direct Government payments and/or subsidies to compete and survive. However, there is still a need for a commercial farm policy.

Criteria for determining what constitutes a large-scale farm are important but somewhat arbitrary. The dividing line developed from this study is about $250,000 in sales for a crop or dairy farm unit under single ownership or control. This level of sales is generally required to achieve most economies of size.[3] Over time, this optimum size has had, and will continue to have, a tendency to increase. As this occurs, criteria for limiting program benefits according to farm size will likewise have to increase.

Creating a Stable Economic Environment

The policy goal of creating a relatively stable economic environment where farmers have an opportunity to sell what they produce implies the following major farm program initiatives:

- Direct Government payments would be eliminated to all farms having over $250,000 in annual sales. This implies the elimination of the target-price concept, at least for this sales class. Elimination of payments to these farms would significantly reduce Government expenditures in agriculture.
- The nonrecourse loan would be converted to a recourse loan. The nonrecourse feature has resulted in the accumulation of large Government commodity stocks. The recourse feature would provide a continuing base of support for the orderly marketing of farm products. It would encourage year-long producer marketings inasmuch as farmers could not avoid interest payments by forfeiting commodities to the Government.

[3]The $250,000 figure is based on census data and the economies of size analysis discussed in previous chapters.

- Government credit to farms having over $250,000 in sales would not be available, except for the recourse price support loan.
- An expanded international development assistance program would be established. Such a program would have to include an optimum balance of commodity aid and economic development aid. Its primary objective would be to help developing countries reach the takeoff phase of economic growth, and thus become better future customers of American agriculture.
- A balanced macroeconomic policy that facilitates growth of export markets and maintains a relatively low real rate of interest would have to be maintained. Reduced deficits, combined with more expansionary monetary policies, would have the effect of expanding the growth of agricultural export markets and would result in reduced interest payments on the record agricultural debt.

Maintaining Technological Competitiveness

The technological competitiveness of American farmers would be assured by continuing a policy that encourages public and private investment in agricultural research. The major thrust of the research and extension programs as they affect large farms would be as follows:

- The trend toward increased public sector emphasis on basic research would be continued. Increased reliance would be placed on the private sector for applied research in the development of new products.
- While the public sector would emphasize basic research, an important problem-solving component would be maintained to adapt new technologies to various agro-ecosystems and to maintain newly achieved productivity from pests and disease, decline in soil fertility, and other factors.
- Extension's role in the direct education of, or consultation with, large farmers would be deemphasized. Private consultants would play an increasing role in technology transfer to the large farm segment.

Policy for Moderate-Size Farms

Policy for moderate-size farms must include not only the elements of policy postulated for large farms, but also additional elements that are specific to the more complex needs of this farm segment. For example, OTA finds that moderate farms having $100,000 to $250,000 in gross sales face major problems of competing and surviving in the biotechnology and information technology era. Some moderate farms will survive and some will not. This latter group should be assisted in their move to other occupations.

The following are specific policy goals for moderate-size farms:

- The risk of moderate farms operating in an open market environment needs to be reduced.
- New and easily adopted technologies should be made available to moderate farms.
- Opportunities for employment outside agriculture should be created for those farmers who are unable to compete.

Diligent enforcement would be needed to assure that the benefits of programs established to assist moderate farms are limited just to those farms.

Risks to Moderate-Size Farms

The most difficult obstacle to survival facing the moderate farm is that of managing risk. The initiation of market-oriented farm policies in the early 1970s greatly increased the amount of price and income risk facing the moderate farm. Large farms are better able to manage risk generally because of the higher level of their management's formal training and because of their greater diversification. The potential advantages of diversification by moderate farms commonly are offset by diseconomies associated with smaller scale, multiple enterprises. Similarly, managers of moderate farms often lack the skills associated with operating in the futures market or understanding various forms of contracting.

Three possible options exist for reducing the risks confronting moderate farms. One involves offering moderate farms a higher level of price and/or income protection than would be available to large farms. It may be argued that such policies foster inefficiency, but this may be a price that must be paid to maintain a decentralized agriculture. The three options are:

1. Income protection could be provided through either a continuation of the current target-price concept for moderate farms only or through a device known as the marketing loan. Like the current nonrecourse loan, the marketing loan is a loan from the Government on commodities in storage. If the commodity is sold for less than the loan value, the farmer pays back only those receipts to the Government in full payment of the loan. The marketing loan, in essence, becomes a guaranteed price to the producer. The level of the marketing loan should be no greater than the average cost of production for moderate farmers.

2. The nonrecourse loan concept could be continued for moderate farms. However, the level of the nonrecourse loan should not be set any higher than the recourse loan suggested previously for large farms; otherwise, the Government could end up acquiring most of the production from moderate farms.

3. The public sector could provide significantly increased assistance as a means of reducing risk to moderate farms. Such assistance could be in the form of, for example, educational programs on risk management, futures markets, contracting, and cooperative marketing. In addition, special assistance could be provided for cooperatives that offer marketing and pooling programs designed to reduce risk. While such programs might also benefit large farms, cooperatives have tended to be institutions used primarily by moderate and small farms.

Technology Availability and Transfer to Moderate Farms

OTA finds that agricultural research generally is not inherently biased against moderate farms. Rather, moderate farms may be seriously disadvantaged either by lags in adoption or by lack of access to competitive markets for the products produced by new technology. The following initiatives could help minimize such problems of technology availability and transfer:

- Extension's evaluation of the increasing number of new products entering the market would be extended. This increased effort would play a dual role of providing a check on the efficacy and the efficiency of new products of biotechnology and information technology, and would eliminate the costs associated with individual farmer experimentation with them. These test results would be available to all farms, regardless of size.
- Extension technology transfer services would be specifically aimed at moderate farms. The primary goal of such programs would be to make technologies available to moderate farms on the same schedule as large farms. Farming systems encompassing new technologies would have to be adapted specifically to moderate farm needs and made available through extension programs. Where this requires special research initiatives, the U.S. Department of Agriculture (USDA) and the Experiment Stations would provide the support. In States where technological change threatens to displace large numbers of moderate farms, such as in Midwest dairying, special initiatives by State and local governments to support research and extension would also be warranted.
- The development of cooperatives that emphasize technology supply and transfer services to moderate farms would have to be undertaken. Unlike private sector agribusiness firms, cooperatives do not appear to conduct or fund any aspect of biotechnology and information technology research. Current financial stress in the cooperative sector suggests that this sector may not be able to marshal the capital needed for such research. At a minimum, there seems to be a need for cooperatives to have a strong applied and developmental program of research in biotechnology and information technology buttressed by land-grant university basic research. To achieve such a research objective cooperatives should consider carefully the formation of a research agency in common (RAC). USDA or land-grant university research along with RAC could receive special public sector Federal and State appropriations and support. Formal links might be encouraged between research, extension, and cooperative institutions to maximize the effectiveness of technology transfer to cooperatives and their moderate-size farm members.
- Ample credit would have to be made available to moderate farms that have the potential to survive. Government credit, in concert with cooperative credit, should be aimed specifically at filling the needs of moderate farms. Emphasis should be placed on credit required to keep moderate farms technologically up-to-date.

Transition Policy to Other Agricultural Enterprises or Nonfarm Employment

Regardless of the effectiveness of the initiatives discussed above, there will be an accelerated need to move farm families either to other agricultural enterprises or out of agriculture into new occupations. The need arises, therefore, for specific public action to facilitate adjustment of resources from the current farm operation into gainful, productive employment elsewhere. Adjustments in rural community business activity and social service will be directly affected by such changes. (The specific nature of these adjustments and potential public policy considerations are treated in greater detail later in this chapter.) Specific initiatives to ease this structural adjustment process include the following:

- As a continuously evolving industry, new opportunities for employment of displaced farmers need to be explored and developed within agriculture. Aquaculture, for example, is becoming an important and viable agricultural industry. A more urbanized society has resulted in the growth of a large horticulture and nursery industry. Changing population demographics, particularly in terms of aging, suggest marked increases in the demand for fruits and vegetables. Land-grant universities and the Extension Service bear an important responsibility for fostering the growth of these industries through education and training. Displaced farmers, having prior experience in agriculture, are logical clientele for such education and training activities.

- Special skills training programs aimed at those areas where significant employment opportunities exist need to be designed to assist with the transition to nonfarm jobs. Jobs in rapidly growing service, health care, or care-for-the aged industries provide contemporary examples.
- Financial assistance from Federal, State, and local governments, similar to the famous G.I. bill, might be established to assist displaced farmers or rural residents during the period of transition while they receive skills training. For example, the Federal Job Training Partnership Act Title III program is a federally funded, State-administered program that assists displaced workers in obtaining vocational retraining and counseling. Such a program could be made available to displaced farmers.
- In areas of severe financial stress, assistance may be provided in the form of Government purchase of land or production rights from displaced farmers at its "long-term fair market value." The returns from the land could be used by the displaced farmer for relocation and retraining. The Government could retain the land in conservation reserve status until it might be needed for future production.
- An alternative program to ease the transition for farmers leaving agriculture is a self-financed agricultural transition loan. Its objective would be to allow a farmer to leave agriculture without having to worry about generating the funds needed to live on while seeking new employment. Such a program could involve the following: 1) farmer terminates the farm operation and becomes eligible for a Federal or State guaranteed living loan, 2) farmer liquidates the farm business over time and ultimately finds other employment, and 3) farmer uses the net proceeds from liquidation and earnings from new job to repay the loan.

Policy for Small/Part-Time Farms

Policy for small/part-time farms includes the elements of policy for large farms plus additional elements.

With few exceptions, small farms having less than $100,000 in sales are not viable economic entities in the mainstream of commercial agriculture—nor can they be made so. However, even a small increase in their farm income could have a significant multiplier effect on the local economy because of the large number of small farms. These farms survive because their operators have substantial outside income (part-time farmers), or because they have found themselves a niche in marketing a unique product with special services attached (often direct to consumers), and/or because they are willing to accept a very low return on resources contributed to the farming operation.

The Government's role would be severely restricted for the small farms who either have substantial outside income or who have found a niche in the market. They are as much able to take care of themselves as large farms are.

However, subsistence farmers who have limited resources, and often limited technical abilities, represent a genuine problem for which public concern is warranted—these indeed are the rural people left behind. Commercial farm programs have done and can do little to solve their problems. These impoverished individuals are a social and economic problem for which only social programs can help. However, while programs such as food stamps, social security, and aid to families with dependent children are important to many subsistence farmers, these programs do not serve the farmers' unique agricultural and related needs. The following suggestions are made for dealing with the problems of subsistence farmers:

- Initiate a special study to identify these individuals and their specific status and needs. Develop social programs to meet those needs.
- USDA and the land-grant universities bear a special burden of responsibility for serving the needs of these subsistence farmers. This responsibility has not generally been realized and, therefore, has not been fulfilled. In the South, this responsibility falls particularly on the 1890 land-grant universities along with the statewide extension education programs and the 1862 land-grant universities. In the North, the responsibility for serving the agricultural educational and research needs of subsistence farmers falls exclusively on the 1862 land-grant universities.
- USDA and the land-grant universities could be directed to develop a joint plan for serving the agricultural research and educational needs of these farmers. Such a plan should include the delivery of farming, credit, and marketing systems designed to maximize the small farms' agricultural production and earning capacity.
- Farming systems must be developed specifically to serve the needs of small subsistence farms. Such systems should, to the extent practicable, encompass the use of new technologies. Special USDA and land-grant research program components must be designed specifically to develop and/or modify technology for use by small subsistence farms.
- Credit delivery systems for small subsistence farmers must be specifically developed by USDA through the Farmers Home Administration. Such systems should consider the unique capital and cash flow limiting factors associated with subsistence farmers who commonly are not in a position to take advantage of other farm programs such as price and income supports.
- Marketing programs geared to subsistence agriculture are essential for providing hope for this farm segment. The difficulty lies in the inability of these farmers to obtain access to the mass markets through which most agricultural production moves. Cooperatives and direct marketing to consumers offer two potentially viable alternatives. USDA and the Extension Service should play a critical role in assisting in the establishment of such markets.

Policy for Rural Communities

The impact of adjustment in agriculture to changing technology will by no means be limited to the farm sector. Rural communities will be at least equally affected by increasing farm size, integration, and moderate-size farm displacement. Although these effects will initially be felt by implement dealers, farm supply and marketing firms, or bankers, the reverberations will extend throughout the community in terms of employment levels, tax receipts, and required services. Rural communities should be assessing these impacts and preparing to make needed adjustments. To ease the pain of adjustment the following actions are suggested:

- Comprehensive programs for community redevelopment and change need to be initiated throughout rural America. Such development plans should be fostered and facilitated by both Federal and State government agencies. Rural community development research and extension programs must be revitalized to serve the needs of communities in transition.
- Increased employment opportunities in rural areas should be fostered by aggressively attracting new business activities to rural communities. Particular emphasis would be placed on attracting those businesses that develop technologies and serve the needs of high-technology agriculture in rural areas.
- Rural communities should be assisted in developing and modernizing the infrastructure needed to be a socially and economically attractive place to live. Some rural communities can serve as an attractive retirement residence for an aging popula-

tion. But a higher level of social services would clearly be required.

- To attract new industry to these areas, rural communities need to play a vital role in skills training for displaced farmers and rural community employees. School and university outreach programs can be modified to serve this important role.

Policy for Technology and Environmental Resource Adjustment

Technological change inherently creates a disruption or imbalance in the allocation of resources. Much of this study has been devoted to analyzing these effects. Some may question whether this degree of change is either necessary or desirable.

One of the major reasons that American agriculture has been so productive is because technological change has been fostered by the public sector and nurtured by a profit-seeking private sector. Consequently, American consumers have enjoyed a plentiful supply of low-cost food and natural fiber. In addition, agricultural exports have made a major contribution to the overall development of export markets, to the benefit of the general economy. Biotechnology and information technology offer more of the same, with the added bonus of using less chemicals in the production of food—whether for the control of pests, disease, and weeds or for the production of commercial fertilizer.

Maintaining the productivity and competitiveness of U.S. agriculture in the public interest requires a delicate balance between public and private sector support for technological change. Yet it would be wrong to imply that there are no risks. The conferring of property rights on discoveries of the agricultural research system has shifted the agricultural research balance to the private sector. While the effects of this shift appear to be positive, concerns exist that a substantial portion of the benefits of even public research could be captured by private firm interests. In addition, no scientifically acceptable methodology exists for weighing the risks or hazards of biotechnology research. To deal with such issues, the following policy suggestions are made:

- Steps should be taken to secure the public interest social contract on which the USDA and land-grant university agricultural research system has been based. Assurance needs to be provided that the benefits of publicly supported research and extension are not inappropriately captured in the form of private monopoly rents. The effect would be to stifle the process of discovery and the dissemination of new knowledge.
- Major investments need to be made to foster the development of human capital that is in a position to cope with the process of rapidly changing agricultural technology. This need extends from the training and development of the most basic biological research scientists, through the extension specialist and county agent, to the farmer who adopts the new technology and the banker who supplies the loan for its purchase. At a time when agriculture is in a low-income crisis state, there may be a tendency not to make such investments in the future. Such a strategy would clearly be counterproductive.
- Biotechnology is not likely to replace land and water as vital agricultural resources. In recent years, soil conservation has taken a back seat from a policy perspective to full-production policies. Such a strategy would appear to be very short-sighted. Likewise, the inability of policymakers to establish a national water policy runs counter to maintaining the competitive edge of U.S. agriculture internationally.
- Little is known about the adverse impacts of potential biotechnology developments on the ecosystem. These risks must be carefully assessed, monitored, and, where necessary, regulated. Care must be taken, as well, not to overregulate and thereby stifle the potential competitiveness and productivity of U.S. agriculture.

CONCLUSIONS

While the biotechnology and information technology revolution will create many adjustment problems, it has the potential for creating benefits in a safer, less expensive, more stable, and more nutritious food supply. The substantial costs of these improvements to farming and rural communities can be minimized by careful policy analysis, planning, and implementation. This study is only the first step in that direction.

Appendixes

Appendix A

Animal and Plant Technology Workshop Methodology and Procedures

To assess the impacts of emerging agricultural production technologies, two workshops—one for animal technology and the other for plant, soil, and water technology—were conducted in April 1984. The objective of the workshops was to obtain information about the development and adoption of emerging technologies so that the information could be used to analyze the economic, social, and environmental impacts of technology adoption.

Since the information needed spanned a wide range on the spectrum of the process of technological innovation—from successful completion of research to widespread commercialization of the technology—participants of the workshops were carefully selected to include expertise in different stages of technological innovation. Participants comprised physical and biological scientists, engineers, economists, extension specialists, agribusiness representatives, and experienced farmers.

The Delphi technique was used to obtain collective judgments from the workshop participants. To facilitate the process of obtaining consensus, an electronic Consensor was employed to tabulate the ratings assigned by each expert. In addition to registering the ratings, the device allowed each expert to weight his rating according to the degree of confidence or expertise he had in his rating. That level of confidence or expertise could be set at zero, 25, 50, 75, or 100 percent.

The Consensor provided an immediate video screen readout of the rating distribution, the weighted average rating, and the average degree of confidence. If the first vote showed a very wide distribution on ratings, those experts with ratings that were outliers were asked to explain their reasons for the ratings assigned. After additional discussion, another vote was taken. Since lack of a consensus after such discussion was, in itself, an indication of considerable uncertainty about the impacts of new technology, no attempt was made to force a consensus beyond a second vote.

The principal tasks accomplished at the workshops were:

a. Estimation of the year that each technology was likely to be introduced for commercial adoption.
b. Estimation of the yield trends for each commodity in 1990 and 2000 under the no-new-technology environment.[1]
c. Packaging of technologies that are likely to be introduced in the production of each commodity in 1990 and 2000.
d. Estimation of the increases in crop and livestock performance measures when the package of technologies is fully adopted by farmers.
e. Estimation of the adoption profile (i.e., the number of years it takes to reach a certain percentage of adoption and the maximum percentage of adoption) of each package of technologies applied to a particular commodity.
f. Discussion of major barriers to the adoption of a particular package of technologies to the production of each commodity.
g. Identification of public policy options that could remove the barriers or facilitate adoption of the packages of technologies.

Information obtained from this workshop was used to assess the economic, social, and environmental impacts of these technologies. Iowa State's CARD econometric and hybrid models were used to simulate the impacts of these emerging technologies on plant and animal production, inventory, demand, supply, prices, gross farm income, production expenses, and net farm income in 1990 and 2000 under alternative technology environments.

Alternative Environments for the Development and Adoption of Technology

Since the information to be obtained at the workshops depended on the environment under which a new technology would be developed and adopted, it was necessary to make certain assumptions about future environments, or scenarios. Four technology environments were developed and used in the workshops: most likely, more-new-technology, less-new-technology, and no-new-technology environments.

[1]Alternative technology environments will be discussed in the next section.

The most likely environment is bordered by the more-new-technology and the less-new-technology environments; both deviate from the position of the most likely environment. It is assumed in the most likely environment that the historical trends will continue into the future. Forces, such as gross national product (GNP), population growth, export demand for U.S. agricultural commodities, trade policy, inflation rates, energy prices, and research and extension expenditures, that have shaped the past would continue to evolve as they had in the past decades. Assumptions made for various economic variables under the three environments are shown in table A-1. Factors underlying the more-new-technology environment are generally more favorable for development and adoption of new technologies than those under the most likely environment, and factors underlying the less-new-technology environment are less favorable than those under the most likely environment.

The assumptions under the no-new-technology environment are the same as that of the less-new-technology environment except for new technologies. It is assumed that all emerging technologies discussed in the two workshops will not be available for commercial introduction before year 2000. Existing technologies will continue to be used. Through education and extension, farmers will learn to use the existing technologies better to increase productivity. Productivity is likely to continue to increase at a decreasing rate and will eventually level off.

Presentation and Discussion of Technologies

In a plenary session, each author of a technology paper made a 10-minute presentation to give the participants essential information about a technology area so that they would be able to make intelligent projections about the development and adoption of that area. The authors' presentations focused on the following:

a. When would a significant technology emerge from each major line of research?
b. What is the output of the new technology? Is it a new product (e.g., a new vaccine for a particular disease) or a new process to produce the same product (e.g., no till)?
c. How will each technology be used by farmers? Can it be used alone or in combination with other technologies? If it has to be combined with other technologies, how will they be packaged?
d. What will it take for farmers to adopt it? Do they have to make a capital investment, such as the purchase of new chemicals, instruments, equipment, or machinery?
e. What specific crops or livestock would be affected by adoption of a specific package of technologies?
f. How would the package of technologies affect the performance of crop and livestock production?
g. How would each package of technologies affect the quality of the environment and resource use?

Table A-1.—Alternative Technology Environments

Factors	More new technology	Most likely	Less new technology
Population growth rate:			
United States	1.0%	0.7%	0.5%
World	1.8	1.6	1.3
GNP growth rate:			
United States	4.0	3.4	3.0
World	5.0	3.5	2.0
Trade policy	Less protectionist, more favorable terms of trade	Continuation of present trends	More protectionist, less favorable terms of trade
Rate of growth of export demand:			
Grain	1.8%	1.4%	0.8%
Oilseeds	2.3	1.8	1.2
Red meat	2.0	1.0	0.0
Energy price growth rate (constant dollars)	5.0	3.0	1.0
Growth rate of research and extension expenditures (constant dollars)	4.0	2.0	0.0
Inflation rate	8.0	5.0	3.0

SOURCE: Office of Technology Assessment.

A checklist was given on the above information to the participants. Based on information obtained from the presentation and on interactions with the authors, the participants collectively packaged the technologies and estimated the impacts and the adoption profile for each package of technologies (see section "Packaging of Technologies").

Timing of Commercial Introduction

Since the impact of a new technology on agriculture at a given time depends on when the technology is introduced for commercial adoption, each author at the workshops was asked to make an initial estimate on the probable year of commercial introduction for each technology. Following each presentation, the entire group evaluated and discussed the author's initial estimate. The entire group then collectively estimated the year of commercial introduction of each technology under the three technology environments. Table A-2 shows the probable years of commercial introduction of animal technologies, and table A-3 shows the same for plant technologies under the three alternative environments.

The years of commercial introduction estimated ranged from the present or possibly earlier—for genetically engineered pharmaceutical products; control of infectious disease in animals; superovulation, embryo transfer, and embryo manipulation of cows; and control of plant growth and development—to 2000 and beyond for genetic engineering techniques for farm animals and cereal crops. Of the 57 potentially available animal technologies, 27 were estimated to be available for commercial introduction before 1990, and the other 30 between 1990 and 2000, under the most likely environment. In plant agriculture, 50 of the 90 technologies examined were projected to be available for commercial introduction by 1990, and the other 40 technologies between 1990 and 2000.

Packaging of Technologies

Since in practice most technologies would be used in combination with other technologies, the 28 areas of technologies were grouped into packages according to their probable impacts on particular commodities under different technology environments. Table A-4 shows different packages of technologies used in producing different commodities under the three alternative environments. In beef production, for example, 12 animal technology areas were grouped into six packages. Since more new technologies would be available for commercial adoption in later years than earliers years, each package of technologies was further categorized as a 1990 package and a 2000 package. For example, package 1990A would include all genetic engineering technologies introduced commercially by 1990, and package 2000A would include all genetic engineering technologies introduced commercially by 2000, including all package 1990A technologies. Thus, there are really a total of 12 packages of technologies for beef production.

Performance Estimates Under the No-New-Technology Environment

To estimate the net impact of emerging technologies on agricultural production, the participants of the workshops were first asked to project the performance measures of crop and livestock production, such as crop yields and livestock feed efficiency, to 1990 and 2000 under the no-new-technology economic environment. Historical trend lines of performance measures of crop and livestock production were provided to the participants as a basis for their discussion. Through the Delphi process, participants collectively projected the performance measures for each of the nine commodities for 1990 and 2000 assuming that all emerging technologies identified and discussed in this study would not be available for commercial adoption by 2000. The performance measures used in this study were as follows:

- Wheat: bushels per acre, percent of planted acreage harvested.
- Corn: bushels per acre, percent of planted acreage harvested.
- Soybeans: bushels per acre, percent of planted acreage harvested.
- Cotton: pounds per acre, percent of planted acreage harvested.
- Rice: bushels per acre, percent of planted acreage harvested.
- Beef: pounds of meat produced per pound of feed, calves per cow per year.
- Swine: pounds of meat produced per pound of feed, pigs per sow per year.
- Dairy: pounds of milk produced per pound of feed, pounds of milk produced per cow per year.
- Poultry: pounds of poultry produced per pound of feed, eggs per layer per year.

The results of the estimates are shown in table A-5. If all the new technologies identified in this study do not become available for commercial adoption

Table A-2.—Timing of Commercial Introduction of Animal Technologies

	Technology environments		
Technology	More new technology	Most likely	Less new technology
Genetic engineering:			
Production of pharmaceuticals	1982	1982	1982
Control of infectious diseases	1983	1983	1983
Improvements in animal production	1990	2000	>2000*
Detection and treatment of genetic abnormalities:			
Detection	1990	1995	>2000
Treatment	1990	2000	>2000
Control of cancer and leukemia	1990	1990	>2000
Animal production:			
Cycle regulation	1985	1989	1995
Superovulation, embryo transfer, and embryo manipulations	1983	1983	1983
Improvement of fertility	1990	1995	1995
Genetic engineering techniques for farm animals	1995	2000	>2000
Regulation of growth and development:			
Muscle and adipose tissue accretion	1987	1992	>2000
Hormone, serum, and tissue factors important to growth	1995	2000	>2000
Immunological attraction of animals	1990	1995	>2000
Measuring body composition and animal identification	1990	1995	>2000
Animal nutrition:			
Animal product consumption and human health	1995	2000	>2000
Alimentary tract microbiology and digestive physiology	1989	2000	>2000
Voluntary feed intake and efficiency of animal production	1989	1995	>2000
Maternal nutrition and progeny development	1984	1984	1984
Aquaculture	1984	1984	1984
Livestock pest control:			
Slow-release insecticides	1984	1984	1984
Vaccines	1986	1986	1991
Integrated systems	1987	1989	1994
Modification of insect habitat	2000	2000	2000
Insect-resistant animals	2000	2000	2000
Utilizing immunity systems	1990	1990	1995
Disease control:			
Data managment and systems analysis	1980	1980	1980
Diagnostic methodologies	1986	1986	1988
Selection for disease resistance	1994	1999	>2000
Genetic engineering of micro-organisms and embryos:			
Embryos	1995	1999	>2000
Micro-organism	1988	1989	1999
Immunobiology	1983	1983	1983
Environment and animal behavior:			
Energy conservation			
Non-integrated system	1985	1990	2000
Integrated system	1995	2000	>2000
Optimizing total stress	1995	2000	>2000
Stress and immunity	1995	2000	>2000
Photoregulation of physiological phenomena	1990	1990	>2000
Utilization of crop residues and animal wastes:			
Energy from manure	1985	1985	1985
Chemicals from crop residues	1990	1990	>2000
Animal feed from crop residue	1990	1990	>2000
Animal feed from manure	1990	1995	>2000
Monitoring and control technologies:			
Sensors	1985	1985	1985
Controllers	1985	1985	1985
Displayers	1985	1985	1985
Actuators	1985	1985	1985
Communication and information management:			
Local communication networks	1985	1985	1985
Data terminals	1985	1985	1985
Software and database systems	1985	1985	1985
Manufacturing management systems	1987	1990	2000
Expert systems	1992	1995	2000
Telecommunications:			
Digital communication	1990	2000	>2000
Fiber optics	1990	2000	>2000
Personal computers	1985	1985	1985
Videotex and teletext	1985	1985	1985
Value-added networks	1985	1985	1985
Integrated services digital network	1987	1990	2000
Remote sensing	1985	1985	1985
Labor-saving technologies:			
Robotic farming	1995	2000	>2000

*> = After.

SOURCE: Office of Technology Assessment.

by 2000, the performance of crops and livestock could continue to improve through 2000 (but at slower rates), primarily because of better applications of existing technologies through education and extension. For example, corn yields are projected to increase from 115 bushels per acre in 1982 to 117 bushels per acre in 1990 and 124 bushels per acres in 2000. Wheat yields are projected to increase from 36 bushels per acre in 1982 to 38 bushels per acre in 1990 and 41 bushels per acre in 2000. And milk production could increase from 12,300 pounds per cow per year in 1982 to 13,700 pounds in 1990 and 15,700 pounds in 2000.

Table A-3.—Timing of Commercial Introduction of Plant Technologies

Technology	Technology environments		
	More new technology	Most likely	Less new technology
Genetic engineering:			
Microbial inoculums	1990	1990	Never
Plant propagation	1983-90	1983-90	>1990*
Genetically engineered plants:			
Vegetable	1990	1990	1995
Soybeans/cotton	1990	1995	2000
Cereals	1995	2000	2010
Enhancement of photosynthetic efficiency:			
Basic process of photosynthesis	1983	1983	1983
Photosynthetic control by internal and external factors	1983-90	1983-90	1983-2000
Photosynthetic molecular biology and genetics	1990-2000	1990-2000	1990-2000
Estimation of photosynthesis and project management needs	1983-90	1983-90	1983-90
Mechanisms of response and adaptation to stress	1990	1983-95	2000
Plant growth regulators:			
Controlling growth and development	1984	1984	1985
Resistance to disease and insect pests	1986	1988	1990
Overcoming environmental stresses	1986	1988	1990
Postharvest preservation	1985	1986	1990
Plant disease and nematode control:			
Breed cultivators	1984	1984	1984
Genetic engineering	2000	2000	2025
Bacteriocides, fungicides, and nematicides	1988	1990	2000
Biocontrol agents	1985	1990	2010
Crop loss assessment	1985	1990	2000
Management of insects and mites:			
Chemical controls	>1995	2000	>2000
New chemicals	1984	1984	1984
Application technology	1988	1990	1995
Genetic engineering:			
Pathogenic chemicals	1995	2000	2005
Plants	2000	2005	2010
Information processing	1984	1984	1984
Weed control:			
Bioregulation through chemical and biological technology	1984-2000	1984-2000	1984-2000
Allelopathic chemicals as bioregulators	1990	1995	2000
Crop tolerance and susceptibility to control agents	1992	1998	>2000
IWMS for conservation tillage and annual multicrop production	1984-2000	1984-2000	1984-2000
Biological nitrogen fixation:			
Improved strains of rhizobia	1984	1984	1984
Stress-tolerant rhizobia	1987	1990-95	1995-2000
Legumes more active in nitrogen fixation (plant breeding)	1990-95	1990-95	1990-95
Root zone of cereals	>2000	>2000	>2000
Nitrogen-fixing cereals	>2000	>2000	>2000
Chemical fertilizers:			
Increasing efficiency of nitrogen use	1990	1995	2000
Decreasing energy required	1980	1980	1980
Processing of lower quality phosphate rock into fertilizers	1990	1990	1990
Ammonia from coal	1995	2000	2000

Technology	Technology environments		
	More new technology	Most likely	Less new technology
Water and soil-water-plant relations:			
Understanding drought resistance/tolerance	2000	2020	2050
Plant breeding	1984	1984	1984
Biotechnology: recombinant DNA:			
Water use efficiency	2010	2030	2050
Water management	1984	1984	1984
Photovoltaic systems	1995	1995	2010
Soil erosion, productivity, and tillage:			
Conservation farming systems	1995	1995	1995
Assessing erosion and its impact	1995	1995	2000
Reclaiming lands	1995	1995	>2000
Use of public for soil conservation projects	1995	1995	1995
Multiple cropping:			
Breeding crops for intensive planting systems	1985	1985	1985
Double cropping/intercropping	1990	1985	1985
Competition by plant species for growth factors	1990	1995	1990**
Plant nutrition through fertilizers and microbiology	1995	2000	>2000
Mechanization for multiple cropping	1987	1990	1987**
Organic farming:			
Reduced use of inputs:			
Biocides	1984	1984	1984
Reduced soil erosion	1984	1984	1984
Self-sufficiency for nutrients	1984	1984	1984
Minimum tillage with minimal biocide use	1990	1990-95	2000
Rotations:			
Use	1984	1984	1984
Knowledge	1990	1990-95	2000
Labor-saving technologies:			
Mechanized fruit and vegetable operations	1985	1985	1985
Robotic farming:			
Fruit and vegetable	1995	2000	2010
Grains	1995	2000	2010
Crop separation, cleaning, and processing:			
New methods for separating and cleaning grain	1995	1995	1995
Infield or onfarm processing of forages and oilseeds:			
Vegetable	1984	1984	1984
Forage	1990	1990	2000
Oilseed	1984	1984	1984
Engine and fuels:			
Adiabatic compression ignition engines with turbocompounding	1990	1990	1990
Electronic engine controls	1985-86	1986	1986
Alternative fuels:			
Grains	1984	1984	1984
Cellulose	1995	2000	2010
Land management:			
Conservation tillage	1984	1984	1984
Controlled traffic farming	1987	1990	1995
Customed-prescribed tillage	2000	2005	2020
Multicropping	1984	1984	1984
Organic farming			

Table A-3.—Timing of Commercial Introduction of Plant Technologies—Continued

Technology	Technology environments: More new technology	Technology environments: Most likely	Technology environments: Less new technology
Communication and information management:			
Local communication networks	1985	1985	1985
Data terminals	1985	1985	1985
Software and database systems	1985	1985	1985
Manufacturing management systems	1987	1990	2000
Expert systems	1990	1992	1997
Monitoring and control:			
Sensors	1984	1984	1984
Controllers	1984	1984	1984
Displayers	1984	1984	1984
Actuators	1984	1984	1984

Technology	Technology environments: More new technology	Technology environments: Most likely	Technology environments: Less new technology
Telecommunications:			
Digital communication	1995	2000	2010
Fiber optics	1990	2000	2010
Personal computers	1985	1985	1985
Videotex and teletext	1985	1985	1985
Value-added networks	1985	1985	1985
Integrated services digital network	1990	1990	>2000
Remote sensing	1985	1985	1985

*> = After.

**May actually accelerate development in this area if there is increased interest in resource efficient/sustainable cropping systems.

SOURCE: Office of Technology Assessment.

Table A-4.—Packages of Technologies

Beef:	
Package A:	Genetic engineering
Package B:	Animal reproduction
	Regulation of growth and development
	Animal nutrition
	Crop residue and animal waste
Package C:	Pest control
	Disease control
Package D:	Environment and animal behavior
Package E:	Monitoring and control
	Communication and information management
	Telecommunications
Package F:	Labor saving
Swine:	
Package A:	Genetic engineering
	Animal reproduction
	Regulation of growth and development
	Animal nutrition
	Pest control
	Disease control
Package B:	Environment and animal behavior
	Monitoring and control
	Communication and information management
	Telecommunications
Package C:	Crop residue and animal waste
Package D:	Labor saving
Dairy:	
Package A:	Genetic engineering
	Animal reproduction
	Pest control
	Disease control
Package B:	Regulation of growth and development
	Animal nutrition
	Environment and animal behavior
	Crop residue and animal waste
Package C:	Monitoring and control
	Communication and information management
	Telecommunications
	Labor saving
Package D:	Bovine growth hormone
Poultry:	
Package A:	Genetic engineering
	Animal reproduction
	Regulation of growth and development
	Animal nutrition
Package B:	Pest control
	Disease control
	Environment and animal behavior
	Crop residue and animal waste
Package C:	Monitoring and control
	Communication and information management
	Telecommunications
	Labor saving
Wheat:	
Package A:	Plant growth regulators
	Plant disease and nematode control
	Management of insects and mites
	Weed control
	Chemical fertilizers
	Water and soil-water-plant relations
	Soil erosion, productivity, and tillage
	Multiple cropping
	Organic farming
	Land management
Package B:	Labor saving
	Crop separation, cleaning, and processing
	Engines and fuels
Package C:	Communication and information management
	Monitoring and control
	Telecommunications
Corn:	
Package A:	Genetic engineering
	Plant disease and nematode control
	Management of insects and mites
	Water and soil-water-plant relations
	Communication and information management
	Monitoring and control
	Telecommunications
Package B:	Weed control
	Chemical fertilizers
	Soil erosion, productivity, and tillage
	Multiple cropping
	Land management
Package C:	Organic farming
Soybean:	
Package A:	Genetic engineering
	Enhancement of photosynthetic efficiency
	Plant growth regulators
	Plant disease and nematode control
	Multiple cropping
Package B:	Management of insects and mites
	Weed control
	Biological nitrogen fixation
	Chemical fertilizers
	Water and soil-water-plant relations
	Soil erosion, productivity, and tillage
	Organic farming
	Labor saving
	Crop separation, cleaning, and processing
Package C:	Communication and information management
	Monitoring and control
	Telecommunications
Rice:	
Package A:	Genetic engineering
	Enhancement of photosynthetic efficiency
	Plant growth regulators
	Plant disease and nematode control
Package B:	Management of insects and mites
	Weed control

Table A-4.—Packages of Technologies—Continued

	Chemical fertilizers
	Water and soil-water-plant relations
	Multiple cropping
	Crop separation, cleaning, and procesing
	Communication and information management
	Monitoring and control
	Telecommunications
Cotton:	
Package A:	Genetic engineering
	Enhancement of photosynthetic efficiency
	Plant growth regulators
	Plant disease and nematode control
	Management of insects and mites
	Weed control
	Chemical fertilizers
	Water and soil-water-plant relations
	Soil erosion, productivity, and tillage
	Multiple cropping
	Labor saving
	Engines and fuels
	Land management
Package B:	Communication and information management
	Monitoring and control
	Telecommunications
Package C:	Biological nitrogen fixation
	Organic farming
	Crop separation, cleaning, and processing

SOURCE: Office of Technology Assessment.

Table A-5.—Performance Projections Under No-New-Technology Environment

Commodity	Unit	1990	2000
Beef	lb. meat per lb. feed	0.071	0.066
	calves per cow	0.940	0.950
Dairy	lb. milk per lb. feed	0.938	0.952
	milk per cow per year (thousand lb)	13.7	15.7
Poultry	lb. meat per lb. feed	0.52	0.53
	eggs per layer per year	255.0	260.0
Swine	lb. meat per lb. feed	0.167	0.17
	pigs per sow per year	14.8	15.7
Corn	bushels/acre	116.5	123.5
Cotton	pounds/acre	502.0	511.0
Rice	bushels/acre	108.6	111.9
Soybean	bushels/acre	32.2	34.8
Wheat	bushels/acre	37.8	40.8

SOURCE: Office of Technology Assessment.

Appendix B

U.S. Regional Agricultural Sales by Sales Class and Commodity

Table B-1.—Sales of Cash Grains by Sales Class and Region, 1982

Sales class definition	Northeast region	North Central region	Southern region	Western region	U.S. total
Small <$19.9K	$ 58,622	$ 768,433	$ 516,810	$ 92,769	$ 1,436,634
Part-time $20 to 99.9K	127,304	6,935,115	1,556,510	572,129	9,191,058
Moderate $100 to 249.9K	128,851	8,222,262	2,086,754	1,017,181	11,455,048
Large $250 to 500K	71,517	3,774,781	1,563,810	727,328	6,137,436
Very large >$500K	46,052	1,899,146	1,269,735	1,180,086	4,395,019
Total	432,346	21,599,737	6,993,619	3,589,493	32,615,195
		Percentage of total regional sales of cash grains, 1982			
Small <$19.9K	13.6%	3.6%	7.4%	2.6%	
Part-time $20 to 99.9K	29.4	32.1	22.3	15.9	
Moderate $100 to 249.9K	29.8	38.1	29.8	28.3	
Large $250 to 500K	16.5	17.5	22.4	20.3	
Very large >$500K	10.7	8.8	18.2	32.9	
Total	100.0	100.0	100.0	100.0	
		Percentage of total national sales of cash grains, 1982			
Small <$19.9K	0.2%	2.4%	1.6%	0.3%	4.4%
Part-time $20 to 99.9K	0.4	21.3	4.8	1.8	28.2
Moderate $100 to 249.9K	0.4	25.2	6.4	3.1	35.1
Large $250 to 500K	0.2	11.6	4.8	2.2	18.8
Very large >$500K	0.1	5.8	3.9	3.6	13.5
Total	1.3	66.2	21.4	11.0	100.0

NOTE: Moderate and large sales class intervals in this table differ from intervals used elsewhere in the report.

SOURCE: Office of Technology Assessment. Compiled from regional data provided by Bureau of the Census, Agriculture Division.

Table B-2.—Sales of Cattle and Calves by Sales Class and Region, 1982

Sales class definition	Northeast region	North Central region	Southern region	Western region	U.S. total
Small <$19.9K	$ 95,569	$ 860,155	$1,491,739	$ 328,404	$ 2,775,867
Part-time $20 to 99.9K	84,425	1,450,508	1,089,760	719,211	3,343,904
Moderate $100 to 249.9K	74,117	2,022,486	726,592	745,354	3,568,549
Large $250 to 500K	57,267	1,825,647	577,156	836,720	3,296,790
Very large >$500K	58,996	6,677,929	4,048,095	4,372,600	15,157,620
Total	370,374	12,836,725	7,933,342	7,002,289	28,142,730
	Percentage of total regional sales of cattle and calves, 1982				
Small <$19.9K	25.8%	6.7%	18.8%	4.7%	
Part-time $20 to 99.9K	22.8	11.3	13.7	10.3	
Moderate $100 to 249.9K	20.0	15.8	9.2	10.6	
Large $250 to 500K	15.5	14.2	7.3	11.9	
Very large >$500K	15.9	52.0	51.0	62.4	
Total	100.0	100.0	100.0	100.0	
	Percentage of total national sales of cattle and calves, 1982				
Small <$19.9K	0.3%	3.1%	5.3%	1.2%	9.9%
Part-time $20 to 99.9K	0.3	5.2	3.9	2.6	11.9
Moderate $100 to 249.9K	0.3	7.2	2.6	2.6	12.7
Large $250 to 500K	0.2	6.5	2.1	3.0	11.7
Very large >$500K	0.2	23.7	14.4	15.5	53.9
Total	1.3	45.6	28.2	24.9	100.0

NOTE: Moderate and large sales class intervals in this table differ from intervals used elsewhere in the report.

SOURCE: Office of Technology Assessment. Compiled from regional data provided by Bureau of the Census, Agriculture Division.

Table B-3.—Sales of Dairy Products by Sales Class and Region, 1982

Sales class definition	Northeast region	North Central region	Southern region	Western region	U.S. total
Small <$19.9K	$ 25,152	$ 109,070	$ 26,613	$ 11,508	$ 172,343
Part-time $20 to 99.9K	1,040,132	2,760,985	437,924	134,948	4,373,989
Moderate $100 to 249.9K	1,381,918	2,600,968	1,085,013	437,406	5,505,305
Large $250 to 500K	467,390	743,530	647,896	550,289	2,409,105
Very large >$500K	178,001	197,635	666,780	2,053,233	3,095,649
Total	3,092,593	6,412,188	2,864,226	3,187,384	15,556,391
	Percentage of total regional sales of dairy products, 1982				
Small <$19.9K	0.8%	1.7%	0.9%	0.4%	
Part-time $20 to 99.9K	33.6	43.1	15.3	4.2	
Moderate $100 to 249.9K	44.7	40.6	37.9	13.7	
Large $250 to 500K	15.1	11.6	22.6	17.3	
Very large >$500K	5.8	3.1	23.3	64.4	
Total	100.0	100.0	100.0	100.0	
	Percentage of total national sales of dairy products, 1982				
Small <$19.9K	0.2%	0.7%	0.2%	0.1%	1.1%
Part-time $20 to 99.9K	6.7	17.7	2.8	0.9	28.1
Moderate $100 to 249.9K	8.9	16.7	7.0	2.8	35.4
Large $250 to 500K	3.0	4.8	4.2	3.5	15.5
Very large >$500K	1.1	1.3	4.3	13.2	19.9
Total	19.9	41.2	18.4	20.5	100.0

NOTE: Moderate and large sales class intervals in this table differ from intervals used elsewhere in the report.

SOURCE: Office of Technology Assessment. Compiled from regional data provided by Bureau of the Census, Agriculture Division.

Table B-4.—Sales of Poultry and Eggs by Sales Class and Region, 1982

Sales class definition	Northeast region	North Central region	Southern region	Western region	U.S. total
Small <$19.9K	$ 7,081	$ 13,634	$ 23,616	$ 4,488	$ 48,819
Part-time $20 to 99.9K	25,334	66,701	386,349	31,085	509,469
Moderate $100 to 249.9K	85,870	241,939	1,719,082	52,012	2,098,903
Large $250 to 500K	153,000	490,911	1,559,160	96,664	2,299,735
Very large >$500K	562,178	845,410	2,373,048	1,009,005	4,789,641
Total	833,463	1,658,595	6,061,255	1,193,254	9,746,567
	Percentage of total regional sales of poultry and eggs, 1982				
Small <$19.9K	0.8%	0.8%	0.4%	0.4%	
Part-time $20 to 99.9K	3.0	4.0	6.4	2.6	
Moderate $100 to 249.9K	10.3	14.6	28.4	4.4	
Large $250 to 500K	18.4	29.6	25.7	8.1	
Very large >$500K	67.5	51.0	39.2	84.6	
Total	100.0	100.0	100.0	100.0	
	Percentage of total national sales of poultry and eggs, 1982				
Small <$19.9K	0.1%	0.1%	0.2%	0.0%	0.5%
Part-time $20 to 99.9K	0.3	0.7	4.0	0.3	5.2
Moderate $100 to 249.9K	0.9	2.5	17.6	0.5	21.5
Large $250 to 500K	1.6	5.0	16.0	1.0	23.6
Very large >$500K	5.8	8.7	24.3	10.4	49.1
Total	8.6	17.0	62.2	12.2	100.0

NOTE. Moderate and large sales class intervals in this table differ from intervals used elsewhere in the report.

SOURCE: Office of Technology Assessment. Compiled from regional data provided by Bureau of the Census, Agriculture Division.

Table B-5.—Sales of Hogs by Sales Class and Region, 1982

Sales class definition	Northeast region	North Central region	Southern region	Western region	U.S. total
Small <$19.9K	$ 18,399	$ 261,398	$ 139,914	$ 19,696	$ 439,407
Part-time $20 to 99.9K	34,735	1,268,807	213,832	33,453	1,550,827
Moderate $100 to 249.9K	49,804	2,596,083	273,900	50,072	2,969,859
Large $250 to 500K	35,215	1,595,321	252,456	35,143	1,918,135
Very large >$500K	36,848	1,158,830	436,821	80,313	1,712,812
Total	175,001	6,880,439	1,316,923	218,677	8,591,040
	Percentage of total regional sales of hogs, 1982				
Small <$19.9K	10.5%	3.8%	10.6%	9.0%	
Part-time $20 to 99.9K	19.8	18.4	16.2	15.3	
Moderate $100 to 249.9K	28.5	37.7	20.8	22.9	
Large $250 to 500K	20.1	23.2	19.2	16.1	
Very large >$500K	21.1	16.8	33.2	36.7	
Total	100.0	100.0	100.0	100.0	
	Percentage of total national sales of hogs, 1982				
Small <$19.9K	0.2%	3.0%	1.6%	0.2%	5.1%
Part-time $20 to 99.9K	0.4	14.8	2.5	0.4	18.1
Moderate $100 to 249.9K	0.6	30.2	3.2	0.6	34.6
Large $250 to 500K	0.4	18.6	2.9	0.4	22.3
Very large >$500K	0.4	13.5	5.1	0.9	19.9
Total	2.0	80.1	15.3	2.5	100.0

NOTE: Moderate and large sales class intervals in this table differ from intervals used elsewhere in the report.

SOURCE: Office of Technology Assessment. Compiled from regional data provided by Bureau of the Census, Agriculture Division.

Table B-6.—Sales of Fruit and Nuts by Sales Class and Region, 1982

Sales class definition	Northeast region	North Central region	Southern region	Western region	U.S. total
Small <$19.9K	$ 25,913	$ 26,709	$ 49,595	$ 139,210	$ 241,427
Part-time $20 to 99.9K	71,720	64,518	132,191	803,335	1,071,764
Moderate $100 to 249.9K	75,058	69,900	135,530	456,103	736,591
Large $250 to 500K	93,410	58,372	137,860	643,407	933,049
Very large >$500K	112,430	66,268	826,993	1,490,301	2,495,992
Total	378,531	285,767	1,282,169	3,532,356	5,478,823
	Percentage of total regional sales of fruit and nuts, 1982				
Small <$19.9K	6.8%	9.3%	3.9%	3.9%	
Part-time $20 to 99.9K	18.9	22.6	10.3	22.7	
Moderate $100 to 249.9K	19.8	24.5	10.6	12.9	
Large $250 to 500K	24.7	20.4	10.8	18.2	
Very large >$500K	29.7	23.2	64.5	42.2	
Total	100.0	100.0	100.0	100.0	
	Percentage of total national sales of fruit and nuts, 1982				
Small <$19.9K	0.5%	0.5%	0.9%	2.5%	4.4%
Part-time $20 to 99.9K	1.3	1.2	2.4	14.7	19.6
Moderate $100 to 249.9K	1.4	1.3	2.5	8.3	13.4
Large $250 to 500K	1.7	1.1	2.5	11.7	17.0
Very large >$500K	2.1	1.2	15.1	27.2	45.6
Total	6.9	5.2	23.4	64.5	100.0

NOTE: Moderate and large sales class intervals in this table differ from intervals used elsewhere in the report.

SOURCE: Office of Technology Assessment. Compiled from regional data provided by Bureau of the Census, Agriculture Division.

Table B-7.—Sales of Vegetables and Melons by Sales Class and Region, 1982

Sales class definition	Northeast region	North Central region	Southern region	Western region	U.S. total
Small <$19.9K	$ 27,070	$ 34,743	$ 50,355	$ 19,196	$ 131,364
Part-time $20 to 99.9K	61,918	61,181	64,749	69,592	257,440
Moderate $100 to 249.9K	68,836	59,792	80,025	90,604	299,257
Large $250 to 500K	57,776	69,208	97,667	218,614	443,265
Very large >$500K	86,165	135,505	734,475	1,901,387	2,857,532
Total	301,765	360,429	1,027,271	2,299,393	3,988,858
	Percentage of total regional sales of vegetables and melons, 1982				
Small <$19.9K	9.0%	9.6%	4.9%	0.8%	
Part-time $20 to 99.9K	20.5	17.0	6.3	3.0	
Moderate $100 to 249.9K	22.8	16.6	7.8	3.9	
Large $250 to 500K	19.1	19.2	9.5	9.5	
Very large >$500K	28.6	37.6	71.5	82.7	
Total	100.0	100.0	100.0	100.0	
	Percentage of total national sales of vegetables and melons, 1982				
Small <$19.9K	0.7%	0.9%	1.3%	0.5%	3.3%
Part-time $20 to 99.9K	1.6	1.5	1.6	1.7	6.5
Moderate $100 to 249.9K	1.7	1.5	2.0	2.3	7.5
Large $250 to 500K	1.4	1.7	2.4	5.5	11.1
Very large >$500K	2.2	3.4	18.4	47.7	71.6
Total	7.6	9.0	25.8	57.6	100.0

NOTE: Moderate and large sales class intervals in this table differ from intervals used elsewhere in the report.

SOURCE: Office of Technology Assessment. Compiled from regional data provided by Bureau of the Census, Agriculture Division.

Table B-8.—Sales of Cotton by Sales Class and Region, 1982

Sales class definition	Northeast region	North Central region	Southern region	Western region	U.S. total
Small <$19.9K	0.0	$ 729	$ 49,604	$ 4,279	$ 54,612
Part-time $20 to 99.9K	0.0	5,967	276,144	32,822	314,933
Moderate $100 to 249.9K	0.0	18,952	459,384	106,598	584,934
Large $250 to 500K	0.0	11,578	558,039	168,870	738,487
Very large >$500K	0.0	5,085	311,511	1,003,784	1,320,380
Total	0.0	42,311	1,654,682	1,316,353	3,013,346
		Percentage of total regional sales of cotton, 1982			
Small <$19.9K	0.0	1.7%	3.0%	0.3%	
Part-time $20 to 99.9K	0.0	14.1	16.7	2.5	
Moderate $100 to 249.9K	0.0	44.8	27.8	8.1	
Large $250 to 500K	0.0	27.4	33.7	12.8	
Very large >$500K	0.0	12.0	18.8	76.3	
Total	0.0	100.0	100.0	100.0	
		Percentage of total national sales of cotton, 1982			
Small <$19.9K	0.0%	0.0%	1.6%	0.1%	1.8%
Part-time $20 to 99.9K	0.0	0.2	9.2	1.1	10.5
Moderate $100 to 249.9K	0.0	0.6	15.2	3.5	19.4
Large $250 to 500K	0.0	0.4	18.5	5.6	24.5
Very large >$500K	0.0	0.2	10.3	33.3	43.8
Total	0.0	1.4	54.9	43.7	100.0

NOTE: Moderate and large sales class intervals in this table differ from intervals used elsewhere in the report.

SOURCE: Office of Technology Assessment. Compiled from regional data provided by Bureau of the Census, Agriculture Division.

Appendix C
Participants in OTA Workshops

Animal Technology Workshop Participants

Michael Phillips *(Co-leader)*
Project Director
Office of Technology Assessment
U.S. Congress
Washington, DC

Yao-chi Lu *(Co-leader)*
Senior Analyst
Office of Technology Assessment
U.S. Congress
Washington, DC

C. Eugene Allen
Professor and Associate Dean
College of Agriculture
University of Minnesota
St. Paul, MN

Howard Bachrach
Consultant
Agricultural Research Service
Plum Island Disease Center
U.S. Department of Agriculture
Greenport, NY

Roger Breeze
Director of Research
College of Veterinary Medicine
Washington State University
Pullman, WA

Keith Brown
Professor
Department of Poultry Science
Ohio State University
Wooster, OH

John Campbell
Professor
North Platte Agricultural Experiment Station
North Platte, NE

Dennis Campion
Research Leader
Animal Physiology Research
Richard B. Russell Research Center
U.S. Department of Agriculture
Athens, GA

B. Joe Conlin
Extension Dairy Specialist
University of Minnesota
St. Paul, MN

Larry Corah
Extension Beef Cattle Specialist
Kansas State University
Manhattan, KS

Nicholas Cross
President
Cattle Code America
Schiller Park, IL

Stanley E. Curtis
Professor
Department of Animal Science
University of Illinois
Urbana, IL

James R. Fischer
Associate Director
Agricultural Experiment Station
Michigan State University
East Lansing, MI

Harold Hafs
Vice President
Agricultural Research
Merck Institute for Therapeutic Research
Rahway, NJ

William Hansel
Liberty Hyde Bailey Professor of Animal Physiology
Department of Physiology
New York State College of Veterinary Medicine
Cornell University
Ithaca, NY

Maynard Hogberg *(Swine group leader)*
Extension Swine Specialist
Michigan State University
East Lansing, MI

Alan Holiday
Vice-President for Research and Development
Farmland Industries
Kansas City, MO

Keith Hora
Farmer
Riverside, IA

Ronald Knutson *(Dairy group leader)*
Professor
Department of Agricultural Economics
Texas A&M University
College Station, TX

Sid E. Kunz
Research Leader
U.S. Livestock Insects Laboratory
Agricultural Research Service
U.S. Department of Agriculture
Kerrville, TX

Juliette Linzer
Research Assistant
Office of Technology Assessment
U.S. Congress
Washington, DC

Charles McGinnis
Farmer
Kinards, SC

John McKnight, Jr.
Farmer
Searcy, AR

A.J. Muehling
Professor
Agricultural Engineering Extension Specialist
Agricultural Engineering Department
University of Illinois
Urbana, IL

John Nye
Professor
Agricultural Engineering Department
Purdue University
West Lafayette, IN

P.A. Oltenacu
Associate Professor of Systems Analysis
Department of Animal Science
Cornell University
Ithaca, NY

B.I. Osburn
Professor of Pathology
Associate Dean of Research
School of Veterinary Medicine
University of California
Davis, CA

John Paterson
Associate Professor
Animal Science Research Center
University of Missouri
Columbia, MO

Wilson G. Pond
Research Leader
Agricultural Research Service
Roman L. Hruska U.S. Meat Animal Research Center
U.S. Department of Agriculture
Clay Center, NE

Rod Preston
Thornston Distinguished Professor
Department of Animal Sciences
Texas Tech University
Lubbock, TX

Hoyle B. Puckett
Research Leader
ARS-USDA
University of Illinois
Urbana, IL

Allan Rahn *(Poultry group leader)*
Professor
Department of Animal Science
Michigan State University
East Lansing, MI

Dave Reed
Director of Clinical Research and Immunology
Molecular Genetics, Inc.
Minnetonka, MN

Robert Reining
Analyst
Office of Technology Assessment
U.S. Congress
Washington, DC

Barry Richmond
Assistant Professor
Dartmouth College
Hanover, NH

William Riddle
Research Leader
Food and Agricultural Economics
Battelle Columbus Laboratories
Columbus, OH

Marvin Rohlf
Vice President
Director of Research and Nutrition
Golden Sun Feeds
Estherville, IA

John V. Shutze
Head
Extension Poultry Science
University of Georgia
Athens, GA

Tom Sporleder *(Beef group leader)*
Professor
Department of Agricultural Economics
Texas A&M University
College Station, TX

R.H. Strickler
President
ROCCO Enterprises, Inc.
Harrisonburg, VA

Roy Van Arsdall
Economist
ERS-USDA
University of Illinois
Urbana, IL

Harold Wilcke
Consultant
Ralston-Purina
St. Louis, MO

Plant, Soil, and Water Technology Workshop Participants

Michael Phillips *(Co-leader)*
Project Director
Office of Technology Assessment
U.S. Congress
Washington, DC

Yao-chi-Lu *(Co-leader)*
Senior Analyst
Office of Technology Assessment
U.S. Congress
Washington, DC

Delmar Akerlund
Farmer
Valley, NE

Martin Alexander
Professor of Soil Microbiology
Department of Agronomy
Cornell University
Ithaca, NY

Joseph A. Berry
Staff Member
Plant Biology Department
Carnegie Institution of Washington
Stanford, CA

Larry L. Boersma
Professor of Soil Physics
Department of Soil Science
Oregon State University
Corvallis, OR

J.A. Browning
Chairman
Department of Plant Sciences
Texas A&M University
College Station, TX

Bill Burrows
Manager
Product Systems Research
Deere & Co.
Moline, IL

John A. Chapman
Vice President
Engineering, Irrigation Division
Valmont Industries
Valley, NE

Louie J. Chapman
Head
Extension Agronomy
Auburn University
Auburn, AL

C.H. Davis
Director of Chemical Division
National Fertilizer Development Center
Tennessee Valley Authority
Muscle Shoals, AL

Lynn Forster *(Corn group leader)*
Associate Professor
Department of Agricultural Economics and Rural Sociology
Ohio State University
Columbus, OH

George R. Foster
Hydraulic Engineer
National Soil Erosion Lab
ARS-USDA
Purdue University
West Lafayette, IN

Robert Fraley
Manager
Plant Molecular Biology Group
Monsanto Agriculture Products Co.
St. Louis, MO

Charles Francis
Professor
Department of Agronomy
University of Nebraska
Lincoln, NE

Nicholas M. Frey
Director, Department of Bio-Technology Research
Pioneer Hi-Bred, Inc.
Johnston, IA

Arthur Gerlow *(Rice group leader)*
Economist
Texas A&M University
Bryan, TX

Don W. Graffis
Professor
Extension Agronomist
University of Illinois
Urbana, IL

W.R. Guthrie
Vice President
Agricultural Research
Butler Manufacturing Co.
Grandview, MO

Robert Hall
Assistant Professor
Extension Agronomist
Plant Science Department
Brookings, SD

Paul Hedin
Research Leader
Chemistry Research Unit
Boll Weevil Research Laboratory
ARS-USDA
Mississippi State, MS

Alan Holiday
Vice President for Research and Development
Farmland Industries
Kansas City, MO

Bobby A. Huey
Extension Agronomist
University of Arkansas
Stuttgart, AR

Robert Jolly *(Soybeans group leader)*
Associate Professor
Department of Economics
Iowa State University
Ames, IA

George Kennedy
Associate Professor
Department of Entomology
North Carolina State University
Raleigh, NC

John Knepler
Manager, Design Engineering
DICKEY-john, Corp.
Auburn, IL

Ron Lacewell *(Cotton group leader)*
Professor
Department of Agricultural Economics
Texas A&M University
College Station, TX

William Liebhardt
Assistant Director
Rodale Research Center
Kutztown, PA

Juliette Linzer
Research Assistant
Office of Technology Assessment
U.S. Congress
Washington, DC

Chester G. McWhorter
Chief, Southern Weed Science Laboratory
Agricultural Research Service
Delta State Research Center
U.S. Department of Agriculture
Stoneville, MS

Thomas A. Miller *(Wheat group leader)*
Research Leader
ERS-USDA
Colorado State University
Fort Collins, CO

Page W. Morgan
Professor of Plant Physiology
Department of Plant Sciences
Texas A&M University
College Station, TX

Louis G. Nickell
Vice President of Research and Development
Velsicol Chemical Co.
Chicago, IL

Paul Nixon
Agricultural Engineer
Soil and Water Conservation Research
Agricultural Research Service
U.S. Department of Agriculture
Weslaco, TX

Bob Nowatzki
Farmer
Langdon, ND

James H. Palmer
Extension Agronomist
Professor of Agronomy and Soils
Clemson University
Clemson, SC

Buford Perry
Farmer
Lake Providence, LA

Robert Reining
Analyst
Office of Technology Assessment
U.S. Congress
Washington, DC

Barry Richmond
Assistant Professor
Dartmouth College
Hanover, NH

William Riddle
Research Leader
Food and Agricultural Economics
Battelle Columbus Laboratories
Columbus, OH

Bill Robertson
Farmer
Holly Ridge, MS

Robert L. Schafer
Research Leader
Tillage Research
National Tillage Machinery Laboratory
U.S. Department of Agriculture
Auburn, AL

Lyle Schertz
Senior Economist
Economic Research Service
U.S. Department of Agriculture
Washington, DC

James W. Searcy
Research Manager
E.I. du Pont de Nemours & Co.
Stine-Haskell Research Center
Newark, DE

B. Stout
Professor
Agricultural Engineering Department
Texas A&M University
College Station, TX

E.D. Threadgill
Associate Professor
Agricultural Engineering Department
Coastal Plain Experimental Station
University of Georgia
Tifton, GA

Larry Williams
Farmer
Hoopeston, IL

Agricultural Structure and Natural Environment and Resources Workshop Participants

Resource Persons

Robert Fraley
Manager
Plant Molecular Biology Group
Monsanto Agriculture Products Co.
St. Louis, MO

Charles Francis
Professor
Department of Agronomy
University of Nebraska
Lincoln, NE

William Hansel
Liberty Hyde Bailey Professor of Animal Physiology
Department of Physiology
New York State College of Veterinary Medicine
Cornell University
Ithaca, NY

B.I. Osburn
Professor of Pathology
Associate Dean of Research
School of Veterinary Medicine
University of California
Davis, CA

William Riddle
Research Leader
Food and Agricultural Economics
Battelle Columbus Laboratories
Columbus, OH

Jeffrey Ruckman
Engineer
Food and Agricultural Economics
Battelle Columbus Laboratories
Columbus, OH

Agricultural Structure Group

Michael Phillips *(Co-leader)*
Project Director
Office of Technology Assessment
U.S. Congress
Washington, DC

Thomas Sporleder *(Co-leader)*
Professor
Department of Agricultural Economics
Texas A&M University
College Station, TX

Michael Boehlje
Professor and Head
Department of Agricultural and Applied Economics
University of Minnesota
St. Paul, MN

Boyd Buxton
Economist
ERS-USDA
University of Minnesota
St. Paul, MN

James Cothern
Extension Economist
Department of Agricultural Economics
University of California
Davis, CA

Lynn Forster
Associate Professor
Department of Agricultural Economics and Rural Sociology
Ohio State University
Columbus, OH

Gene Futrell
Professor
Department of Economics
Iowa State University
Ames, IA

Harold Harris
Professor
Department of Agricultural Economics
Clemson University
Clemson, SC

Lester Myers
Branch Chief
Food Economics
Economic Research Service
U.S. Department of Agriculture
Washington, DC

George Mew
Industry Analyst
Farmland Industries
Kansas, City, MO

Allan Rahn
Professor
Department of Animal Science
Michigan State University
East Lansing, MI

Lyle Schertz
Senior Economist
Economic Research Service
U.S. Department of Agriculture
Washington, DC

Lee Schrader
Professor
Department of Agricultural Economics
Purdue University
West Lafayette, IN

Ronald Ward
Professor
Department of Food and Resource Economics
University of Florida
Gainesville, FL

Natural Environment and Resources Group

James Hite *(Co-leader)*
Professor
Department of Agricultural Economics
Clemson University
Clemson, SC

Yao-chi-Lu *(Co-leader)*
Senior Analyst
Office of Technology Assessment
U.S. Congress
Washington, DC

Robert Fraley
Manager
Plant Molecular Biology Group
Monsanto Agricultural Products Co.
St. Louis, MO

Charles Francis
Professor
Department of Agronomy
University of Nebraska
Lincoln, NE

Dellworth Gardner
Professor
Department of Agricultural Economics
University of California
Davis, CA

Clyde Kiker
Associate Professor
Department of Food and Resource Economics
University of Florida
Gainesville, FL

Randall Kramer
Associate Professor
Department of Agricultural Economics
Virginia Polytechnic Institute
Blacksburg, VA

James Moore
Associate Professor
Agricultural Engineering Department
Oregon State University
Corvallis, OR

Ford Runge
Assistant Professor
Department of Agricultural and Applied Economics
University of Minnesota
St. Paul, MN

Ray Supalla
Associate Professor
Department of Agricultural Economics
University of Nebraska
Lincoln, NE

Robert Todd
Professor and Head
Department of Microbiology
South Dakota State University
Brookings, SD

Appendix D

Analysis of Size Economies and Comparative Advantage in Crop Production in Various Areas of the United States

This appendix provides the detailed analysis of size economies and comparative advantage by area for crop production. A summary of this analysis was presented in chapter 8. The analysis was conducted at the University of Minnesota by Steve Cooke under the guidance of Burt Sundquist. Cooke's manuscript "Size Economies and Comparative Advantage in the Production of Corn, Soybean, Wheat, Rice, and Cotton in Various Areas of the United States" is published in a second volume to this OTA report.

The following analysis is organized by commodity. Each section follows the same format. First, there is a discussion of size economies by selected producing areas. Second, there is a discussion of comparative advantage, including the relationship between comparative advantage and size economies. Each of these commodity sections includes a summary of the size economy indices and "scoring table."

Corn

The corn-producing areas selected are:

1. Illinois area 300—Corn for grain
2. Indiana area 101—Corn for grain
3. Iowa area 201—Corn for grain
4. Nebraska area 400—Irrigated corn for grain

Size Economies in Corn Production

The four measures, or indicators, of size economies are estimated according to the procedures outlined by Cooke (1985). These indicators include production cost, use of harvesting equipment, and static and dynamic Herfindahl production concentration indices.[1] A summary of these indices is presented in a table for each commodity by enterprise size for each of the selected production areas (table D-1). There is an element of judgment required in using these indices.

To clarify and facilitate the judgment used, a scoring table was set up and marked for each of the measurement categories (table D-2). Each enterprise within a category was given a plus or minus for the presence or absence, respectively, of a "clear advantage" that enterprise size exhibited relative to the others within a given production area. Each measurement category was weighted equally so that the overall or total measure of size economies was expressed as the sum of the phases. The range in scoring was from 0 to 4. Zero implies no advantage for that enterprise size. Four implies a clear advantage for that enterprise size relative to one or both of the other enterprise sizes within a production area. A table for each commodity presents the results of the scoring procedure (table D-3).

In Illinois the very large corn enterprises have a cost advantage both in cost per bushel and in capital ownership costs per acre (table D-1). Very large corn enterprises in this area can fully use one to five self-propelled, six-row harvesters. These harvesters are assumed to have an annual harvesting capacity of about 450 acres per harvester. The static Herfindahl index (which is the measure of relative production concentration) in 1982 was greatest for large and very large enterprises. The dynamic Herfindahl index (which is the change in relative concentration) from 1978 to 1982 was positive only for the very large enterprise size. The scoring results are 4, 1, and 0 for the very large, large, and moderate-size enterprises, respectively (table D-2). There is strong evidence to argue for the existence of size economies for very large enterprises relative to large and moderate enterprises in corn production in Illinois.

In Indiana large and very large enterprises have nearly identical costs per bushel. In ownership cost per acre, very large enterprises have a cost advantage. Very large enterprises can fully use two, and large enterprises can fully use one, self-propelled,

[1] For simplicity, linear production possibilities curves and homogeneous commodity price ratios were assumed in the analysis.

Table D-1.—Indices Used to Determine Size Economies in Selected Corn-Producing Areas, 1983

State, area, and enterprise size	Total cost[a] per bushel (percent)	Ownership cost[a] per acre (percent)	Harvest machinery full utilization		Herfindahl indices	
			Maximum	Minimum	Static (percent)	Dynamic (percent)
IL 300						
VL	100	100	5.4	0.9	100	35
L	105	105	0.9	0.7	96	0
M	119	110	0.6	0.5	43	−15
IN 101						
VL	100	100	2.1	1.7	100	32
L	99	105	1.4	0.9	95	21
M	105	105	0.9	0.4	79	11
IA 201						
VL	100	100	1.8	1.0	100	48
L	107	128	0.6	0.5	118	11
M	105	155	0.5	0.3	85	−11
NE 400[b]						
VL	100	100	5.2	2.6	100	42
L	107	100	2.1	0.9	62	11
M	113	105	0.8	0.5	43	−2

[a]Excluding land charge.
[b]Irrigated.

SOURCE: Office of Technology Assessment.

Table D-2.—Scoring Table Used to Determine Size Economies in Selected Corn-Producing Areas, 1983

State, area, and enterprise size	Production cost	Harvester utilization	Herfindahl indices		Total
			Static	Dynamic	
IL 300					
VL	+	+	+	+	4
L	−	−	+	−	1
M	−	−	−	−	0
IN 101					
VL	+	+	+	+	4
L	+	+	+	+	4
M	−	−	+	+	2
IA 201					
VL	+	+	+	+	4
L	−	−	+	+	2
M	−	−	+	−	1
NE 400[a]					
VL	+	+	+	+	4
L	−	+	−	+	2
M	−	−	−	−	0

[a]Irrigated.

SOURCE: Office of Technology Assessment.

Table D-3.—Production Costs and Yield by Enterprise Size in Selected Corn-Producing Areas, 1983

State, area, and enterprise size	Total cost[a] $/bu.	Total cost[a] Percent	Yield[b] Bu/acre	Yield[b] Percent	Total cost[a] $/acre	Total cost[a] Percent
IL 300						
VL	1.67	100	130.3	100	217	100
L	1.75	105	128.6	99	225	103
M	1.99	119	123.1	94	245	113
IN 101						
VL	1.69	100	125.6	100	212	100
L	1.67	99	125.3	100	209	98
M	1.77	105	122.4	97	217	102
IA 201						
VL	1.67	100	119.0	100	199	100
L	1.80	107	117.4	99	211	106
M	1.75	105	113.0	95	198	99
NE 400[c]						
VL	2.83	100	118.6	100	336	100
L	3.03	107	112.6	95	341	102
M	3.21	113	106.2	90	341	102

[a]Excluding land charges.
[b]State level yields per harvested acre for irrigated and nonirrigated in 1982.
[c]Irrigated.

SOURCE: Office of Technology Assessment.

six-row harvesters. The relative production concentration in 1982 was nearly uniform across enterprise sizes within 21 percent. The change in relative production concentration from 1978 to 1982 was positive and nearly uniform across enterprise sizes in this area within 21 percent. The scoring results are 4, 4, and 2 for the very large, large, and moderate enterprises, respectively. There is evidence to argue that size economies exist for large and very large enterprises relative to moderate enterprises in corn production in Indiana.

In Iowa very large corn enterprises have a cost advantage both in total cost per bushel and in capital ownership cost per acre. Very large enterprises can fully use one to two self-propelled, six-row corn harvesters. The relative production concentration in 1982 was nearly uniform across enterprise size within 33 percent. The change in relative production concentration from 1978 to 1982 was positive for large and very large enterprises in this area. The scoring results are 4, 2, and 1 for very large, large, and moderate enterprises, respectively. There is evidence to argue that size economies exist for very large enterprises relative to large and moderate enterprises in corn production in Iowa.

In Nebraska very large irrigated corn enterprises have a production cost advantage in terms of total cost per bushel. However, ownership costs per acre are equal for large and very large enterprises and less than those for medium enterprises. Very large enterprises can fully use three to five, and large enterprises one to two, self-propelled, six-row harvesters. The relative production concentration from 1978 to 1982 was substantially higher for very large enterprises relative to large and moderate enterprises. The change in relative production concentration from 1978 to 1982 was positive for large and very large enterprises in this area. The scoring results are 4, 2, and 0 for very large, large, and moderate enterprises, respectively. There is clear evidence to argue that size economies exist for very large enterprises in irrigated corn production in Nebraska.

The source of size economies can be found by examining the components of the production cost measures (table D-3). Very large enterprises in general tend to have the lowest total cost per bushel. Large and very large enterprises all have at least slightly higher yields per acre relative to moderate enterprises. Yield is a source of size economies in corn production. Total cost per acre is relatively uniform across enterprises in which very large enterprises have a slight cost advantage.

In Illinois yield is a source of size economies. The very large enterprises in this area have lower expenditures per acre for fertilizer, fuel lubrication, repairs, and labor relative to large and moderate enterprises. In Indiana yield is also a source of size economy. Preharvest and capital ownership costs are not a source of size economies in this area. In Iowa yield is again a source of size economy, as is custom harvesting. In Nebraska yield is a source of size economies. Purchased irrigation water represents a potential for size diseconomies in Nebraska.

Very large enterprises in each of the selected producing areas consistently have slightly lower variable and ownership costs associated with machinery and equipment. This implies that very large corn enterprises tend to use some combination of fewer and/or smaller machines and tractors, and ones that go over the field fewer times. This is a constant source of size economies in the selected corn-producing areas.

Comparative Advantage in Corn Production

The overall objective for including a discussion on comparative advantage is to provide a context within which to analyze size economies and to determine the source or explain the absence of comparative advantage between the selected production areas.

In Illinois the total cost of corn production is about 14 percent higher than that in Iowa. There are size economies for very large corn enterprises only. The relative lack of comparative advantage is due to higher expenditures on phosphate, potash, herbicide, and pesticide in conjunction with a 6-percent lower yield compared with that in Iowa. The trends in relative yield and land prices indicate that the competitive position in corn production will decrease in this area. The absolute measure of production concentration in this area is low compared with the other selected producing areas. In addition, corn production is not particularly concentrated in any one enterprise size category, which implies that producers are not beginning to exploit size economies to increase their competitive position.

In Indiana the total cost of producing corn is about 10 percent more than in Iowa, and size economies exist for large and very large enterprises relative to moderate enterprises. The lack of comparative advantage is due to the relative price of nitrogen and additional expenditures on pesticides compared with those in Iowa. The absolute measure of production concentration in this area is high compared with that of the other selected areas. This implies that corn production is concentrated in one or more enterprise size categories and that producers are moving toward larger enterprise sizes to exploit size economies in this area so as to increase their competitive position.

In Iowa the total cost of producing corn is the lowest of the selected corn-producing areas. Size economies exist in this area for very large enterprises relative to large and moderate enterprises. Iowa's comparative advantage is related to higher yields relative to fertilizer, herbicide, and pesticide use. The absolute measure of production concentration is also the lowest of the selected producing areas, implying that corn production is not concentrated in any one particular enterprise size in this area. However, size economies exist for very large enterprises and can be exploited to improve Iowa's comparative advantage.

In Nebraska the total cost of producing irrigated corn is 8 percent higher than in Iowa, and size economies exist for very large enterprises relative to large and moderate enterprises. The lack of comparative advantage is due to the additional cost of irrigation water pumped from wells. The trends in yield and land prices indicate that the competitive position of this area will substantially decrease. The absolute measure of production concentration in this area is the highest of the selected corn-producing areas. This implies that production is concentrated in one or more enterprise size categories and that producers are moving toward larger enterprise sizes to exploit size economies and to enhance or maintain their competitive positions.

Soybeans

The soybean-producing areas are:

1. Illinois area 300
2. Iowa area 201
3. Mississippi area 100
4. Ohio area 101

Size Economies in Soybean Production

In Illinois very large soybean enterprises have about a 5-percent cost disadvantage relative to large enterprises in total cost per bushel (table D-4). Large and very large enterprises in this area have nearly equal capital ownership costs per acre. Large and very large soybean enterprises in this area can fully use one and two self-propelled, six-row harvesters, respectively. This size harvester has an annual harvesting capacity of about 380 acres. The static Herfindahl index (or measure of relative production concentration) in 1982 was greatest for the large enterprises. Finally, the dynamic Herfindahl index (or the measure of the change in relative production concentration) between 1978 and 1982 was positive only for the large and very large enterprises in this production area. The scoring results are 2, 4, and 0 for the very large, large, and moderate enterprises, respectively (table D-5). Thus, there is evidence to argue that size economies exist for the large enterprises

Table D-4.—Indices Used to Determine Size Economies in Selected Soybean-Producing Areas, 1983

State, area, and enterprise size	Total cost per bushel (percent)	Ownership cost per acre (percent)	Harvest machinery full utilization		Herfindahl indices	
			Maximum	Minimum	Static (percent)	Dynamic (percent)
IL 300						
VL	100	100	2.4	1.4	100	65
L	95	102	1.3	0.9	142	16
M	102	106	0.9	0.6	80	−4
IA 201						
VL	100	100	2.7	1.3	100	38
L	104	86	1.2	0.7	202	19
M	108	90	0.7	0.4	198	−5
MS 100						
VL	100	100	2.1	1.2	100	9
L	116	98	1.2	1.1	24	11
M	99	72	1.1	1.0	24	11
OH 101						
VL	100	100	2.5	2.2	100	71
L	84	90	1.6	1.0	115	21
M	86	90	0.9	0.3	73	−3

SOURCE: Office of Technology Assessment.

Table D-5.—Scoring Table Used to Determine Size Economies in Selected Soybean-Producing Areas, 1983

State, area, and enterprise size	Production cost	Harvester utilization	Herfindahl indices		Total
			Static	Dynamic	
IL 300					
VL	−	+	−	+	2
L	+	+	+	+	4
M	−	−	−	−	0
IA 201					
VL	+	+	−	+	3
L	−	+	+	+	3
M	−	−	+	−	1
MS 100					
VL	+	+	+	−	3
L	−	−	−	−	0
M	+	+	−	−	2
OH 101					
VL	−	−	+	+	2
L	+	+	+	+	4
M	+	−	−	−	1

SOURCE: Office of Technology Assessment.

relative to very large and moderate enterprises in soybean production in Illinois.

In Iowa very large soybean enterprises have about a 4- to 8-percent cost advantage relative to large and moderate enterprises in total cost per bushel. However, very large enterprises in this area have about a 10- to 14-percent cost disadvantage relative to large and moderate enterprises in ownership costs per acre. Large and very large enterprises can fully use one and two self-propelled, six-row harvesters, respectively. The relative production concentration in 1982 was greatest for large and moderate enterprises. Finally, the change in relative production concentration between 1978 and 1982 was positive only for very large and large enterprises in this area. The scoring results are 3, 3, and 1 for very large, large, and moderate enterprises, respectively. There is evidence to argue that size economies exist for very large and large enterprises relative to moderate enterprises in soybean production in Iowa.

In Mississippi very large and moderate soybean enterprises have about a 16-percent cost advantage relative to large enterprises in total cost per bushel. However, very large enterprises have a 26- to 28-

percent cost disadvantage in ownership costs per acre in this area relative to large and moderate enterprises, respectively. Moderate and very large soybean enterprises can fully use one and two self-propelled, eight-row harvesters, respectively. This size harvester has an annual harvesting capacity of about 800 acres. The relative production concentration in 1982 was greatest for very large enterprises. Finally, the change in relative production concentration between 1978 and 1982 was relatively low and uniform across enterprise sizes in this area. The scoring results are 3, 0, and 2 for very large, large, and moderate enterprises, respectively. The evidence suggests that there is no clear advantage for any enterprise size relative to another in soybean production in Mississippi.

In Ohio very large soybean enterprises have between a 14- to 16-percent cost disadvantage relative to large and moderate enterprises in total cost per bushel. Similarly, very large enterprises have about a 10-percent cost disadvantage in ownership cost per acre relative to large and moderate enterprises in this area. Large and very large soybean enterprises can fully use one and two self-propelled, six-row harvesters, respectively. The relative production concentration in 1982 was greatest for very large and large enterprises. Finally, the change in relative production concentration between 1978 and 1982 was positive for very large and large enterprises in this area. The scoring results are 2, 4, and 1 for very large, large, and moderate enterprises, respectively. Thus, there is evidence to argue that size economies exist for the large enterprises relative to very large and moderate enterprises in soybean production in Ohio.

The source or absence of size economies can be found by examining the components of the production cost measure. Total cost per unit of output is equal to the total cost per acre divided by the yield per acre. Table D-6 summarizes the production costs and yield by enterprise size for the selected soybean-producing areas. In Illinois soybean yields for large and very large enterprises are equal and 2 percent higher than those of moderate enterprises. In Iowa soybean yields are nearly uniform across enterprise sizes, with larger enterprises having 1 percent higher yields. In Mississippi soybean yields are 6 percent higher for very large enterprises relative to large and moderate enterprises. In Ohio soybean yields are nearly uniform across enterprise sizes, with larger enterprises having 2 to 3 percent higher yields. Yield is only a slight source of size economies in soybean production (table D-6).

In Illinois size diseconomies for very large enterprises are associated with the substantially higher fertilizer, herbicide, and pesticide expenditures, without corresponding higher yield. In Iowa the modest size economies for very large and large enterprises relate to lower costs of owner-provided relative to custom-provided durable services. In Mississippi the absence of size economies relate to the diseconomies of additional horsepower used by very large enterprises. In Ohio size diseconomies for very

Table D-6.—Production Costs and Yield by Enterprise Size in Selected Soybean-Producing Areas, 1983

State, area, and	Total cost[a]		Yield[b]		Total cost[a]	
enterprise size	$/bu.	Percent	Bu/acre	Percent	$/acre	Percent
IL 300						
VL	3.56	100	38.2	100	136	100
L	3.38	95	38.2	100	129	95
M	3.64	102	37.4	98	136	100
IA 201						
VL	3.32	100	36.8	100	122	100
L	3.44	104	36.6	99	126	103
M	3.58	108	36.3	99	130	107
MS 100[c]						
VL	5.20	100	25.0	100	130	100
L	6.02	116	23.6	94	142	109
M	5.17	99	23.6	94	122	94
OH 101						
VL	4.27	100	35.6	100	152	100
L	3.59	84	34.8	98	125	82
M	3.66	86	34.4	97	126	83

[a]Excluding land charges.
[b]State level yields per harvested acre for irrigated and nonirrigated in 1982.
[c]Production year data for 1982.

SOURCE: Office of Technology Assessment.

large enterprises relate to the diseconomies of additional horsepower used in land preparation and to the substantially higher expenditure on fertilizer, herbicide, and pesticide, without corresponding higher yields.

Comparative Advantage in Soybean Production

In Illinois total cost per bushel of soybeans is about 3 percent higher than in Iowa. Size economies exist for large enterprises only. The slight lack of comparative advantage is due to higher expenditures on herbicides, pesticides, and land. Trends in relative yield and land value indicate that the competitive position of Illinois will substantially improve. Soybean production is not concentrated in one or more enterprise sizes; however, size economies exist for large enterprises and can be exploited to improve comparative advantage.

In Iowa total cost per bushel of soybeans is the lowest of the selected soybean-producing areas. This comparative advantage is related to the level of yield, which is high relative to seed, herbicide, pesticide, and fertilizer expenditures. Trends in relative yields and land values indicate that Iowa's comparative position will decrease in the future. The measure of production concentration in Iowa is the lowest of the selected States, implying that soybean production is not concentrated in one or more enterprise sizes. Size economies exist for the large and very large enterprises and can be exploited to improve comparative advantage in this area.

Total cost per bushel of soybeans in Mississippi is about 24 percent higher than in Iowa. The substantial lack of comparative advantage relative to Iowa is a result of low yields and high expenditures on herbicides, pesticides, and ownership costs. Trends in yield and land values indicate that the competitive position of Mississippi will substantially decrease in the future. Production is concentrated in one or more enterprise sizes, and no size economies remain to be exploited in this area.

In Ohio total cost per bushel of soybeans is about 5 percent higher than in Iowa. The slight lack of comparative advantage is the result of lower yields and higher expenditures on herbicides, pesticides, potash, and phosphate. Trends in yield and land values indicate that the competitive position of Ohio will improve slightly in the future. Production is concentrated in one or more enterprise size categories. Size economies exist for large soybean enterprises only and can be exploited to improve comparative advantage in this area.

Wheat

For wheat the selected producing areas, type of wheat grown, and cultural practices followed are:

1. Kansas area 100—Hard red winter wheat following fallow
2. Montana area 200—Hard red spring wheat following fallow
3. North Dakota area 200—Hard red spring wheat following crop
4. Washington area 400—White wheat following fallow

Size Economies in Wheat Production

In Kansas very large wheat enterprises have a substantial cost advantage both in cost per unit of output and in ownership cost per acre (table D-7). Very large, large, and moderate wheat producers in this area can fully use 4 to 20, 2 to 3, and 1 to 2 self-propelled, 20-foot-wide harvesters, respectively. This size harvester has an annual capacity of harvesting about 500 acres. The static Herfindahl index (or measure of relative production concentration) in 1982 was greatest for the large and very large producers. This difference in static concentration was less pronounced in this area than in other wheat-producing areas in this study, however. Finally, the dynamic Herfindahl index (or the change in relative concentration) from 1978 to 1982 was positive for the large and very large enterprises. The scoring results are 4, 3, and 2 for very large, large, and moderate enterprises, respectively (table D-8). There is evidence to argue for the existence of size economies advantages for very large enterprises relative to the large and moderate wheat enterprise in Kansas.

In Montana very large wheat enterprises have a cost advantage both in cost per unit of output and ownership cost per acre. The very large, large, and moderate-size enterprises can fully use one to six self-propelled, 20-foot-wide harvesters. The static measure of relative production concentration in 1982 was greatest for the very large enterprises. Finally, the change in the relative concentration from 1978 to 1982 was positive only for the very large enterprise size. The scoring results are 4, 1, and 1 for the very large, large, and moderate categories, respectively. There is strong evidence to argue for the existence of size economies for very large enterprises relative to large and moderate sizes in the production of wheat in Montana.

In North Dakota large wheat enterprises have a cost advantage when measured either in cost per unit of output or ownership cost per acre. Large and

Table D-7.—Indices Used to Determine Size Economies in Selected Wheat-Producing Areas, 1983

State, area, and enterprise size	Total cost per bushel (percent)	Ownership cost per acre (percent)	Harvest machinery full utilization Maximum	Minimum	Herfindahl indices Static (percent)	Dynamic (percent)
KS 100						
VL	100	100	20.6	4.1	100	17
L	112	109	3.9	2.0	100	17
M	118	113	2.0	1.1	75	−7
MT 200						
VL	100	100	6.9	1.5	100	20
L	106	115	1.5	1.0	23	−12
M	110	108	1.0	0.6	6	−31
ND 200						
VL	100	100	3.5	1.6	100	47
L	95	90	1.6	0.9	72	11
M	103	98	0.9	0.4	36	−8
WA 400						
VL	100	100	6.2	3.7	100	3
L	118	117	2.7	1.9	100	3
M	85	91	1.9	1.3	31	−2

SOURCE: Office of Technology Assessment.

Table D-8.—Scoring Table Used to Determine Size Economies in Selected Wheat-Producing Areas, 1983

State, area, and enterprise size	Production cost	Harvestor utilization	Herfindahl Static	Herfindahl Dynamic	Total
KS 100					
VL	+	+	+	+	4
L	−	+	+	+	3
M	−	+	+	−	2
MT 200					
VL	+	+	+	+	4
L	−	+	−	−	1
M	−	+	−	−	1
ND 200					
VL	−	+	+	+	3
L	+	+	+	+	4
M	−	−	−	−	0
WA 400					
VL	−	+	+	−	2
L	−	+	+	−	2
M	+	+	−	−	2

SOURCE: Office of Technology Assessment.

very large enterprises can fully use one and two to three self-propelled, 20-foot-wide harvester(s), respectively. The static measure of relative production concentration in 1982 was greatest for large and very large enterprises. Finally, the change in the relative concentration from 1978 to 1982 was positive for both large and very large enterprises. The scoring results are 3, 4, and 0 for the very large, large, and moderate categories, respectively. There is evidence to argue for the existence of size economies for large and very large enterprises relative to moderate enterprises in the production of wheat in North Dakota. The production data suggests size economies for large enterprises in particular.

In Washington moderate wheat enterprises have a substantial cost advantage when measured in cost per unit of output or ownership cost per acre. Moderate, large, and very large producers can fully use one to two and four to six self-propelled, 19.4-foot-wide harvesters. Washington wheat producers in this area typically use a combination of "regular" and "hillside" harvesters in approximately a 70:30

ratio in harvesting their crop. A "composite" harvester is assumed to have an annual harvesting capacity of 495 acres per harvester. The static measure of relative production concentration in 1982 was greatest for large and very large enterprises. Finally, the change in the relative production concentration from 1978 to 1982 was positive for both large and very large enterprises, but only by 3 percent. There was virtually no change in production concentration between enterprise sizes from 1978 to 1982. The scoring results are 2, 2, and 1 for very large, large, and moderate enterprise categories, respectively. There is no clear evidence on which to argue for size economies in the production of wheat in Washington.

Yield is only a slight source of size economies in wheat production (table D-9). Wheat yields in Kansas are nearly uniform across enterprise size. In Montana and North Dakota the very large enterprise has the greatest yield per acre, by about 3 to 6 percent. In Washington large and very large enterprises have the same yield. The moderate enterprises, however, have substantially (20 percent) higher yields per acre than do the large and very large enterprises in this area. In fact, 1982 data reveals that small and very small enterprises have substantially higher yields than do moderate enterprises in this area. In Washington wheat yield is inversely related to enterprise size. Otherwise, yield is only a slight source of size economies in wheat production.

In Kansas the important factor for size economies relates to economies associated with custom harvest rates. In Montana size economies are the result of the combination of slightly higher yields and lower costs, again related to the use of custom harvesting. In North Dakota size economies for large enterprises relate to higher yield and lower ownership and harvest costs relative to those of moderate and very large enterprises. In Washington size economies do not exist for very large enterprises relative to moderate enterprises because of the substantial diseconomies associated with yield and the slightly higher ownership and harvesting costs. In Washington size economies for very large enterprises relative to large enterprises exist because of the substantially lower ownership costs of the very large enterprises in this area, which are related to the differences in horsepower tractors used particularly in land preparation.

Comparative Advantage in Wheat Production

In Kansas and Washington the comparative advantages in producing wheat are nearly equal and are the greatest of the areas studied. Kansas also has the potential for increasing its comparative advantage relative to unexploited size economies that exist for large and very large enterprises. This area has a relatively low level of production concentration. Finally, there is little change (about 1 percent per year) in production concentration in this area. The relatively large average size of the wheat enterprise in Kansas implies that enterprises are larger on average across size categories and that no one enterprise size dominates production in the area. (A similar set of characteristics exists for Iowa corn enterprises

Table D-9.—Production Costs and Yield by Enterprise Size in Selected Wheat-Producing Areas, 1983

State, area, and enterprise size	Total cost $/bu.	Total cost Percent	Yield Bu/acre	Yield Percent	Total cost $/acre	Total cost Percent
KS 100						
VL	2.05	100	33.1	100	68	100
L	2.30	112	33.1	100	76	112
M	2.41	118	33.2	100	80	118
MT 200						
VL	2.77	100	31.1	100	86	100
L	2.94	106	29.9	96	88	102
M	3.05	110	29.2	94	89	103
ND 200						
VL	3.79	100	31.7	100	100	120
L	3.60	95	30.8	97	111	93
M	3.91	103	29.7	94	116	97
WA 400						
VL	3.26	100	39.9	100	130	100
L	3.86	118	39.9	100	154	118
M	2.76	85	47.8	120	132	102

SOURCE: Office of Technology Assessment.

except that the enterprises there tend to be smaller, on average.)

It is only about 3 percent more costly to produce a bushel of wheat in Montana than in Kansas. This slight lack of comparative advantage is due to high ownership costs associated with more and larger machines being used on the land more times in land preparation. Unlike those in Kansas, size economies in Montana exist for very large enterprises only. This implies that producers in Montana can exploit size economies as a strategy to remain competitive.

It is about 9 percent more costly to produce a bushel of wheat in North Dakota than in Kansas. The increased costs are due to additional expenditures on seed, fertilizer, and chemicals associated with increasing relative yields, spring planting, and continuous cropping. Size economies exist for very large and large enterprises in North Dakota. Production concentration, though higher than in Kansas, is still low. Thus, no one enterprise size dominates production in this area. Size economies can be exploited to improve North Dakota's comparative advantage.

Washington's comparative advantage in wheat production is nearly identical to that of Kansas. It is less than 1 percent more costly to produce a bushel of wheat in Washington than in Kansas. All size economies within Washington are nearly fully exploited. The average enterprise size is quite large, about 1,600 acres. The level of production concentration is the highest of any of the wheat-producing areas studied. This implies that one or more enterprise sizes dominate production in this area.

Rice

For rice the selected producing areas and type of rice grown include:

1. California area 400—medium- and short-grain rice
2. Texas area 1001—long-grain rice
3. Delta (Mississippi 100 and Arkansas 300—long-grain rice
4. Arkansas area 200—long-grain rice

Size Economies in Rice Production

In California the total cost per hundredweight of rice for all three enterprise sizes is nearly identical, with moderate enterprises having a slight advantage of about 3 percent (table D-10). On the basis of ownership cost per acre, the moderate enterprise has a cost advantage of about 8 percent relative to very large enterprises. Very large, large, and moderate enterprise sizes can fully use 5 to 11, 3 to 4, and 2 self-propelled, 16-foot-wide harvesters, respectively. This size harvester has an annual harvesting capacity of 465 acres. The static Herfindahl index (or measure of relative production concentration) in 1982 was greatest for the large and very large producers. The dynamic Herfindahl index (or the change in relative concentration) from 1978 to 1982 was negative for all enterprise sizes in this rice-producing area. Unfortunately, the data associated with the Herfindahl indices are not sufficiently disaggregated at the large and very large rice enterprise sizes to allow

Table D-10.—Indices Used to Determine Size Economies in Selected Rice-Producing Areas, 1979

State, area, and enterprise size	Total cost per cwt[a] (percent)	Ownership cost per acre (percent)	Harvest machinery full utilization		Herfindahl indices	
			Maximum	Minimum	Static (percent)	Dynamic (percent)
CA 400[b]						
VL	100	100	11.7	4.7	100	−4
L	99	103	4.6	2.4	100	−4
M	97	92	2.4	1.3	25	−19
TX 1001[b]						
VL	100	100	6.2	3.7	100	28
L	96	96	2.8	1.8	80	24
M	97	95	1.8	1.1	61	20
DLT 100 and 300[b]						
VL	100	100	6.5	2.2	100	−35
L	94	109	3.3	1.0	85	−19
M	92	113	1.5	0.6	58	11
AR 200[b]						
VL	100	100	6.3	2.1	100	21
L	100	105	2.1	1.0	84	−20
M	96	102	1.0	0.6	67	−7

[a]Hundredweight.
[b]Irrigated.

SOURCE: Office of Technology Assessment.

for more detailed analysis. The scoring results are 2, 2, and 2 for very large, large, and moderate enterprises, respectively (table D-11). There is evidence to argue that no size economies exist in California rice production, given the 1979 configuration of enterprise sizes.

In Texas large and moderate rice enterprises have a slight cost advantage (3 to 4 percent), both in cost per unit of output and ownership cost per acre. The very large, large, and moderate enterprises can fully use four to six, and one to two self-propelled, 16-foot-wide harvesters. The static measure of relative production concentration in 1982 was relatively uniform across enterprise sizes in this area. Finally, the change in the relative concentration from 1978 to 1982 was also relatively uniform and positive across enterprise sizes in this area. The scoring results are 2, 2, and 2 for the very large, large, and moderate categories, respectively. There is evidence to argue that no size economies exist in Texas rice production, given the 1979 configuration of enterprise sizes.

In the Delta large and moderate enterprises have a cost advantage, when measured in cost per unit of output, by about 6 to 8 percent. However, very large enterprises have a cost advantage in capital ownership cost per acre. Very large, large, and moderate enterprises in this area can fully use three to six, two to three, and one self-propelled, 17-foot-wide harvesters, respectively. This size harvester has an annual harvesting capacity of 495 acres per harvester. The static measure of relative production concentration in 1982 was nearly uniform across enterprise sizes. Finally, the change in the relative concentration from 1978 to 1982 was positive for the moderate enterprise size only. The scoring results are 1, 2, and 3 for the very large, large, and moderate enterprises, respectively. There is evidence to argue that no size economies exist in the Delta rice production, given the 1979 configuration of enterprise sizes.

In Arkansas moderate enterprises have a cost advantage, when measured in cost per unit of output, by about 4 percent. However, very large enterprises have a cost advantage in capital ownership costs per acre. Very large, large, and moderate enterprises can fully use three to six, one to two, and one self-propelled, 17-foot-wide harvesters, respectively. The static measure of relative production configuration in 1982 was nearly uniform across enterprise sizes. Finally, the change in the relative production concentration from 1978 to 1982 was positive for the very large enterprise size only. The scoring results are 2, 1, and 2 for the very large, large, and moderate enterprises, respectively. There is evidence to argue that no size economies exist in Arkansas rice production, given the 1979 configuration of enterprise sizes.

The absence of size economies in rice production can be explained by examining the components of the production cost measures (table D-12). Rice yield in all the production areas studied is inversely related to enterprise size, except in Texas. In the case of the Delta, large and moderate rice enterprises have a substantial yield advantage over very large enter-

Table D-11.—Scoring Table Used to Determine Size Economies in Selected Rice-Producing Areas, 1979

State, area, and enterprise size	Production cost	Harvester utilization	Herfindahl: Static	Herfindahl: Dynamic	Total
CA 400[a]					
VL	–	+	+	–	2
L	–	+	+	–	2
M	+	+	–	–	2
TX 1001[a]					
VL	–	+	+	–	2
L	+	+	–	–	2
M	+	–	+	–	2
DLT 100 and 300[a]					
VL	–	+	+	–	2
L	+	+	–	–	2
M	+	+	–	+	3
AR 200[a]					
VL	–	+	–	+	2
L	–	+	–	–	1
M	+	+	–	–	2

[a]Irrigated.

SOURCE: Office of Technology Assessment.

Table D-12.—Production Costs and Yield by Enterprise Size in Selected Rice-Producing Areas, 1979

State, area, and enterprise size	Total cost		Yield		Total cost	
	$/cwt	Percent	Cwt/acre	Percent	$/acre	Percent
CA 400[a]						
VL	6.34	100	51.3	100	325	100
L	6.29	99	52.6	103	331	102
M	6.12	97	52.1	102	319	98
TX 1001						
VL	7.70	100	47.4	100	365	100
L	7.39	96	46.3	98	342	94
M	7.46	97	46.4	98	346	95
DLT 100-300[a]						
VL	6.78	100	39.8	100	270	100
L	6.36	94	42.6	107	271	100
M	6.26	92	43.6	110	273	101
AR 200[a]						
VL	6.33	100	43.1	100	273	100
L	6.31	100	44.7	104	282	103
M	6.09	96	44.5	103	271	99

[a]Irrigated.

SOURCE: Office of Technology Assessment.

prises by about 7 to 10 percent. Total cost per acre for very large enterprises is less than or equal to total cost per acre for large and moderate enterprises in all the production areas again except Texas. In general, rice production has diseconomies of size relative to yield and no economies of size in relation to total costs per acre. Yield diseconomies are related in large part to timeliness of fertilizer and water application, which can be managed better at smaller enterprise sizes than at larger ones.

Size diseconomies in rice production exist uniformly across the selected production area. These diseconomies are primarily the result of yield diseconomies of size assistance in California, the Delta, and Arkansas. Yield diseconomies are related to timeliness of fertilizer, herbicide, pesticide, fungicide, and water application. In Texas size diseconomies are associated with purchased canal water used for irrigation, which in turn allows for lower ownership costs associated with producers' well-pumped irrigation.

Comparative Advantage in Rice Production

The comparative advantage of California in rice production is the greatest of the areas studied. It is the result of high yields, relatively inexpensive irrigation water, and reduced herbicide and fungicide costs relative to those of other selected producing areas. This comparative advantage is not extendable through size economies, since size economies have been more than fully exploited in this area. California has the largest average size rice acreage per enterprise, at 1,071 acres. This implies that production is concentrated in one or more size categories. The combination of size economy and comparative advantage information shows that rice enterprises in California should not increase their size as a means of reducing cost and thereby improving comparative advantage.

The comparative disadvantage of Texas in producing rice is the greatest of the areas studied. The data indicate that in 1979 it was about 25 percent more costly to produce a hundredweight of rice in Texas than in California. This substantial lack of comparative advantage is related to relatively low yields and high irrigation, herbicide, and fungicide costs. It is not correctable by increasing enterprise size in attempting to be more competitive, since size economies have been more than fully exploited. If trends in relative yield and land values continue, Texas will decline from its already marginal competitive position.

It is about 14 percent more costly to produce a hundredweight of rice in the Delta than in California. This lack of comparative advantage is related to additional expenditures on herbicides, fungicides, and irrigation water. This comparative disadvantage is not correctable by simply increasing enterprise size in an attempt to be more competitive, since size economies have been more than fully exploited. The average enterprise size in this area is about 700 acres. Production is very highly concentrated in one or more of the enterprise size categories, and size economies have been more than fully exploited. The dynamic Herfindahl index shows that there was a substantial decrease in rice production concentration between 1978 and 1982 in this area. Rice production in the absence of size economies is becoming less concentrated and may continue to be so into the future. The combination of size economy and comparative advantage information implies that rice enterprises in the Delta could not increase enterprise size as a means of enhancing comparative advantage.

The comparative disadvantage in Arkansas in 1979 was such that it was 8 percent more costly to produce a hundredweight of rice in Arkansas than in California. This lack of comparative advantage is due to additional expenditures on herbicides, fungicides, and irrigation water pumped from wells. This comparative disadvantage is not correctable by increasing enterprise size in an attempt to be more competitive, since size economies do not exist. The average enterprise size in this area is about 485 acres, the smallest of the selected rice-producing areas. Production is distributed relatively uniformly across enterprise sizes, and there is a modest trend toward resource dispersion, or deconcentration. The combined information on size economies and comparative advantage implies that rice enterprises in Arkansas could not increase enterprise size to enhance comparative advantage. In fact, the current size distribution of rice enterprises is well suited for rice production by being small on average and yet capable of fully using a single rice harvester.

Cotton

For cotton the selected upland cotton-producing areas and cultural practices include:

1. Alabama area 600—dryland
2. California area 500—irrigated
3. Mississippi area 100—mixed
4. Texas area 200—irrigated
5. Texas area 200—dryland

Size Economies in Cotton Production

In Alabama very large cotton enterprises have the lowest total cost per bale and lowest ownership costs per bale by about 7 to 8 percent and lowest cost per acre relative to large and moderate enterprises by about 20 to 26 percent (table D-13). Very large enterprises in this area can fully use three to six self-propelled, two-row cotton pickers. This size harvester has an annual harvesting capacity of about 400 acres. The static Herfindahl index (or measure of relative production concentration) in 1982 was greatest for the very large producers. Finally, the

Table D-13.—Indices Used to Determine Size Economies in Selected Cotton-Producing Areas, 1982

State, area, and enterprise size	Total cost per bale (percent)	Ownership cost per acre (percent)	Harvest machinery full utilization		Herfindahl indices	
			Maximum	Minimum	Static (percent)	Dynamic (percent)
AL 600						
VL	100	100	6.7	2.9	100	63
L	107	120	2.6	2.1	64	8
M	108	126	1.8	1.3	64	8
CA 500[a]						
VL	100	100	9.3	5.2	100	39
L	91	86	4.2	2.3	100	39
M	98	98	2.0	0.9	6	9
MS 100						
VL	100	100	11.3	5.2	100	24
L	103	101	5.1	2.3	100	24
M	100	97	2.3	1.6	48	21
TX 200[a]						
VL	100	100	4.7	2.5	100	77
L	94	88	2.4	1.3	70	80
M	99	94	1.1	0.6	35	80
TX 200						
VL	100	100	29.5	5.8	100	6
L	114	143	5.0	2.7	100	6
M	117	124	2.7	1.4	51	5

[a]Irrigated.

SOURCE: Office of Technology Assessment.

dynamic Herfindahl index (or the change in relative concentration) from 1978 to 1982 was positive for each enterprise size, particularly for very large enterprises. The scoring results are 4, 1, and 1 for very large, large, and moderate enterprises, respectively (table D-14). There is strong evidence to argue that size economies exist for very large cotton enterprises in Alabama.

In California large cotton enterprises have the lowest total cost per bale and ownership costs per acre relative to very large and moderate enterprises. Very large, large, and moderate enterprises can fully use six to nine, three to four, and one to two self-propelled, two-row cotton pickers, respectively. The static Herfindahl index in 1982 was greatest for the large and very large enterprises. Finally, the dynamic Herfindahl index for 1978 to 1982 was positive for each enterprise size, particularly for the large and very large enterprise sizes. The scoring results are 3, 4, and 1 for very large, large, and moderate enterprises, respectively. There is evidence to argue that size economies do exist for very large enterprises in the production of irrigated cotton in California, given the enterprise configuration in 1982.

In Mississippi the total cost per bale and the ownership cost per acre is nearly equal across cotton enterprise size categories. Very large, large, and moderate enterprises can fully use 5 to 11, 3 to 5, and 2 self-propelled, two-row cotton pickers. The measure of relative production concentration in 1982 was greatest for very large and large enterprise sizes. The change in relative production concentration from 1978 to 1982 was nearly constant across all enterprise sizes. The scoring results are 4, 4, and 3 for very large, large, and moderate enterprises, respectively. There is evidence to argue that size economies do not exist for large and very large cotton enterprises relative to moderate enterprises in Mississippi, given the enterprise configuration in 1982.

In Texas (irrigated) large cotton enterprises have the lowest total cost per bale, by about 6 percent. Large and moderate enterprises have the lowest ownership cost per acre. Moderate, large, and very large enterprises can fully use one, two, and three to four, "composite" cotton strippers with an annual harvesting capacity of about 525 acres per harvester. The measure of relative production concentration in 1982 was greatest for large and very large enterprises. The change in relative production concentration from 1978 to 1982 was nearly constant across enterprise size categories. The scoring results are 3, 4, and 2 for very large, large, and moderate enterprises, respectively. There is evidence to argue that size economies do not exist for very large enterprises in irrigated cotton in Texas, given the 1982 configuration of enterprises.

In Texas (dryland) very large cotton enterprises have the lowest total cost per bale, by about 14 to

Table D-14.—Scoring Table Used to Determine Size Economies in Selected Cotton-Producing Areas, 1982

State, area, and enterprise size	Production cost	Harvester utilization	Herfindahl		Total
			Static	Dynamic	
AL 600					
VL	+	+	+	+	4
L	–	–	+	–	1
M	–	–	+	–	1
CA 500[a]					
VL	–	+	+	+	3
L	+	+	+	+	4
M	–	+	–	–	1
MS 100					
VL	+	+	+	+	4
L	+	+	+	+	4
M	+	+	–	+	3
TX 200[a]					
VL	–	+	+	+	3
L	+	+	+	+	4
M	–	+	–	+	2
TX 200					
VL	+	+	+	–	3
L	–	+	+	–	2
M	–	+	–	–	1

[a]Irrigated.

SOURCE: Office of Technology Assessment.

17 percent, and lowest ownership cost per acre, by about 24 to 43 percent. Moderate, large, and very large enterprises can fully use 2, 3 to 5, and 6 to 30 "composite" cotton strippers. The measure of relative production concentration in 1982 was greatest for large and very large enterprises. The change in relative production concentration from 1978 to 1982 was positive, nearly constant, and quite small across enterprise sizes. The scoring results are 3, 2, and 1 for very large, large, and moderate enterprises, respectively. There is evidence to argue that size economies exist for very large enterprises in dryland cotton production in Texas.

Circumstances regarding size economies in the selected cotton-producing areas can be explained by examining the components of the production cost measures (table D-15). Cotton yields tend to be related to enterprise size, as in Alabama, California, and Mississippi, by about 3 to 7 percent. In Texas, yields tend to be inversely related to enterprise size by about 2 to 3 percent. Total cost per acre tends to be nearly uniform across enterprises in Alabama, Mississippi, and Texas (irrigated). In California total cost per acre is directly related to enterprise size, whereas in Texas (dryland) the total cost per acre is inversely related to enterprise size.

In summary, size economies exist for very large cotton enterprises in Alabama because these enterprises incur lower machinery and tractor-related expenses for a given field operation and still manage to obtain a slightly higher yield. In California, size diseconomies are primarily related to the pecuniary diseconomies of purchased irrigation water for cotton production. In Mississippi the lack of size economies in cotton production for large and very large enterprises relative to moderate enterprises relates to similar preharvest and ownership costs in conjunction with slightly higher yields. In Texas (irrigated) the lack of size economies is related to the combination of size diseconomies in harvesting and cultivation, along with slightly higher yields enjoyed by large and moderate enterprises in this area. In Texas (dryland) size economies for very large enterprises relate to the substantial preharvest and ownership cost advantages associated with lower machinery and tractor-related expenses for a given field operation, without substantial loss in yield.

Comparative Advantage in Cotton Production

In Alabama the average total cost per bale in producing cotton is about 23 percent higher than in Mississippi. This comparative disadvantage is due to low yields and high fertilizer, lime, insecticide, and harvesting costs. Size economies exist in cotton pro-

Table D-15.—Production Costs and Yield by Enterprise Size in Selected Cotton-Producing Areas, 1982

State, area, and enterprise size	Total cost[a]		Yield		Total cost[a]	
	$/bale	Percent	Bales/acre	Percent	$/acre	Percent
AL 600						
VL	279	100	1.52	100	424	100
L	298	107	1.47	97	438	103
M	301	108	1.47	97	443	104
CA 500[b]						
VL	298	100	2.28	100	680	100
L	271	91	2.28	100	619	91
M	291	98	2.11	93	613	90
MS 100						
VL	230	100	1.79	100	412	100
L	237	103	1.79	100	424	103
M	229	100	1.74	97	399	97
TX 200[b]						
VL	319	100	0.67	100	214	100
L	299	94	0.69	103	206	96
M	315	99	0.68	102	214	100
TX 200						
VL	259	100	0.44	100	114	100
L	298	114	0.45	103	134	118
M	302	117	0.45	102	136	119

[a]Excluding land charge.
[b]Irrigated.

SOURCE: Office of Technology Assessment.

duction in this area for very large enterprises. The average enterprise size is about 1,200 acres and is one of the lowest of the cotton areas studied. Production concentration, on the other hand, is the highest of the areas studied, and production and resources were concentrating at a substantial rate of 36 percent between 1978 and 1982, or 9 percent per year. Since cotton production in this area is at a substantial competitive disadvantage, producers appear to be adopting a strategy of increasing enterprise size as a means of exploiting size economies and increasing competitiveness. This strategy will continue to work in the future, as well.

In California the average total cost per bale of cotton is about 9 percent higher than in Mississippi. This comparative disadvantage is due to high irrigation costs. Size economies do not exist for very large enterprises in California, given the 1982 enterprise configuration. The average cotton enterprise is about 2,100 acres in size and is the second most concentrated cotton-producing area in the selected areas. The diseconomies associated with the rates of purchased irrigation water limit the extent to which other size economies can be used to decrease production costs.

In Mississippi the average total cost per bale of cotton is the lowest of the areas studied. Size economies do not exist in this area for very large and large enterprises relative to moderate enterprises. The average cotton enterprise in this area is about 1,800 acres. Production concentration is "moderate" relative to the other areas in the study. Since cotton production is the most competitive, producers have adopted a strategy of moderate enterprise expansion as a strategy to increase total revenue rather than to increase comparative advantage.

In Texas (irrigated) the average total cost per bale of cotton is about 1 percent higher than in Mississippi. In part, this is because of a comparative advantage in soil. Size economies in Texas (irrigated) do not exist for very large enterprise sizes under the 1982 enterprise configuration. The average enterprise size is about 1,224 acres per enterprise. Production concentration is low relative to the other areas. Producers in this area seem to be adopting irrigation as a means of decreasing variability as well as increasing average yield. However, size economies do not appear to exist for very large irrigated cotton enterprises and, therefore, do not exist as an additional means of improving comparative advantage.

In Texas (dryland) the average total cost per bale of cotton is about 4 percent higher than in Mississippi. Again, this is due in part to the inherent quality of the soil. Size economies remain to be exploited by very large enterprises in this area. The average cotton enterprise size is about 3,300 acres. The production concentration is the lowest of the cotton-producing areas studied. The change in production concentration was about a 6-percent increase from 1978 to 1982, or about 2 percent per year. This is very low relative to the other cotton-producing areas, particularly when compared to irrigated cotton. In this area yields are subject to wide variation, owing to climate and the absence of irrigation water. Therefore, producers seem to have adopted a strategy of nonexpansion in the face of size economies for very large enterprises. The future success of this strategy will depend in part on relative yields and land prices compared with those of other cotton-producing areas.

Appendix E

Methodology and Detailed Results of Microeconomic Impacts of Technology and Public Policy for Crop Farms

Chapter 8 presented the summary results of the microeconomic impacts of public policies and technology on the viability of crop farms. This appendix discusses in more detail the methodology used for the analysis and the specific results by area. For further information the reader is advised to read the individual commissioned papers published in a separate volume to this report.

The first step for each production area was to describe representative farms that included moderate, large, and very large farms. The second step involved a simulation of the representative farms using a Monte-Carlo, whole-farm simulation model (FLIPSIM V) under alternative farm policy, income tax, finance, and technology scenarios.

Simulation Model

The current version of the General Firm Level Policy Simulator—FLIPSIM V, developed by James Richardson and Clair Nixon at Texas A&M University—was used to simulate the three representative farms for selected policy and technology scenarios. The model is capable of simulating the annual functions of a crop farm, i.e., production, marketing, financial growth and decay, machinery depreciation and replacement, family consumption, fixed and variable costs, and participation in farm programs.

Each representative farm was simulated over the 10-year planning horizon beginning in 1982 and extending through 1992. The planning horizon was then repeated 50 times, using a different set of random cotton prices and yields for each iteration. At the end of each iteration, values for key output variables were calculated.

The model began each year of the planning horizon by determining the production costs for the current size of the farm, based on information provided for larger farms. Because the representative farms were permitted to grow over time, crop mix and per-acre production costs were forced to change to correspond to those for larger representative farms.

After determining the relevant crop mix and costs for the farm, the model selected the random crop prices and yields for that year. Random yields were drawn to reflect the historical variability typical of the study area.

FLIPSIM V simulated variable production costs for each crop by multiplying the per-acre input costs by planted acreages for the respective crops. Labor costs were calculated as the sum of full-time labor charges plus the cost of part-time labor. Part-time labor needs were based on the difference between hours of monthly labor available from full-time employees and nonpaid family members, and the monthly labor needs for all crops. Harvesting costs were the product of the per-unit harvest costs, random yield, and harvested acreage. Each farm's initial production and harvesting costs were expressed in 1982 dollars.

Annual crop yields were selected at random, based on the historical yield variability observed for the study area subject to the technology scenario being evaluated, the year of the planning horizon, and the size of farm. Under the base technology scenario it was assumed that the very largest farm would adopt the new technology first. The next smallest farm was assumed to adopt this technology in a similar pattern during subsequent years, and the smallest farm would make the adoption even later. The specific lag years for each commodity were based on the results of technology workshops discussed in chapter 3.

The model calculated property taxes based on the price of land and the property tax rate for the study area. Other fixed costs were determined by the analyst. The model amortized all outstanding loans under the assumption that they were simple interest mortgages. Annual interest rates for existing debt on land, machinery, and operating loans were, respectively, 8.5, 13.4, and 14.4 percent. Annual interest rates for new debts and refinanced loans (on long-term and intermediate-term assets) were 11.4 and 13.4 percent, respectively. Cash reserves and off-farm investments were allowed to earn 10 percent interest annually. The market value of farm machinery was updated under the assumption that the real market value of used equipment decreased 1 percent per year. The market value of cropland was estimated using the historical relationship between the capital gains rate for cropland and the rate of returns for farms. The capital gains rate was a function of the capital gains rate for land in the previous year and the rate of return to production assets for the farm in the previous year.

The model next depreciated each piece of equipment on the farm for income tax purposes. Equipment purchased prior to 1981 was depreciated using the double-declining balance method and a 5-year to 7-year life. Equipment placed into use after 1980 was cost recovered assuming a 5-year life and the Accelerated Cost Recovery System (ACRS) rules. Regular-purpose and special-purpose buildings were depreciated using ACRS rules, or the double-declining balance method, where applicable. Equipment that had passed its economic life was traded for a replacement, if sufficient cash was available to cover the required downpayment. The cost of replacement equipment, expressed in 1982 dollars, was held constant throughout the planning horizon. First-year expensing and maximum investment tax credit (ITC) were calculated for all equipment purchases.

The fraction of each crop marketed in the current tax year was estimated internally, based on the operator's desired taxable income ($7,400), estimated cash receipts, and income tax deductions. If the market price was less than the effective loan rate for a crop, it was placed in a Commodity Credit Corporation (CCC) loan when available, rather than being sold. Stocks were released from the loan if the market price in the following year exceeded the loan rate plus interest. Deficiency payments were paid if the season average price was less than the target price.

The deficiency payment is a function of the payment rate, farm program yield, and harvested acreage. When an acreage set-aside or diversion program was simulated, the model reduced planted acreage the specified amount and accounted for increases in production on the more productive land left in production (slippage).

After simulating the farm policies specified by the user, the model determined the farm operator's year-end financial position, calculated family cash withdrawals, and calculated income taxes payable in the following year. Cash surpluses were deposited in an interest-bearing account at 10 percent interest. Year-end cash flow deficits were handled in the following order: 1) grant a lien on crops in storage at the operating loan interest rate, 2) refinance long-term equity, 3) refinance intermediate-term equity, and/or 4) sell cropland. If the operator was unable to cover the deficit in one of these ways, the farm was declared insolvent and the model proceeded to the next iteration after calculating the operator's accrued income and self-employment taxes.

Personal income taxes and self-employment taxes were calculated with the assumption that the operator was married, filed a joint income tax return, and itemized personal deductions. The regular income tax liability was computed using income averaging (if qualified) and the standard tax tables. The model selected the tax strategy that resulted in the lower income tax liability.

The farm was permitted to grow at the end of each year by purchasing cropland if the operator had cash available (after meeting all expenses) to cover a 30-percent downpayment for land and a 35-percent downpayment for any additional machinery necessary for the proposed larger farm. The operator was permitted to borrow against equity in land to meet up to 50 percent of the downpayment for land. The farm operation could also grow by leasing land if the operator had sufficient cash available to cover the 35-percent downpayment required for purchasing additional machinery needed to operate the larger farm. If machinery was purchased because of growth, the machinery was depreciated, the investment tax credit was calculated, and the operator's income taxes were recomputed.

After checking the farm's prospects for growth, the model updated the farm operator's balance sheet and cash flow statement and prepared to simulate the next year of the planning horizon. The steps in the simulation process described above are repeated for 10 years, or until the farm is declared insolvent. After completing each iteration, the model summarized the information for numerous key output variables and returned the farm to its initial economic situation (year one). This insured that the farm faced the same economic, policy, and physical relationships for each of 50 iterations analyzed.

Policy and Technology Scenarios

The three representative farms for each area were simulated for 10 years under the alternative scenarios described below. Seven farm policy scenarios (including a continuation of the 1981 farm bill), one income tax provision scenario, three financial bailout scenarios, and three alternative technology scenarios were simulated for each farm. All policy values associated with each scenario were held constant across farm sizes to allow direct comparison of their impacts on different farm sizes. Each scenario is described in detail in this section.

Farm Policy Scenarios

1. Base Policy.—The base policy scenario involves continuation of the 1981 farm bill through 1992 and continuation of the income tax provisions under the 1982 Tax Act through 1992. Annual mean crop yields were assumed to increase based on expected adoption of new technology, as indicated in the previous section. For this scenario it was assumed the following farm policies were in effect:

- CCC loan program is available to producers.
- An acreage diversion/set-aside program in effect for 1983-85, was used, excluding payment in kind. No acreage diversion/set-aside program was in effect for 1986 through 1992.
- A target price-deficiency payment program is available for cotton in all years.
- The $50,000-payment limitation for deficiency and diversion payments is in effect.
- Farms of all sizes are eligible to participate in these farm program provisions.

Values for loan rates, target prices, diversion rates, and diversion payment rates for 1983, 1984, and 1985 were set at their actual values. Loan rates and target prices for 1986 through 1992 were held constant at their 1985 levels.

It was assumed that the following options for depreciating machinery and calculating income taxes are used for the base scenario:

- Machinery and buildings placed in use prior to 1981 are depreciated using the double-declining balance method.
- Machinery and buildings placed in use after 1980 are depreciated using an ACRS method.
- The operator elects to claim first-year expensing for all depreciable items.
- The operator elects to take maximum ITC and reduce the basis.
- The operator adjusts crop sales across tax years to reduce current-year taxes.
- The operator may use either the regular income tax computation or income averaging to calculate Federal income tax liabilities.
- There is no maximum interest deduction for calculating taxable income.
- The actual self-employment tax rates and maximum income levels subject to this tax for 1983 and 1984 are used. Announced values for these variables in 1985 through 1986 were used, and the 1986 values were held constant through 1992.
- The operator elects to trade in old machinery on new replacements at the end of each item's economic life.

2. A Twenty-Percent Acreage Reduction.—The provisions of the base policy scenario were modified by adding a 15-percent set-aside with a 5-percent paid diversion for cotton in 1986 through 1992. Reasonable slippage (70 percent for cotton) and program participation rates were used to estimate the resulting increase in mean prices in 1986 through 1992. All other provisions of the base scenario were used without change.

3. No Farm Program Payment Limitation.—All provisions of the base scenario were used except that there was no limitation on diversion and deficiency payments.

4. No Price Supports and No Deficiency Payments.—The CCC loan and target price provisions under the base scenario were assumed to have been eliminated for all years in the planning horizon (1983-92). Annual mean prices were decreased based on the expected impact of removing the price and income support programs. Relative variability in prices about their means was increased based on the work of Morton, Devadoss, and Heady as to the effects of no farm program on U.S. agriculture. To isolate the impact of price and income supports on the representative farms, the acreage diversion and set-aside programs in the base policy for 1983 through 1985 were assumed to remain in effect.

5. No Target Price/Deficiency Payment.—The target price and deficiency payment provision was assumed to be eliminated for all years of the planning horizon 1983 through 1992. All other provisions of the base scenario were used without change to isolate the effects of removing only the deficiency payment.

6. Target Farm Program Benefits.—All farm program and income tax provisions of the base scenario were used except that farms with more than $300,000 of sales were not eligible to participate in farm program provisions. This program restriction excluded the very large farms from participating directly in the program provisions (CCC loan, target price/deficiency payments, and set-aside/diversions). Mean prices and relative variability in prices were not adjusted because sufficient "smaller" farms were assumed to be participating in the farm program for the price support actions of the CCC loan to function normally.

7. No-Farm Program.—All farm program provisions outlined for the base scenario were eliminated for all 10 years of the planning horizon. Mean annual prices and relative variance in prices for the no-price and income supports scenario (4) were used due to eliminating provisions of the CCC loan.

Income Tax Scenarios

8. Reduced Income Tax Benefits and Base Farm Program.—The Federal income tax provisions in place for the base policy scenario were made more restrictive. All farm policy provisions of the base scenario were left unchanged. The more restrictive Federal income tax provisions included the following:

- Machinery and buildings were depreciated using the straight-line cost recovery method.
- First-year expensing provisions were eliminated for all depreciable items.
- ITC provisions were continued, but the maximum ITC provision was eliminated.
- The maximum annual interest expense that could be used to reduce taxable income was $15,600. This value represented the annual interest expense deductions a consumer might have for a home, automobiles, and the like.
- The operator must sell obsolete machinery upon disposition rather than trading it in on new replacements, thus forcing recapture of excess depreciation deductions.

All other Federal income tax provisions for the base scenario were used as outlined earlier.

Financial Stress Scenarios

9. Base Finance Scenario.—Each farm's long-term debt-to-asset ratio was increased to 0.55, and its intermediate-term debt-to-asset ratio was increased to 0.60, to represent a highly leveraged farm. Annual long-term and intermediate-term interest rates were increased to their average values (0.1139 and 0.1343, respectively) for 1980 to 1983 to represent a farm that had been forced to refinance its assets during the past 4 years. The farm program provisions associated with the base policy scenario were continued for this scenario.

10. Debt Restructure.—The length of intermediate-term loans was increased by 1 year, and a portion of intermediate debt was converted to long-term debt. The conversion of intermediate-term debt was permitted as long as the long-term debt-to-asset ratio did not exceed 0.65. For some farms, this allowed all intermediate-term debt to be converted to long-term debt, while for other farms this constraint substantially restricted debt conversion. Total debt loads and farm program provisions were the same as those used for the base finance scenario (9).

11. Interest Subsidy.—The annual interest rates, debt levels, and farm program provisions in the base finance scenario (9) were simulated, but an interest subsidy was provided during the first 2 years. The interest subsidy took the form of an interest rate reduction equal to 3.4 percent for long-term interest rates and 5.4 percent for intermediate-term interest rates. These interest rate reductions were the amounts necessary to reduce the respective interest rates (0.1137 and 0.1343) to a 4-percent rate of interest.

No-New-Technology Scenarios

12. No-New-Technology and Base Farm Policy.—The Federal income tax and farm program provisions in the base policy scenario were simulated assuming no increase in mean yields over the planning horizon. For the no-new-technology scenarios, mean irrigated and dryland cotton yields for all 10 years were set equal to their respective means observed over the period 1974-83.

13. No-New-Technology and No Deficiency Payments.—The farm program provisions in the no-target-price/deficiency payments scenario (5) were simulated assuming the same average annual cotton yields used for the base no-new-technology scenario (12).

14. No-New-Technology and No-Farm Program.—All farm program provisions were eliminated (scenario 7), and annual average crop yields used for the base no new technology scenario (12) were assumed.

Evaluation Criteria

The FLIPSIM V model provides considerable detail about the viability of a representative farm at the end of each iteration, e.g., ending leverage ratio, ending net worth, ending farm size, total assets, total debt, net present value, and the solvency of the farm over 10 years. By repeating each scenario for 50 iterations, the model generates the information necessary for estimating values for key output variables. The means of these key output values are used to compare the economic impacts of selected policy and technology scenarios on representative farms. The following output variables for the model were selected to compare the impacts of the scenarios described in the previous section:

- *Probability of survival* is defined as the probability that the representative farm will remain solvent for 10 years. In other words, it is the probability that the farm operator will maintain at least the minimum financial ratios required by bankers in the local area for all 10 years of the planning horizon.
- *Probability of a positive net present value* is the probability that the representative farm will have a positive after-tax net present value. An

after-tax, real discount rate of 3 percent was used to calculate the farm's net present value. Thus this statistic indicates the probability of the representative farm providing at least a 3-percent real rate of return to the operator's initial net worth.

- *After-tax net present value (NPV)* is the present value of the operator's annual cash withdrawals (CW) plus the present value of the change in net worth (NW) minus the present value of annual off-farm income (OF):

$$NPV = \sum_{t=1}^{T} \frac{CW_t - OF_t}{(1.03)^t} + \frac{NW_T}{(1.03)^T} - NW_o$$

  Cash withdrawals equal family living expenses plus State and Federal income taxes and self-employment taxes. Initial net worth (NW_o) and ending net worth (NW_T) explicitly consider the value of off-farm investments and accrued taxes. A 3-percent after-tax, real discount rate was used to calculate net present value for all representative farms.
- *Present value of ending net worth* is used to indicate the change in the farm's real net worth over the planning horizon. Net worth is affected by increases (or decreases) in asset (land, machinery, and livestock) value and retained earnings. This value can be compared directly with initial net worth to indicate the relative magnitude of real financial growth.
- *Acres owned, leased, and controlled* at the end of the planning horizon for each iteration indicate the impacts of alternative scenarios on the rate of growth for representative farms. These three statistics provide an indication of how the farm grew through either the purchase or lease of land.
- *Total long-term and intermediate-term debts* at the end of the planning horizon provide an insight into the financial stress of the farm over the planning horizon. Increases in average ending debt from one scenario to another can be due either to rapid growth through purchasing land and machinery or to the farm operator being forced to refinance large cash flow deficits. When surplus cash is available, the operator is permitted to prepay intermediate-term debts first and then prepay new long-term debts. Therefore, large ending intermediate-term debts indicate insufficient cash was available to reduce intermediate-term debt through prepayment of principal.
- *Ending equity ratio* is the farm's ending ratio of total net worth to total assets. This ratio provides a "bottomline" measure for comparing the representative farm's ending financial position across scenarios.
- *Average annual net farm income* is the average net farm income received by the operator over all years simulated. Net farm income equals total farm receipts plus total Government payments minus all cash production expenses, interest payments, labor costs, fixed cash costs, and depreciation. This value excludes all nonfarm income and interest earned on cash reserves.
- *Average annual Government payment* is the average annual Government payment (deficiency and diversion payments) received over all years simulated.

Results of Analysis

Texas Southern High Plains Cotton Farms

The results indicate that under the most likely technology scenario and continuation of the provisions of the 1981 farm bill, all three representative cotton farms had a high probability of remaining solvent through 1992 (table E-1). Additionally, all three farms had an 88-percent or greater chance of receiving a reasonable return to equity. All three farms were able to grow over the 10-year planning horizon. The greatest percentage of increase in ending farm size was for the 1,088-acre farm, followed by the 3,383-acre farm and the 5,570-acre farm.

Imposing an acreage reduction program (acreage diversion and set-aside) increased net farm incomes and average net present value for all three farms. Acreage reduction programs increased the annual rate of growth more for the 1,088-acre farm than for the two larger farms.

Removing the deficiency payment program (income supports) reduced the probability of survival, net farm incomes, and annual growth rates for all three farms. The greatest percentage decrease in annual net farm income was experienced by the 1,088-acre farm, followed by the 3,383-acre farm. Similarly, the two smaller farms experienced greater reductions in their annual growth rates.

Removing both price supports (CCC loan) and deficiency payments reduced the probability of survival the most for the 1,088-acre farm (36 percent), whereas the probability of survival for the 5,570-acre farm fell only 2 percent. All three farms had slower rates of growth in the absence of price and income supports. The annual rate of growth for the 1,088-

Table E-1.— Comparison of Selected Farm Commodity and Income Tax Policy Scenarios on Representative Texas Southern High Plains Cotton Farms

Criteria	Alternative scenarios[a]							
	1	2	3	4	5	6	7	8
Moderate size (1,088 acres):								
Probability of survival	92.0	94.0	94.0	56.0	68.0	92.0	42.0	88.0
Present value of ending net worth ($1,000)	564.0	648.0	601.0	242.0	301.0	564.0	167.0	516.0
Ending farm size (acres)	1,558.0	1,635.0	1,648.0	1,216.0	1,274.0	1,558.0	1,213.0	1,565.0
Annual net farm income ($1,000)	8.3	13.3	11.9	−28.9	−21.7	8.2	−40.6	−6.0
Annual Government payment ($1,000)	26.0	22.2	29.5	1.3	1.1	25.9	0.0	25.8
Large size (3,383 acres):								
Probability of survival	90.0	94.0	94.0	72.0	82.0	86.0	62.0	88.0
Present value of ending net worth ($1,000)	1,412.0	1,697.0	1,853.0	931.0	1,055.0	1,191.0	801.0	1,226.0
Ending farm size (acres)	4,289.0	4,455.0	4,577.0	3,748.0	3,857.0	3,985.0	3,649.0	3,965.0
Annual net farm income ($1,000)	33.4	53.6	83.3	−14.8	3.6	12.9	−39.7	−7.2
Annual Government payment ($1,000)	38.0	35.1	83.3	3.2	3.0	16.8	0.0	37.9
Very large size (5,570 acres):								
Probability of survival	94.0	96.0	98.0	92.0	96.0	88.0	78.0	94.0
Present value of ending net worth ($1,000)	3,027.0	3,489.0	4,047.0	2,367.0	2,645.0	2,287.0	2,066.0	2,583.0
Ending farm size (acres)	6,002.0	6,047.0	6,514.0	5,781.0	5,848.0	5,727.0	5,736.0	5,746.0
Annual net farm income ($1,000)	66.6	100.6	170.6	−3.2	31.0	−13.9	−40.5	−15.6
Annual Government payment ($1,000)	40.2	39.1	135.8	4.8	4.6	0.0	0.0	40.4

[a]The scenarios are:
1. Continuation of the 1981 farm bill and 1983 Federal income tax provisions.
2. A 20-percent acreage reduction in 1986-92.
3. No farm program payment limitation in 1983-92.
4. No price supports and no deficiency payment in 1983-92.
5. No target price/deficiency payment in 1983-92.
6. Target farm program benefits to farms that produce less than $300,000 in program crops.
7. No farm program in 1983-92.
8. Reduced income tax benefits and the base farm program.

The impact of price supports can be derived by subtracting scenario 5 from scenario 6.
The impact of income supports can be derived by subtracting scenario 6 from scenario 1.
The impact of income supports with a $50,000 payment limitation can be found by subtracting scenario 6 from scenario 4.

SOURCE: Office of Technology Assessment.

acre farm was reduced five times more than for the 5,570-acre farm.

Removing all farm program provisions reduced the probability of survival for all three farms. The probability of survival declined from 92 to 42 percent for the 1,088-acre farm and from 90 to 62 percent for the 3,383-acre farm. The probability of survival for the 5,570-acre farm remained above 75 percent. Average annual net farm incomes for all three farms were substantially less than zero, and average net present values were considerably lower than under the current farm program.

Imposing a more restrictive set of Federal income tax provisions on the three representative farms reduced the average annual rate of growth more for the two larger farms. Net farm incomes were also reduced more for the two larger farms. Growth occurred by leasing because higher taxes reduced available cash for land purchases (downpayments).

The results of analyzing the three farms, assuming they were highly leveraged, reveal that debt restructuring would not greatly help these farms (table E-2). Although their probabilities of survival would not be increased, the farms would be able to remain solvent 1 to 3 years longer. A 2-year interest subsidy would provide greater benefits to net present value, net farm income, and ending net worth than would a debt restructure program.

Yield-enhancing technology anticipated over the next 10 years for cotton did not significantly change the average annual growth rates of the representative farms (table E-3). Changing the farm program or Federal income tax provisions had a greater impact on farm growth.

In conclusion, the results indicate that moderate (1,088-acre) cotton farms in the Texas Southern High Plains depend more on farm program provisions for their continued growth and economic viability than do larger farms. Larger farms are better able to survive without farm program benefits because of lower production costs (dollars/lb), higher average cotton lint prices, and a greater asset base from which to meet cash flow deficits. The loss of any farm program provision reduces the economic viability and

Table E-2.—Comparison of Selected Financial Bailout Scenarios for Three Representative Texas Southern High Plains Cotton Farms[a]

	Alternative scenarios for 1,088-acre farm			Alternative scenarios for 3,383-acre farm			Alternative scenarios for 5,570-acre farm		
Criteria	9	10	11	9	10	11	9	10	11
Probability of survival	64.0	66.0	72.0	56.0	50.0	60.0	66.0	64.0	66.0
Present value of ending net worth ($1,000)	304.0	314.0	343.0	604.0	600.0	733.0	1,310.0	1,356.0	1,619.0
Ending farm size (acres)	1,414.0	1,434.0	1,443.0	3,770.0	3,841.0	3,821.0	5,733.0	5,976.0	5,772.0
Annual net farm income ($1,000)	−5.4	−6.4	1.3	−9.1	−21.2	6.9	−41.8	−57.3	−6.3
Annual Government payment ($1,000)	24.4	24.8	24.7	36.8	36.4	37.2	41.1	41.3	41.6

[a]The scenarios are:
9. Continuation of the 1981 farm bill and the 1983 Federal tax provisions for a highly leveraged farm.
10. Restructure of debt for a highly leveraged farm.
11. Interest rate subsidy (buy-down) in the first 2 years for a highly leveraged farm.

SOURCE: Office of Technology Assessment.

Table E-3.—Comparison of Selected Policy Scenarios Assuming No New Technology for Three Representative Texas Southern High Plains Cotton Farms[a]

	Alternative scenarios for 1,088-acre farm			Alternative scenarios for 3,383-acre farm			Alternative scenarios for 5,570-acre farm		
Criteria	12	13	14	12	13	14	12	13	14
Probability of survival	92.0	68.0	42.0	88.0	78.0	60.0	94.0	90.0	76.0
Present value of ending net worth ($1,000)	552.0	290.0	161.0	1,325.0	966.0	738.0	2,807.0	2,322.0	1,843.0
Ending farm size (acres)	1,590.0	1,280.0	1,206.0	4,273.0	3,818.0	3,633.0	5,960.0	5,816.0	5,724.0
Annual net farm income ($1,000)	7.0	−22.2	−41.0	25.4	−3.6	−45.5	47.0	0.2	−65.9
Annual Government payment ($1,000)	26.3	1.1	0.0	37.9	3.0	0.0	40.5	4.8	0.0

[a]The scenarios are:
12. Continuation of the 1981 farm bill and the 1983 Federal tax provisions, assuming no-new-technology scenario.
13. No target price deficiency payment program, assuming no new technology scenario.
14. Deficiency plus diversion payments and any other Government payments received for Government loans and storage costs.

SOURCE: Office of Technology Assessment.

growth rate of the 1,088-acre farm more than the larger farms; however, all size farms are negatively affected.

Southern High Plains Wheat Farms

Three different size wheat farms in the Southern High Plains, representative of a majority of the commercial agricultural production for the region, were analyzed. The farms initially operating 1,280 acres, 1,920 acres, and 3,200 acres, reflected debt-to-asset ratios typical of farms in the area, owned the necessary machinery complement, and farmed both owned and leased cropland.

Analysis results indicate that under the most likely technology scenario and a continuation of the provisions of the 1981 farm bill (base scenario), all three representative wheat farms had a high probability of remaining solvent through 1992 (table E-4). Additionally, all three farms had a high probability of generating a reasonable return on equity and were able to grow over the 10-year planning horizon. The greatest percentage increase in average ending farm size was for the 1,280-acre farm, followed by the 1,920-acre operation.

Imposing an acreage reduction program (acreage set-aside and paid diversion) increased net farm incomes and average net present value for all three farms. Acreage reduction programs increased average ending farm size slightly more for the 3,200-acre farm than for the 1,280-acre farm.

Removing the deficiency payment program (income supports) reduced the probability of survival for only the 1,920-acre farm. Although each farm suffered a reduction in average annual net farm income, the reduction was significantly greater for the 1,280-acre farm. Average annual growth rates declined more for the two smaller farms.

Removing both price (CCC loan and farmer-owned reserve) and income supports (deficiency payments) reduced the probability of survival for both the 1,280-acre and 1,920-acre farms. All three farms experi-

Table E-4.—Comparison of Selected Farm Commodity and Income Tax Policy Scenarios on Representative Southern Plains Wheat Farms[a]

Criteria	Alternative scenarios[a] 1	2	3	4	5	6	7	8
Moderate size (1,280 acres):								
Probability of survival	100.0	100.0	100.0	76.0	100.0	100.0	48.0	100.0
Present value of ending net worth ($1,000)	803.0	1,032.0	811.0	283.0	426.0	761.0	189.0	710.0
Ending farm size (acres)	1,901.0	1,955.0	1,901.0	1,565.0	1,648.0	1,910.0	1,478.0	1,757.0
Annual net farm income ($1,000)	2.6	18.3	3.1	−33.6	−21.4	−0.9	−41.6	−8.3
Annual Government payment ($1,000)	30.9	31.5	31.6	2.5	2.5	27.7	0.0	29.4
Large size (1,920 acres):								
Probability of survival	100.0	100.0	100.0	50.0	90.0	96.0	32.0	100.0
Present value of ending net worth ($1,000)	1,028.0	1,359.0	1,117.0	294.0	475.0	696.0	179.0	833.0
Ending farm size (acres)	2,765.0	2,890.0	2,755.0	2,234.0	2,339.0	2,618.0	2,093.0	2,499.0
Annual net farm income ($1,000)	9.0	28.5	17.3	−52.5	−34.9	−17.6	−67.9	−21.8
Annual Government payment ($1,000)	39.0	39.1	44.7	4.2	3.7	16.2	0.0	37.3
Very large size (3,200 acres):								
Probability of survival	100.0	100.0	100.0	100.0	100.0	100.0	92.0	100.0
Present value of ending net worth ($1,000)	1,936.0	2,204.0	2,231.0	1,096.0	1,412.0	1,087.0	925.0	1,657.0
Ending farm size (acres)	4,218.0	4,365.0	4,483.0	3,552.0	3,834.0	3,494.0	3,472.0	3,805.0
Annual net farm income ($1,000)	48.9	59.5	78.4	−7.8	15.6	−13.6	−25.1	28.1
Annual Government payment ($1,000)	44.2	45.0	76.9	5.8	5.9	0.0	0.0	44.1

[a]The scenarios are:
1. Continuation of the 1981 farm bill and 1983 Federal income provisions.
2. A 20-percent acreage reduction in 1986-92.
3. No farm program payment limitation in 1983-92.
4. No price supports and no deficiency payment in 1983-92.
5. No target price/deficiency payment in 1983-92.
6. Target farm program benefits to farms that produce less than $300,000 in program crops.
7. No farm program in 1983-92.
8. Reduced income tax benefits and the base farm program.

The impact of price supports can be derived by subtracting scenario 5 from scenario 6.
The impact of income supports can be derived by subtracting scenario 6 from scenario 1.
The impact of income supports with a $50,000 payment limitation can be found by subtracting scenario 6 from scenario 4.

SOURCE: Office of Technology Assessment.

enced slower rates of growth as a result of eliminating price and income supports. Average ending farm size ranged from 19 to 16 percent less than under the base scenario. All three farms experienced negative annual net farm incomes on the average.

Removing all farm program provisions reduced the probability of survival for all three wheat farms. Probability of survival (100 percent under the base scenario) declined to 48 percent for the 1,280-acre farm, 32 percent for the 1,920-acre farm, and 92 percent for the 3,200-acre farm. Average ending net worth for the farms declined over the period, owing to a decline in land values.

Imposing a more restrictive set of Federal income tax provisions on the three representative wheat farms slowed the average annual growth rate more for the two larger farms. Farm growth occurred more by leasing cropland than by purchasing land, owing to reduced cash reserves. The 3,200-acre farm experienced the greatest absolute reduction in annual net farm income (about $20,000), followed by the 1,920-acre farm.

A 2-year interest rate subsidy program would provide greater benefits to highly leveraged wheat farms than would a debt restructure program. Probability of survival for a highly leveraged 1,920-acre wheat farm was increased from 40 to 80 percent by an interest rate subsidy (table E-5).

Yield-enhancing technology anticipated over the next 10 years will likely contribute to farm growth. The greatest benefit will accrue to those farms initially adopting the new technology (table E-6).

In conclusion, the results of this analysis indicate that moderate wheat farms—1,280 to 1,920 acres—in the Southern High Plains depend more on farm program provisions than do larger farms for their continued growth and economic viability. The loss of any farm program provision, however, negatively affects farms of all sizes.

Corn-Soybean Farms in the Corn Belt

All three Illinois farms had a survival probability at or near 100 percent under the entire range of farm program (and no program) alternatives considered here (table E-7). But the probability of positive after-tax net present value dropped dramatically (particularly for the medium [640-acre] and large [982-acre]

Table E-5.—Comparison of Selected Financial Bailout Scenarios for Three Representative Southern Plains Wheat Farms[a]

	Alternative scenarios for 1,280-acre farm			Alternative scenarios for 1,920-acre farm			Alternative scenarios for 3,200-acre farm		
Criteria	9	10	11	9	10	11	9	10	11
Probability of survival	86.0	98.0	100.0	40.0	70.0	80.0	100.0	100.0	100.0
Present value of ending net worth ($1,000)	289.0	408.0	383.0	258.0	399.0	406.0	1,248.0	1,373.0	1,348.0
Ending farm size (acres)	1,434.0	1,549.0	1,552.0	1,994.0	2,058.0	2,118.0	3,779.0	3,978.0	3,891.0
Annual net farm income ($1,000)	−22.5	−21.2	−14.3	−37.9	−35.1	−24.1	17.1	12.4	27.5
Annual Government payment ($1,000)	25.2	26.4	26.8	34.8	35.2	35.6	43.9	44.1	44.0

[a]The scenarios are:

9. Continuation of the 1981 farm bill and the 1983 Federal tax provisions for a highly leveraged farm.
10. Restructure of debt for a highly leveraged farm.
11. Interest rate subsidy (buy-down) in the first 2 years for a highly leveraged farm.

SOURCE: Office of Technology Assessment.

Table E-6.—Comparison of Selected Policy Scenarios Assuming No New Technology for Three Representative Southern Plains Wheat Farms[a]

	Alternative scenarios for 1,280-acre farm			Alternative scenarios for 1,920-acre farm			Alternative scenarios for 3,200-acre farm		
Criteria	12	13	14	12	13	14	12	13	14
Probability of survival	100.0	90.0	32.0	100.0	44.0	10.0	100.0	82.0	28.0
Present value of ending net worth ($1,000)	726.0	325.0	134.0	780.0	229.0	81.0	1,131.0	562.0	220.0
Ending farm size (acres)	1,859.0	1,632.0	1,430.0	2,605.0	2,304.0	2,048.0	3,699.0	3,542.0	3,322.0
Annual net farm income ($1,000)	−1.3	−28.9	−46.8	−10.9	−52.9	−77.1	−2.1	−45.4	−85.8
Annual Government payment ($1,000)	30.7	2.5	0.0	38.1	3.9	0.0	43.7	5.9	0.0

[a]The scenarios are:

12. Continuation of the 1981 farm bill and the 1983 Federal tax provisions, assuming no-new-technology scenario.
13. No target price deficiency payment program, assuming no new technology scenario.
14. Deficiency plus diversion payments and any other Government payments received for Government loans and storage costs.

farms) when farm program benefits were reduced or removed. In fact, even the loss of target price/deficiency payment programs dropped the probability of positive after-tax net present value into the range of 4 to 6 percent for these two representative farms. As a general rule, the largest farm fared the best with the loss of farm programs because it operates with a substantial acreage of rented land and suffers relatively less from the economic drag of servicing a high real estate debt load. Moreover, this very large unit had much less economic incentive to grow in size than do the two smaller farms.

All three Nebraska farms also had a survival probability at or near 100 percent under the entire range of program and no-program alternatives (table E-8). The loss of farm program benefits had its greatest adverse impact on the very large (2,085-acre) farm, probably because this unit has large machinery investments and uses much more full-time hired labor. In fact, the probability of realizing positive after-tax net present value dropped to the 8 to 12 percent range for this very large operation when program benefits were withdrawn or dramatically reduced. Economic performance measures for the medium (672-acre) and large (920-acre) farm also deteriorated under the latter condition. Overall, the generally stronger economic viability of the Nebraska farms (compared with Illinois farms) was attributable to much lower land prices and the lower debt servicing costs that result.

A modest reduction in income tax benefits did not have major economic impacts on the moderate and large farms in either Illinois or Nebraska. It has its greatest impact (a reduction of $5,800 in net farm income compared with that of the base scenario) for the very large (2,085-acre) farm in Nebraska. Even here, however, the impact was very small compared with the loss of economic benefits from either the target price/deficiency payment program or the entire complement of existing price and income supports.

Increasing debt loads to a level of 60 percent of machinery value and 55 percent of land value resulted in a heavy economic drag on all three Illinois farms, but somewhat less so for the Nebraska farms (tables E-9, E-10). This difference results mainly from

Table E-7.—Comparison of Selected Farm Commodity and Income Tax Policy Scenarios on Representative Corn-Soybean Farms in East Central Illinois[a]

Criteria	Alternative scenarios[a] 1	2	3	4	5	6	7	8
Moderate size (640 acres):								
Probability of survival	100.0	100.0	100.0	100.0	100.0	100.0	100.0	100.0
Present value of ending net worth ($1,000)	703.0	743.0	703.0	568.0	593.0	669.0	563.0	719.0
Ending farm size (acres)	902.0	904.0	902.0	824.0	837.0	907.0	834.0	893.0
Annual net farm income ($1,000)	23.2	29.9	23.2	10.2	11.8	19.1	11.1	19.0
Annual Government payment ($1,000)	11.6	9.8	11.6	0.7	0.7	8.6	0.0	11.7
Large size (982 acres):								
Probability of survival	100.0	100.0	100.0	100.0	100.0	100.0	100.0	100.0
Present value of ending net worth ($1,000)	975.0	970.0	991.0	645.0	693.0	801.0	622.0	852.0
Ending farm size (acres)	1,374.0	1,364.0	1,388.0	1,139.0	1,180.0	1,355.0	1,134.0	1,217.0
Annual net farm income ($1,000)	24.3	22.9	26.4	14.3	5.2	8.0	1.1	24.9
Annual Government payment ($1,000)	22.6	16.6	24.3	1.0	1.0	7.8	0.0	21.9
Very large size (1,630 acres):								
Probability of survival	100.0	100.0	100.0	100.0	100.0	100.0	100.0	100.0
Present value of ending net worth ($1,000)	1,267.0	1,348.0	1,266.0	991.0	1,033.0	1,056.0	1,036.0	1,044.0
Ending farm size (acres)	1,945.0	1,932.0	1,942.0	1,856.0	1,859.0	1,908.0	1,876.0	1,784.0
Annual net farm income ($1,000)	51.8	62.2	52.4	31.1	35.1	34.7	34.8	54.4
Annual Government payment ($1,000)	23.6	19.3	25.3	1.7	1.7	0.0	0.0	23.3

[a]The scenarios are:
1. Continuation of the 1981 farm bill and 1983 Federal income provisions.
2. A 20-percent acreage reduction in 1986-92.
3. No farm program payment limitation in 1983-92.
4. No price supports and no deficiency payment in 1983-92.
5. No target price/deficiency payment in 1983-92.
6. Target farm program benefits to farms that produce less than $300,000 in program crops.
7. No farm program in 1983-92.
8. Reduced income tax benefits and the base farm program.

The impact of price supports can be derived by subtracting scenario 5 from scenario 6.
The impact of income supports can be derived by subtracting scenario 6 from scenario 1.
The impact of income supports with a $50,000 payment limitation can be found by subtracting scenario 6 from scenario 4.

SOURCE: Office of Technology Assessment.

the much higher land prices on Illinois farms. Whereas survival probabilities dropped to as low as 72 percent for the 640-acre Illinois farm, they dropped only to 86 percent for the 672-acre Nebraska unit. Similarly, the probabilities for positive after-tax net present values dropped to as low as 16 percent for the 640-acre Illinois farm, but remained at 100 percent for all three Nebraska farms.

Because of the heavy real estate debt load on Illinois farms, these farms continued to have severe economic problems with either a debt restructuring or an interest rate subsidy type of financial bailout. Of the two, however, the interest rate subsidy was the most beneficial alternative, particularly for the smaller (640-acre and 982-acre) farms. Similarly, the interest rate subsidy was preferable (as compared with debt restructuring) for the two smallest Nebraska farms. Faced with substantial incentives for additional growth in size, these farms were not in a position to profit appreciably from debt restructuring. A financial bailout in the form of an interest rate subsidy improves net farm incomes and provides a "margin of safety" in the event of unexpected economic adversities.

The impact of eliminating new technology fell mainly on the very large farms (tables E-11 and E-12). These farms tend to be the early adopters of new technology, which generally results in a very favorable benefit/cost ratio. One should keep in mind, however, that the simulation analysis conducted here did not permit feedback on the price effects from increased output levels.

Because of high land and machinery costs, the survival probability for new entrants in Illinois was very low (0 to 4 percent). It was much higher (84 percent) for the base scenario on the 672-acre Nebraska farm. But this probability dropped to only 6 percent with the loss of all farm programs. Thus the economic survival of new entrants was particularly dependent on price and income benefits from farm programs (or of some other type of financial assistance).

As a practical matter, new entrants to farming can probably survive with high land prices and high interest rates only if they are able to lease most of their land resources or arrange for a postponement of a portion of their "early year" debt repayment obligations.

Table E-8.—Comparison of Selected Farm Commodity and Income Tax Policy Scenarios on Representative Irrigated Row Crop Farms in South Central Nebraska

Criteria	Alternative scenarios[a] 1	2	3	4	5	6	7	8
Moderate size (672 acres):								
Probability of survival	100.0	100.0	100.0	92.0	100.0	100.0	90.0	100.0
Present value of ending net worth ($1,000)	670.0	736.0	670.0	260.0	476.0	670.0	264.0	628.0
Ending farm size (acres)	921.0	909.0	921.0	882.0	870.0	921.0	808.0	917.0
Annual net farm income ($1,000)	26.8	31.0	26.8	−9.8	10.6	26.8	−11.4	26.8
Annual Government payment ($1,000)	17.3	14.5	17.3	1.0	1.0	17.3	0.0	17.9
Large size (920 acres):								
Probability of survival	100.0	100.0	100.0	100.0	100.0	100.0	100.0	100.0
Present value of ending net worth ($1,000)	1,349.0	1,377.0	1,369.0	739.0	1,084.0	1,180.0	750.0	1,269.0
Ending farm size (acres)	1,257.0	1,253.0	1,257.0	1,242.0	1,240.0	1,257.0	1,243.0	1,234.0
Annual net farm income ($1,000)	58.4	60.9	57.4	0.1	35.7	37.4	−0.5	58.9
Annual Government payment ($1,000)	24.1	19.3	23.9	1.3	1.3	15.3	0.0	24.4
Very large size (2,085 acres):								
Probability of survival	100.0	100.0	100.0	100.0	100.0	100.0	100.0	100.0
Present value of ending net worth ($1,000)	2,259.0	2,374.0	2,407.0	1,013.0	1,863.0	1,270.0	1,007.0	2,072.0
Ending farm size (acres)	2,375.0	2,383.0	2,384.0	2,167.0	2,280.0	2,197.0	2,128.0	2,330.0
Annual net farm income ($1,000)	118.6	127.3	134.6	1.3	88.0	10.8	−0.1	112.8
Annual Government payment ($1,000)	35.9	31.5	49.6	3.0	3.0	0.0	0.0	35.9

[a]The scenarios are:
1. Continuation of the 1981 farm bill and 1983 Federal Income tax provisions.
2. A 20-percent acreage reduction in 1986-92.
3. No farm program payment limitation in 1983-92.
4. No price supports and no deficiency payment in 1983-92.
5. No target price/deficiency payment in 1983-92.
6. Target farm program benefits to farms that produce less than $300,000 in program crops.
7. No farm program in 1983-92.
8. Reduced income tax benefits and the base farm program.

The impact of price supports can be derived by subtracting scenario 5 from scenario 6.
The impact of income supports can be derived by subtracting scenario 6 from scenario 1.
The impact of income supports with a $50,000 payment limitation can be found by subtracting scenario 6 from scenario 4.

SOURCE: Office of Technology Assessment.

Table E-9.—Comparison of Selected Financial Bailout Scenarios for Three Representative Corn-Soybean Farms in East Central Illinois[a]

Criteria	Alternative scenarios for 640-acre farm			Alternative scenarios for 982-acre farm			Alternative scenarios for 1,630-acre farm		
	9	10	11	9	10	11	9	10	11
Probability of survival	80.0	72.0	84.0	88.0	80.0	90.0	100.0	100.0	100.0
Present value of ending net worth ($1,000)	271.0	291.0	299.0	579.0	588.0	654.0	822.0	872.0	831.0
Ending farm size (acres)	653.0	689.0	662.0	1,046.0	1,062.0	1,073.0	1,795.0	1,740.0	1,712.0
Annual net farm income ($1,000)	−0.9	−3.3	3.8	2.0	−3.5	7.8	30.6	27.9	36.9
Annual Government payment ($1,000)	8.9	8.9	9.1	19.2	18.9	19.0	23.0	22.8	22.8

[a]The scenarios are:
9. Continuation of the 1981 farm bill and the 1983 Federal tax provisions for a highly leveraged farm.
10. Restructure of debt for a highly leveraged farm.
11. Interest rate subsidy (buy-down) in the first 2 years for a highly leveraged farm.

SOURCE: Office of Technology Assessment.

Table E-10.—Comparison of Selected Financial Bailout Scenarios for Three Representative Irrigated Row Crop Farms in South Central Nebraska[a]

	Alternative scenarios for 672-acre farm			Alternative scenarios for 920-acre farm			Alternative scenarios for 2,083-acre farm		
Criteria	9	10	11	9	10	11	9	10	11
Probability of survival	96.0	86.0	98.0	100.0	100.0	100.0	100.0	100.0	100.0
Present value of ending net worth ($1,000)	353.0	334.0	387.0	871.0	876.0	893.0	1685.0	1820.0	1714.0
Ending farm size (acres)	822.0	822.0	854.0	1,195.0	1,146.0	1,205.0	2,399.0	2,392.0	2,421.0
Annual net farm income ($1,000)	5.9	2.9	11.3	22.6	16.7	28.2	58.9	77.2	72.1
Annual Government payment ($1,000)	16.7	16.8	17.0	23.0	22.6	22.9	36.0	36.0	36.1

[a]The scenarios are:
9. Continuation of the 1981 farm bill and the 1983 Federal tax provisions for a highly leveraged farm.
10. Restructure of debt for a highly leveraged farm.
11. Interest rate subsidy (buy-down) in the first 2 years for a highly leveraged farm.

SOURCE: Office of Technology Assessment.

Table E-11.—Comparison of Selected Policy Scenarios Assuming No New Technology for Three Representative Corn-Soybean Farms in East Central Illinois[a]

	Alternative scenarios for 640-acre farm			Alternative scenarios for 982-acre farm			Alternative scenarios for 1,630-acre farm		
Criteria	12	13	14	12	13	14	12	13	14
Probability of survival	100.0	100.0	100.0	100.0	100.0	100.0	100.0	100.0	98.
Present value of ending net worth ($1,000)	699.0	589.0	561.0	862.0	604.0	540.0	915.0	694.0	672.0
Ending farm size (acres)	902.0	837.0	850.0	1,392.0	1,190.0	1,116.0	1,899.0	1,801.0	1,796.0
Annual net farm income ($1,000)	23.0	11.7	10.8	23.9	3.3	-0.8	25.3	9.8	6.1
Annual Government payment ($1,000)	11.6	0.7	0.0	22.9	1.0	0.0	22.9	1.7	0.0

[a]The scenarios are:
12. Continuation of the 1981 farm bill and the 1983 Federal tax provisions for a highly leveraged farm.
13. Restructure of debt for a highly leveraged farm.
14. Interest rate subsidy (buy-down) in the first 2 years for a highly leveraged farm.

SOURCE: Office of Technology Assessment.

Table E-12.—Comparison of Selected Policy Scenarios Assuming No New Technology for Three Representative Irrigated Row Crop Farms in South Central Nebraska[a]

	Alternative scenarios for 672-acre farm			Alternative scenarios for 920-acre farm			Alternative scenarios for 2,085-acre farm		
Criteria	12	13	14	12	13	14	12	13	14
Probability of survival	100.0	100.0	90.0	100.0	100.0	100.0	100.0	100.0	100.0
Present value of ending net worth ($1,000)	670.0	475.0	263.0	1,230.0	985.0	671.0	1,812.0	1,388.0	680.0
Ending farm size (acres)	921.0	870.0	808.0	1,257.0	1,221.0	1,226.0	2,402.0	2,240.0	2,107.0
Annual net farm income ($1,000)	26.7	10.6	-11.4	53.9	30.3	-2.6	77.5	51.0	-10.9
Annual Government payment ($1,000)	17.3	0.9	0.0	23.9	1.3	0.0	35.7	3.0	0.0

[a]The scenarios are:
12. Continuation of the 1981 farm bill and the 1983 Federal tax provisions, assuming no-new-technology scenario.
13. No target price deficiency payment program, assuming no-new-technology scenario.
14. Deficiency plus diversion payments and any other Government payments received for Government loans and storage costs.

SOURCE: Office of Technology Assessment.

General Crop Farms in the Delta Region of Mississippi

All three representative farms had a 100-percent probability of survival under the entire range of policy alternatives in that equity in land and machinery did not fall below 30 and 35 percent, respectively (table E-13). One of the principal reasons for the solvency of these farms over the 10-year planning horizon was the availability of off-farm income to meet some of the cash flow needs. The remaining criteria in table E-13 are indicative of farm size, wealth, and financial characteristics that are projected to occur on these representative farms over the 10-year simulation under each policy alternative.

The present value of ending net worth is one measure of real wealth accumulation. In comparing the policy scenarios for each size of farm, substantial greater growth in real net worth occurs on the representative farms under conditions that continue current farm commodity policy and income tax provisions with and without acreage reductions and farm program payments limitation (scenarios 1 to 3) and with a more restrictive set of income tax provisions (scenario 8). For the 1,443-acre farm real net worth increases by 105 to 151 percent under these program alternatives. The largest rate of growth in real net worth (a 151-percent increase from the initial situation) occurs for the alternative that continues the 1981 farm bill provisions, but with no farm program payments limitations (scenario 3). A policy that continues the current farm program but with a 20-percent acreage reduction in 1986 to 1992 results in a 135-percent growth in real net worth. Much lower growth rates in real net worth occur for the policy alternatives that eliminate various provisions of the current farm program, withdraws all farm program support, or targets the benefits to farms producing less than $300,000 of program crops. Similar patterns are evident in the effects of the policy alternatives on rates of growth in real net worth of the 3,119-acre farm and the 6,184-acre farm.

A second noticeable pattern is the decline in the growth rate in real wealth as the size of the representative farm increases from the 1,443-acre farm to the 6,184-acre farm for each of the policy alternatives. Comparisons among the different farm sizes must be made with caution because the initial total

Table E-13.—Comparison of Selected Farm Commodity and Income Tax Policy Scenarios on Representative General Crop Farms in the Delta of Mississippi

Criteria	Alternative scenarios[a]							
	1	2	3	4	5	6	7	8
Moderate size (1,443 acres):								
Probability of survival	100.0	100.0	100.0	100.0	100.0	100.0	100.0	100.0
Present value of ending net worth ($1,000)	1,651.0	1,757.0	1,881.0	1,106.0	1,134.0	1,059.0	1,070.0	1,533.0
Ending farm size (acres)	2,009.0	2,057.0	2,093.0	1,625.0	1,645.0	1,581.0	1,590.0	1,913.0
Annual net farm income ($1,000)	38.9	40.4	64.6	−14.2	−6.9	−16.3	−17.6	29.9
Annual Government payment ($1,000)	48.2	45.2	75.4	1.9	1.9	0.0	0.0	47.9
Large size (3,119 acres):								
Probability of survival	100.0	100.0	100.0	100.0	100.0	100.0	100.0	100.0
Present value of ending net worth ($1,000)	2,940.0	3,280.0	4,418.0	2,482.0	2,537.0	2,433.0	2,454.0	3,139.0
Ending farm size (acres)	3,327.0	3,340.0	3,877.0	3,119.0	3,135.0	3,119.0	3,119.0	3,135.0
Annual net farm income ($1,000)	38.3	65.1	148.0	−20.6	−8.2	−28.9	−25.1	21.8
Annual Government payment ($1,000)	49.9	49.1	160.6	4.7	4.8	0.0	0.0	49.9
Very large size (6,184 acres):								
Probability of survival	100.0	100.0	100.0	100.0	100.0	100.0	100.0	100.0
Present value of ending net worth ($1,000)	5,450.0	6,116.0	7,728.0	5,135.0	5,175.0	4,964.0	5,079.0	5,902.0
Ending farm size (acres)	6,248.0	6,254.0	6,530.0	6,270.0	6,245.0	6,242.0	6,267.0	6,203.0
Annual net farm income ($1,000)	41.9	118.2	227.1	−19.7	−0.6	−42.9	−32.4	5.9
Annual Government payment ($1,000)	49.9	49.8	278.0	7.9	0.0	0.0	0.0	49.9

[a]The scenarios are:

1. Continuation of the 1981 farm bill and 1983 Federal income provisions.
2. A 20-percent acreage reduction in 1986-92.
3. No farm program payment limitation in 1983-92.
4. No price supports and no deficiency payment in 1983-92.
5. No target price/deficiency payment in 1983-92.
6. Target farm program benefits to farms that produce less than $300,000 in program crops.
7. No farm program in 1983-92.
8. Reduced income tax benefits and the base farm program.

The impact of price supports can be derived by subtracting scenario 5 from scenario 6.
The impact of income supports can be derived by subtracting scenario 6 from scenario 1.
The impact of income supports with a $50,000 payment limitation can be found by subtracting scenario 6 from scenario 4.

SOURCE: Office of Technology Assessment.

equity to asset ratios differ. However, the results indicate that the policy alternatives involving farm program payments (scenarios 1 to 3 and scenario 8) induced a greater growth rate in real wealth on the moderate-size farm as compared with the two larger farms.

This pattern of growth is even more evident when examining changes in farm acreage. The 1,443-acre farm experienced considerable growth in both owned land acreage and/or acreage leased under scenarios 1 to 3 and scenario 8. In contrast the two larger farms exhibited less than 7-percent growth in farm size under these scenarios, with the exception of the 3,119-acre farm under scenario 3 wherein payments limitations are removed. The 1,443-acre farm experienced a 10- to 14-percent increase in acreage whereas the two larger farms exhibited virtually no growth in farm acreage for the policy alternatives involving elimination of some or all the program payments provisions and when program payments are targeted to farms with less than $300,000 of program commodity sales. These results indicate that farm program payments are an important inducement to growth of moderate-size general crops farms in the Delta of Mississippi Region.

The two largest representative farms reduced a substantial portion of the long-term real estate debt under all scenarios. The 1,443-acre farm had a much lower rate of long-term debt payback, principally because growth in farm size occurred through purchase of additional cropland under scenarios 1 to 3, and the use of accumulated cash to purchase machinery and equipment for expansion on leased land under scenarios 4 to 8. The 1,443-acre farm generally exhibited a larger liquidation of its intermediate-term debt than the two larger farms for each of the policy alternatives. Each of the representative farms tended to use income from both farm and nonfarm sources to pay back existing debts, and the ratio of total equity to total assets increased appreciably on each farm for all of the policy alternatives.

The three representative general crops farms in the Delta of Mississippi Region are very dependent on farm program payments in maintaining net farm income. Policy alternatives involving relatively little or no Government payments (scenarios 4 to 7) resulted in negative average annual net farm incomes.

All three farms had a 100-percent chance of remaining solvent and having a positive after-tax net present value over the 10-year planning horizon for each financial bailout alternative (table E-14). Present value of ending net worth increased substantially on each farm with the largest rate of growth occurring under the debt restructuring alternative. Each representative farm expanded its acreage, both through purchasing and leasing, with the smallest farm exhibiting the most rapid rate of growth.

The highly leveraged crops farms in this region exhibit characteristics that indicate survival and growth under financial bailout policies. The implementation of debt restructuring and interest rate subsidy policy alternatives would appear to stimulate substantial growth in farm acreage in this production region.

The no-new-technology scenarios had little effect on the probability of having a positive after-tax net present value on each farm (table E-15). It reduced slightly the probability under the policy of "No Farm Program." The probabilities of having a positive after-tax net present value did not change from the most likely technology situation on the 1,443-acre farm. The impacts of these modest technology driven yield increases on product prices were not evaluated. Consequently, in the base farm policy scenario, the moderate-size and large-size farms show small improvement in annual net farm income as a result of technological advance. The very large farm shows a substantial increase in net farm income since the technology adoption rate was much faster on this size of farm.

Rates of growth in cropland purchases, leasing and total farm acreage were almost identical under the two technology situations for a given representative farm. However, the 1,443-acre farm exhibited substantially higher growth rates in farm acreage than the two larger farms. Also, the rates of payback on long-term and intermediate-term loans under the two technology situations were nearly identical for a given representative farm.

These results indicate that the most likely technology changes projected for the Delta of Mississippi Region are expected to have the greatest impact on growth in real wealth and farm acreage of the 1,443-acre farm. The 3,119-acre farm and the 6,184-acre farm are expected to exhibit little growth in farm acreage over the 10-year simulation period. The economic impact expected from new technology is rather minimal compared with the economic impact from changing the farm commodity price and income support programs.

Table E-14.—Comparison of Selected Financial Bailout Scenarios for Three Representative General Crop Farms in the Delta of Mississippi

	Alternative scenarios for 1,443-acre farm			Alternative scenarios for 3,119-acre farm			Alternative scenarios for 6,184-acre farm		
Criteria	9	10	11	9	10	11	9	10	11
Probability of survival	100.0	100.0	100.0	100.0	100.0	100.0	100.0	100.0	100.0
Present value of ending net worth ($1,000)	1,563.0	1,656.0	1,545.0	3,237.0	3,431.0	2,968.0	5,259.0	5,840.0	4,990.0
Ending farm size (acres)	2,109.0	2,115.0	2,025.0	3,845.0	4,719.0	3,685.0	6,606.0	7,656.0	6,453.0
Annual net farm income ($1,000)	35.5	29.4	37.7	30.1	20.4	33.8	3.7	−14.8	5.4
Annual Government payment ($1,000)	48.4	48.4	48.3	49.9	49.9	49.9	49.9	49.9	49.9

[a]The scenarios are:
9. Continuation of the 1981 farm bill and the 1983 Federal tax provisions for a highly leveraged farm.
10. Restructure of debt for a highly leveraged farm.
11. Interest rate subsidy (buy-down) in the first 2 years for a highly leveraged farm.

SOURCE: Office of Technology Assessment.

Table E-15.—Comparison of Selected Policy Scenarios Assuming No New Technology for Three Representative General Crop Farms in the Delta of Mississippi

	Alternative scenarios for 1,443-acre farm			Alternative scenarios for 3,119-acre farm			Alternative scenarios for 6,104-acre farm		
Criteria	12	13	14	12	13	14	12	13	14
Probability of survival	100.0	100.0	100.0	100.0	100.0	100.0	100.0	100.0	100.0
Present value of ending net worth ($1,000)	1,513.0	1,104.0	1,043.0	2,786.0	2,451.0	2,354.0	2,286.0	4,915.0	4,715.0
Ending farm size (acres)	2,006.0	1,638.0	1,587.0	3,343.0	3,148.0	3,119.0	6,322.0	6,277.0	6,261.0
Annual net farm income ($1,000)	38.6	−7.3	−18.3	34.0	−11.9	−29.9	15.1	−27.5	−57.7
Annual Government payment ($1,000)	48.2	1.9	0.0	49.9	4.8	0.0	49.9	7.9	0.0

[a]The scenarios are:
12. Continuation of the 1981 farm bill and the 1983 Federal tax provisions, assuming no-new-technology scenario.
13. No target price deficiency payment program, assuming no-new-technology scenario.
14. Deficiency plus diversion payments and any other Government payments received for Government loans and storage costs.

SOURCE: Office of Technology Assessment.

Appendix F

Detailed Results of Microeconomic Impacts of Technology and Public Policy on Dairy Farms

Chapter 9 presented the summary results of the microeconomic impacts of public policies and technology on the viability of various size dairy farms. This appendix discusses in more detail the specific results of the analysis by area and size of farm. For more detail than is provided in this appendix the reader is advised to read the background paper on which this analysis is based. The paper is published in a separate volume to this report.

As with the crop farms analyses, the first step was to describe representative farms that included moderate, large, and very large farms in each production area. The second step involved a simulation of the representative farms using a Monte-Carlo, whole farm simulation model (FLIPSIM V) under alternative farm policy, income tax, finance, and technology scenarios. This model is described in appendix E.

Results for Alternative Policy and Technology Scenarios

Minnesota Dairy—52 Cows

Given the base policy scenario, the 52-cow Minnesota dairy had a 74-percent chance for survival after 10 years, but only a 26-percent chance of having a positive after-tax, net present value (table F-1). Average present value of ending net worth decreased from the initial $417,000 to $240,000 after 10 years. Total debt increased $199,000, and the equity-to-asset ratio declined from 0.71 to 0.44 over the 10-year period. Average annual net farm income was a −$22,000.

Both policy alternatives involving crop programs that increase feed prices (scenario 1) and the no-

Table F-1.—Comparison of Selected Policy Scenarios on a 52-Cow Minnesota Dairy

		Alternative scenarios[a]								
Criteria	Initial situation	Base	1	2	3	4	5	6	7	8
Probability of survival (percent)	NA	74	62	74	58	22	74	92	50	62
Probability of positive net present value (percent)	NA	26	22	26	18	8	24	38	20	22
After-tax net present value mean ($1,000)	NA	−61	−89	−61	−100	−198	−62	13	−103	−88
Average present value of ending net worth ($1,000)	417	240	213	240	202	114	238	310	191	214
Total debts after 10 years ($1,000)	168	367	392	367	403	443	370	309	400	391
Average ending equity ratio (fraction)	0.71	0.44	0.39	0.44	0.37	0.22	0.43	0.55	0.35	0.39
Average internal rate of return (fraction)	NA	−0.02	−0.03	−0.02	−0.04	−0.09	−0.02	0.01	−0.04	−0.03
Average annual net farm income ($1,000)	NA	−22	−25	−22	−27	−38	−27	−14	−26	−25

[a]The scenarios are:
Base—see chapter 9.
1—A 20-percent acreage reduction crop program—9-percent higher feed costs.
2—No crop program
3—Fifty cents-per-hundredweight lower milk price.
4—No dairy price suport program.
5—Reduce income tax benefit program.
6—Milk supply control program.
7—No information technology.
8—No bovine growth hormone technology.
NA=Not applicable.

SOURCE: Office of Technology Assessment.

Table F-2.—Comparison of Selected Policy Scenarios on a 125-Cow Minnesota Dairy

		Alternative scenarios[a]								
Criteria	Initial situation	Base	1	2	3	4	5	6	7	8
Probability of survival (percent)	NA	100	100	100	100	98	100	100	100	100
Probability of positive net present value (percent)	NA	96	88	96	86	44	94	98	54	82
After-tax net present value mean ($1,000)	NA	369	312	369	283	59	360	461	104	235
Average present value of ending net worth ($1,000)	969	1,120	1,072	1,119	1,049	835	1,083	1,190	869	1,007
Total debts after 10 years ($1,000)	302	154	208	154	235	518	197	93	459	289
Average ending equity ratio (fraction)	0.76	0.89	0.86	0.89	0.85	0.67	0.87	0.93	0.70	0.81
Average internal rate of return (fraction)	NA	0.05	0.04	0.05	0.04	0.01	0.05	0.05	0.02	0.03
Average annual net farm income ($1,000)	NA	20	12	20	8	−21	3	33	−15	2

[a]The scenarios are:
Base—see chapter 9.
1—A 20-percent acreage reduction crop program—9 percent higher feed costs.
2—No crop program
3—Fifty cents-per-hundredweight lower milk price.
4—No dairy price suport program.
5—Reduce income tax benefit program.
6—Milk supply control program.
7—No information technology.
8—No bovine growth hormone technology.
NA=Not applicable.

SOURCE: Office of Technology Assessment.

crop program that increases feed price variability (scenario 2) had little impact on the 52-cow Minnesota dairy because most of the feed is raised at the dairy rather than purchased. Crop yields were fixed, not variable, from year to year. However, simulations where crop yields were stochastic did not significantly change the results.

A support price 50 cents below the price in the base scenario reduced the probability of survival to 58 percent (scenario 3). The decline in the present value of ending net worth was more adversely affected, and debt was $36,000 higher compared with the results from the base scenario after 10 years.

Eliminating the price-stabilizing aspects of the dairy price support program (scenario 4) resulted in a probability of survival of 22 percent for the 52-cow Minnesota dairy. The equity-to-asset ratio was 0.22 after 10 years.

Eliminating income tax benefits (scenario 5) did not change the probability of survival but did adversely affect the present value of ending net worth and total debt when compared with the base scenario.

A supply control program (scenario 6) increased the probability of survival to 92 percent, and the equity-to-asset after 10 years increased from 0.44 to 0.55. Present value of ending net worth declined to $310,000 after 10 years. Total debt increased but less than that under the base scenario. The supply control scenario generally was more favorable than the base scenario for the 52-cow Minnesota dairy.

The probability of survival for the 52-cow Minnesota dairy would decline to 50 percent after 10 years if productivity gains from information and nutrition technology do not materialize. The probability of survival without bovine growth hormone would decline to only 62 percent because this size dairy would adopt this technology in 1989, near the end of the 10-year period.

Minnesota Dairy—125 Cows

Given the base policy scenario, the 125-cow Minnesota dairy had a 100-percent chance of surviving 10 years and a 96-percent chance of having a positive after-tax net present value (table F-2). The present value of ending net worth increased slightly from an initial $969,000 to $1,120,000 at the end of the 10-year period. Both long-term and intermediate-term debt were reduced, and the equity ratio increased from an initial 0.76 to 0.89 by the end of the 10 years.

Like the effects on the 52-cow Minnesota dairy, policies that increase the level or variability of feed prices have little impact on the financial performance (scenarios 1 and 2) of the 125-cow dairy. Because this dairy produces most of its own feed requirement, it is insulated from short-run variations in feed costs.

A support price lower by 50 cents per hundredweight reduced the present value of ending net worth from $1,120,000 under the base scenario to $1,049,000 (scenario 3). The total equity ratio was 0.85.

A dairy support program (scenario 4) reduced the probability of a positive after-tax present value to 44 percent. Both long-term and intermediate-term debt was $216,000 higher than under the base scenario.

Eliminating the income tax benefits (scenario 5) for the 125-cow Minnesota dairy resulted in higher debts and lower present value of ending net worth, equity ratio, and average annual net farm income after 10 years, compared with the results under the base scenario.

A supply control program (scenario 6) increased the probability of a positive after-tax net present value to 98 percent, compared with 96 percent under the base scenario. Under both the base and supply control scenarios, the 125-cow dairy had a 100 percent chance of surviving the 10-year period. The dairy showed good financial progress under the supply control program as the present value of ending net worth was $1,190,000. The total equity ratio was 0.93 at the end of the 10-year period. Average annual net farm income for the 10-year period was $33,000, up from $20,000 under the base scenario.

The probability of survival would remain at 100 percent without either information or bovine growth hormone technology. However, the probability of a positive net present value, average present value of ending net worth, and other financial performance measures were more adversely affected than under the base scenario.

Arizona Dairy—359 Cows

Given the base policy scenario, the 359-cow Arizona dairy had a 96 percent chance of survival and a 96 percent change of a positive after-tax net present value for the 10-year period (table F-3). The dairy showed good financial improvement over the 10 years as present value of ending net worth was $1,296,000 compared with $744,000 at the beginning of the period, most debt was paid, and total equity ratio increased from 0.71 to 0.93. Average annual net farm income was $14,000.

All feed is purchased by the 359-cow Arizona dairy, and the 9 percent increase in feed cost (scenario 1) reduced the probability of survival to 80 per-

Table F-3.—Comparison of Selected Policy Scenarios on a 359-Cow Arizona Dairy

Criteria	Initial situation	Base	1	2	3	4	5	6	7	8
		Alternative scenarios[a]								
Probability of survival (percent)	NA	96	80	96	94	42	96	96	92	94
Probability of positive net present value (percent)	NA	96	72	96	90	26	96	96	90	90
After-tax net present value mean ($1,000)	NA	829	172	822	593	−326	812	1,134	543	592
Average present value of ending net worth ($1,000)	744	1,296	768	1,288	1,120	276	1,247	1,486	1,028	1,107
Total debts after 10 years ($1,000)	298	41	254	43	72	603	42	26	82	78
Average ending equity ratio (fraction)	0.71	0.93	0.71	0.93	0.91	0.26	0.93	0.93	0.90	0.91
Average internal rate of return (fraction)	NA	0.09	0.01	0.09	0.06	−0.15	0.09	0.11	0.06	0.07
Average annual net farm income ($1,000)	NA	14	−64	13	−14	−121	6	54	−21	−19

[a]The scenarios are:
Base—see chapter 9.
1—A 20-percent acreage reduction crop program—9-percent higher feed costs.
2—No crop program
3—Fifty cents-per-hundredweight lower milk price.
4—No dairy price suport program.
5—Reduce income tax benefit program.
6—Milk supply control program.
7—No information technology.
8—No bovine growth hormone technology.
NA=Not applicable.

SOURCE: Office of Technology Assessment.

cent, compared with 96 percent under the base scenario. However, the dairy still made good financial progress as debts were reduced, and the present value of ending net worth was $768,000, compared with $744,000 at the beginning of the period. The no-crop-program scenario (scenario 2) increased the variability of feed prices but reduced the financial progress of the dairy relatively little.

A support price 50 cents below the base scenario (scenario 3) prices reduced the probability of survival and resulted in a lower present value of ending net worth. Total equity-to-asset ratio still increased to 0.91, 2 percentage points less than that of the base scenario.

Eliminating the dairy support program (scenario 4) reduced the probability of survival to 42 percent. The after-tax net present value was −$326. The present value of ending net worth was $1,020,000 less than under the base and more than 60 percent lower than at the beginning of the period. Debt was tripled, and the total equity-to-asset ratio decreased from 0.71 at the beginning of the period to 0.26 at the end of the 10 years. Average annual net farm income was −$121,000, compared with $14,000 under the base scenario.

Eliminating income tax advantages (scenario 5) slightly reduced the present value of ending net worth but had little effect on the probability of survival, remaining debt, and ending total equity-to-asset ratio.

With supply control (scenario 6) the probability of survival was 96 percent. Present value of ending net worth was $190,000 higher at the end of the 10-year period. Average annual net farm income was $54,000.

The financial progress under either the no information technology (scenario 7) or no bovine growth hormone (scenario 8) was somewhat less than under the base scenario that included both these technologies. The average annual net farm income became negative, and the probability of survival declined to 92 percent for the no information scenario and to 94 percent for the no bovine growth hormone scenario.

California Dairy—550 Cows

Given the base policy scenario, the 550-cow California dairy had a 96 percent probability of survival and positive net present value (table F-4). The dairy showed good financial improvement over the 10-year period under the base scenario. Present value of ending net worth increased from $1,261,000 at the beginning to $2,055,000 at the end of the 10-year

Table F-4.—Comparison of Selected Policy Scenarios on a 550-Cow California Dairy

		Alternative scenarios[a]									
Criteria	Initial situation	Base	1	2	3	4	5	6	7	8
Probability of survival (percent)	NA	96	80	96	94	62	96	96	88	94
Probability of positive net present value (percent)	NA	96	58	96	88	32	96	96	80	96
After-tax net present value mean ($1,000)	NA	1,178	157	1,169	796	−292	1,157	1,682	367	659
Average present value of ending net worth ($1,000)	1,261	2,055	1,267	2,045	1,792	799	1,971	2,349	1,360	1,672
Total debts after 10 years ($1,000)	464	105	405	110	157	739	109	92	300	185
Average ending equity ratio (fraction)	0.71	0.92	0.73	0.92	0.90	0.52	0.93	0.92	0.81	0.89
Average internal rate of return (fraction)	NA	0.08	0.00	0.08	0.06	−0.06	0.08	0.10	0.02	0.05
Average annual net farm income ($1,000)	NA	10	−117	9	−35	−16	−5	76	−86	−57

[a]The scenarios are:
Base—see chapter 9.
1—A 20-percent acreage reduction crop program—9-percent higher feed costs.
2—No crop program
3—Fifty cents-per-hundredweight lower milk price.
4—No dairy price suport program.
5—Reduce income tax benefit program.
6—Milk supply control program.
7—No information technology.
8—No bovine growth hormone technology.
NA=Not applicable.

SOURCE: Office of Technology Assessment.

period. Total debts were greatly reduced, and total equity-to-asset ratio increased from 0.71 to 0.92 over the 10-year period.

The crop acreage reduction policy (scenario 1), resulting in a 9 percent increase in feed costs, reduced the probability of survival to 80 percent and the probability of positive after-tax net present value to 58 percent. The no-crop-program scenario (scenario 2) had relatively little impact on the 550-cow California dairy.

A milk support price 50 cents lower (scenario 3) than under the base scenario reduced the probability of survival to 94 percent and the present value of ending net worth to $1,792,000. The total equity-to-asset ratio after 10 years was 0.90.

The no dairy price support program (scenario 4) reduced the probability of survival to 62 percent, and reduced the present value of ending net worth to $799,000. The total equity-to-asset ratio was 0.52 after 10 years.

Eliminating income tax advantages (scenario 5) had little impact on the financial performance of the 550-cow California dairy relative to the base scenario. The total debts and the total equity-to-asset ratio was about the same as under the base scenario.

A mandatory supply control program (scenario 6) resulting in a milk price $1 per hundredweight higher than in the base scenario had a favorable impact on the 550-cow California dairy. Compared with the base scenario the present value of ending net worth was $2,349,000, or 14 percent higher, and the after-tax net present value was $1,682,000, or 43 percent higher. The average annual net farm income increased to $76,000.

The average present value of ending net worth and the equity ratio increased from the beginning of the period to the end even without information (scenario 7) or bovine growth hormone (scenario 8) technology. Also, total debt was reduced. However, both the probability of survival and a positive net present value declined, and average annual net farm income became negative.

California Dairy—1,436 Cows

Given the base policy scenario, the 1,436-cow California dairy had a 98 percent probability of survival and showed strong financial progress from the beginning to the end of the 10-year period (table F-5). The present value of ending net worth was $7,332,000, compared with $2,538,000 at the beginning of the 10-year period. The total equity-to-asset ratio increased from 0.69 at the beginning of the period to

Table F-5.—Comparison of Selected Policy Scenarios on a 1,436-Cow California Dairy

		Alternative scenarios[a]								
Criteria	Initial situation	Base	1	2	3	4	5	6	7	8
Probability of survival (percent)	NA	98	96	98	98	96	98	100	96	98
Probability of positive net present value (percent)	NA	98	96	98	98	92	98	100	96	98
After-tax net present value mean ($1,000)	NA	6,473	3,923	6,454	5,375	2,523	6,415	8,103	3,246	4,450
Average present value of ending net worth ($1,000)	2,538	7,332	5,477	7,316	6,509	4,418	7,142	8,543	4,648	5,747
Total debts after 10 years ($1,000)	1,131	145	220	148	150	307	145	96	201	145
Average ending equity ratio (fraction)	0.69	0.94	0.92	0.94	0.94	0.91	0.94	0.95	0.94	0.95
Average internal rate of return (fraction)	NA	0.15	0.08	0.16	0.15	0.07	0.16	0.19	0.10	0.12
Average annual net farm income ($1,000)	NA	449	171	447	325	7	421	653	101	207

[a]The scenarios are:
Base—see chapter 9.
1—A 20-percent acreage reduction crop program—9-percent higher feed costs.
2—No crop program
3—Fifty cents-per-hundredweight lower milk price.
4—No dairy price suport program.
5—Reduce income tax benefit program.
6—Milk supply control program.
7—No information technology.
8—No bovine growth hormone technology.
NA=Not applicable.

SOURCE: Office of Technology Assessment.

0.94 at the end. Total debt was cut by 87 percent from the beginning to the end of the period.

The acreage reduction scenario, resulting in 9 percent higher feed prices, reduced the probability of survival to 96 percent (scenario 1). Total equity-to-asset ratio was 0.92.

Eliminating the crop program (scenario 2) had little impact on the financial position of the 1,436-cow California dairy. Debt, equity-to-asset ratio, and present value of ending net worth were all about the same as under the base scenario.

A milk support price 50 cents lower (scenario 3) than under the base scenario resulted in an 11 percent lower present value of ending net worth after 10 years. Total debt increased slightly, and the total equity-to-asset ratio remained the same as the base.

Eliminating the dairy price support program (scenario 4) reduced the probability of survival to 96 percent. The present value of ending net worth was $4,418,000, or 40 percent lower than under the base scenario.

Eliminating the income tax advantages (scenario 5) did not affect the probability of survival, total debt, or equity-to-asset ratio. However, the present value of ending net worth decreased 2.6 percent from the base scenario.

With supply control (scenario 6), the probability of survival of the 1,436-cow California dairy increased to 100 percent. The present value of ending net worth was $8,543,000, or about 17 percent higher than under the base scenario. All intermediate-term debt was paid, and only $96,000 of long-term debt remained after 10 years under the supply control program. Average annual net farm income increased 45 percent, to $653,000.

If the productivity gains associated with information on bovine growth hormone technologies do not materialize, the financial performance will be adversely affected, but the probability of survival for the 1,436-cow California dairy would remain about the same. Average present value of ending net worth was 37 percent lower without information and nutrition technology (scenario 7) and 22 percent lower without bovine growth hormone technology (scenario 8).

Florida Dairies

Given the base policy scenario, the probability of survival was 98 percent for the 350-cow and 100 percent for the 600-cow and 1,436-cow Florida dairies (tables F-6, F-7, and F-8). Debt was reduced over the 10-year period on all three dairies. The total equity-to-asset ratio after 10 years was at least 0.84. The present value of ending net worth increased 33 per-

Table F-6.—Comparison of Selected Policy Scenarios on a 350-Cow Florida Dairy

		Alternative scenarios[a]								
Criteria	Initial situation	Base	1	2	3	4	5	6	7	8
Probability of survival (percent)	NA	96	58	96	86	36	94	98	86	88
Probability of positive net present value (percent)	NA	88	34	88	78	28	86	96	66	78
After-tax net present value mean ($1,000)	NA	448	−185	445	235	−305	425	663	154	270
Average present value of ending net worth ($1,000)	757	1,004	463	1,002	825	317	936	1,164	706	846
Total debts after 10 years ($1,000)	304	198	577	198	300	679	226	130	376	287
Average ending equity ratio (fraction)	0.71	0.84	0.46	0.84	0.74	0.32	0.81	0.89	0.68	0.76
Average internal rate of return (fraction)	NA	0.07	−0.04	0.07	0.03	−0.12	0.07	0.10	0.03	0.05
Average annual net farm income ($1,000)	NA	−6	−83	−6	−31	−97	−9	25	−40	−28

[a]The scenarios are:
Base—see chapter 9.
1—A 20-percent acreage reduction crop program—9-percent higher feed costs.
2—No crop program
3—Fifty cents-per-hundredweight lower milk price.
4—No dairy price suport program.
5—Reduce income tax benefit program.
6—Milk supply control program.
7—No information technology.
8—No bovine growth hormone technology.
NA=Not applicable.

SOURCE: Office of Technology Assessment.

Table F-7.—Comparison of Selected Policy Scenarios on a 600-Cow Florida Dairy

		Alternative scenarios[a]								
Criteria	Initial situation	Base	1	2	3	4	5	6	7	8
Probability of survival (percent)	NA	100	96	100	100	72	100	100	98	100
Probability of positive net present value (percent)	NA	100	80	100	100	56	100	100	86	98
After-tax net present value mean ($1,000)	NA	1,617	602	1,612	1,281	85	1,576	2,011	701	1,151
Average present value of ending net worth ($1,000)	1,465	2,453	1,778	2,455	2,255	1,268	2,306	2,681	1,748	2,164
Total debts after 10 years ($1,000)	468	116	280	116	124	587	116	116	233	129
Average ending equity ratio (fraction)	0.76	0.94	0.86	0.94	0.94	0.66	0.94	0.93	0.89	0.94
Average internal rate of return (fraction)	NA	0.10	0.06	0.10	0.09	−0.04	0.11	0.12	0.06	0.08
Average annual net farm income ($1,000)	NA	83	−28	82	44	−97	68	137	−24	23

[a]The scenarios are:
Base—see chapter 9.
1—A 20-percent acreage reduction crop program—9-percent higher feed costs.
2—No crop program
3—Fifty cents-per-hundredweight lower milk price.
4—No dairy price suport program.
5—Reduce income tax benefit program.
6—Milk supply control program.
7—No information technology.
8—No bovine growth hormone technology.
NA=Not applicable.

SOURCE: Office of Technology Assessment.

Table F-8.—Comparison of Selected Policy Scenarios on a 1,436-Cow Florida Dairy

		Alternative scenarios[a]								
Criteria	Initial situation	Base	1	2	3	4	5	6	7	8
Probability of survival (percent)	NA	100	100	100	100	100	100	100	100	100
Probability of positive net present value (percent)	NA	100	100	100	100	100	100	100	100	100
After-tax net present value mean ($1,000)	NA	8,396	6,020	8,387	7,501	4,712	8,182	9,413	5,046	6,667
Average present value of ending net worth ($1,000)	3,343	9,257	7,560	9,263	8,591	6,625	8,966	10,038	6,650	7,874
Total debts after 10 years ($1,000)	1,053	195	195	195	195	195	195	195	195	195
Average ending equity ratio (fraction)	0.76	0.95	0.95	0.95	0.95	0.95	0.95	0.95	0.96	0.96
Average internal rate of return (fraction)	NA	0.16	0.13	0.16	0.15	0.11	0.16	0.17	0.12	0.14
Average annual net farm income ($1,000)	NA	635	404	634	536	242	607	769	272	433

[a]The scenarios are:
Base—see chapter 9.
1—A 20-percent acreage reduction crop program—9-percent higher feed costs.
2—No crop program
3—Fifty cents-per-hundredweight lower milk price.
4—No dairy price suport program.
5—Reduce income tax benefit program.
6—Milk supply control program.
7—No information technology.
8—No bovine growth hormone technology.
NA=Not applicable.

SOURCE: Office of Technology Assessment.

cent on the 350-cow dairy, 67 percent on the 600-cow dairy, and 177 percent on the 1,436-cow dairy.

The crop acreage reduction policy (scenario 1) reduced the probability of surviving the 10-year period to 58 percent for the 350-cow dairy and to 96 percent for the 600-cow dairy. The probability of survival remained at 100 percent on the large 1,436-cow Florida dairy.

Eliminating the crop program (scenario 2) had relatively little impact on all three Florida dairies. The probability of survival remained unchanged at 100 percent for the 600-cow and 1,436-cow dairies and at 96 percent on the 350-cow dairy.

A milk support price 50 cents below the base (scenario 3) did not affect the probability of survival for the 1,436-cow dairy but reduced the probability of survival for the 350-cow dairy to 86 percent. The present value of ending net worth was about 18 percent lower for the 350-cow dairy, 8 percent lower for the 600-cow dairy, and 7 percent lower for the 1,436-cow dairy.

The no-dairy-program scenario (scenario 4) reduced the probability of survival to 36 percent for the 350-cow dairy and to 72 percent for the 600-cow dairy. The probability of survival remained at 100 percent for the 1,436-cow dairy. The present value of ending net worth was at least 28 percent less under the no-dairy-program scenario than under the base scenario for all three Florida dairies.

Eliminating income tax advantages (scenario 5) had relatively little impact on all three Florida dairies. The present value of ending net worth was about 7 percent less for the 350-cow dairy but only 3 percent less for the 1,436-cow dairy. Total debt remained the same for the 1,436-cow dairy but was somewhat higher for the smaller dairies.

A supply control policy (scenario 6) improved the financial position of each of the dairies. Compared with the base scenario, the present value of ending net worth increased about 16 percent for the 350-cow dairy, 9 percent for the 600-cow dairy, and 8 percent for the 1,436-cow dairy.

Like dairies in other regions, both no information technology (scenario 6) and no bovine growth hormone technology (scenario 8) adversely affected the financial positions of the Florida dairies after 10 years.

Results for Financial Stress Scenarios

Minnesota Dairies

Given the high debt (scenario 9), the 52-cow Minnesota dairy had a zero-percent probability of survival over the 10-year period (table F-9). The probability of survival remained zero even with subsidized interest (scenario 10) and restructuring debt (scenario 11). The same result was obtained for new entrant operators (scenarios 12, 13, and 14).

The 125-cow Minnesota dairy in an initial high debt position had a 24 percent probability of survival (scenario 9) (table F-10). A new entrant with high debt also had a zero probability of survival (scenario 12). Therefore, there was a zero probability of survival under both higher feed costs (scenario 13) and no dairy programs (scenario 14).

Table F-9.—Comparison of Selected Policy Scenarios on a 52-Cow Minnesota Dairy, Assuming High Debt and New Entrant Conditions

Criteria	High debt				New entrants			
	Initial situation	Scenarios 9	10	11	Initial situation	Scenarios 12	13	14
Probability of survival (percent)	NA	0	0	0	NA	0	0	0
Probability of positive net present value (percent)	NA	0	0	0	NA	0	0	0
After-tax net present value mean ($1,000)	NA	−103	−108	−99	NA	−103	−103	−102
Average present value of ending net worth ($1,000)	246	104	96	109	264	143	144	145
Total debts after 10 years ($1,000)	340	423	424	419	466	505	503	504
Average ending equity ratio (fraction)	0.42	0.21	0.20	0.22	0.36	0.23	0.23	0.23
Average internal rate of return (fraction)	NA	−0.17	−0.19	−0.16	NA	−0.11	−0.10	−0.10
Average annual net farm income ($1,000)	NA	−46	−42	−44	NA	−90	−90	−91

[a]The scenarios are:
9—Base—continuation of present dairy policy and assuming high debt.
10—Subsidize interest rate so that effective rate on all loans is 8 percent.
11—Restructure debt.
12—Base policy and new entrant.
13—New entrant and no price support for dairy.
14—New entrant and a 9-percent increase in feed costs.
NA=Not applicable.

SOURCE: Office of Technology Assessment.

Table F-10.—Comparison of Selected Policy Scenarios on a 125-Cow Minnesota Dairy, Assuming High Debt and New Entrant Conditions

	High debt				New entrants			
	Initial	Scenarios			Initial	Scenarios		
Criteria	situation	9	10	11	situation	12	13	14
Probability of survival (percent)	NA	24	16	28	NA	0	0	0
Probability of positive net present value (percent)	NA	24	16	28	NA	0	0	0
After-tax net present value mean ($1,000)	NA	−80	−147	−59	NA	−237	−256	−241
Average present value of ending net worth ($1,000)	554	341	280	351	575	291	274	292
Total debts after 10 years ($1,000)	718	910	943	906	963	1,016	1,026	1,021
Average ending equity ratio (fraction)	0.44	0.30	0.25	0.31	0.37	0.23	0.22	0.24
Average internal rate of return (fraction)	NA	−0.06	−0.09	−0.06	NA	−0.15	−0.16	−0.16
Average annual net farm income ($1,000)	NA	−51	−58	−49	NA	−117	−121	−121

[a]The scenarios are:
9—Base—continuation of present dairy policy and assuming high debt.
10—Subsidize interest rate so that effective rate on all loans is 8 percent.
11—Restructure debt.
12—Base policy and new entrant.
13—New entrant and no price support for dairy.
14—New entrant and a 9-percent increase in feed costs.
NA=Not applicable.

SOURCE: Office of Technology Assessment.

Table F-11.—Comparison of Selected Policy Scenarios on a 359-Cow Arizona Dairy, Assuming High Debt and New Entrant Conditions

	High debt				New entrants			
	Initial	Scenarios			Initial	Scenarios		
Criteria	situation	9	10	11	situation	12	13	14
Probability of survival (percent)	NA	66	70	68	NA	52	6	16
Probability of positive net present value (percent)	NA	66	70	68	NA	52	6	16
After-tax net present value mean ($1,000)	NA	417	466	440	NA	270	−320	−173
Average present value of ending net worth ($1,000)	471	709	746	730	528	583	64	217
Total debts after 10 years ($1,000)	570	326	297	325	715	459	787	700
Average ending equity ratio (fraction)	0.45	0.63	0.66	0.64	0.42	0.52	0.04	0.23
Average internal rate of return (fraction)	NA	0.01	0.03	0.02	NA	−0.01	−0.22	−0.08
Average annual net farm income ($1,000)	NA	−59	−46	−56	NA	−109	−191	−191

[a]The scenarios are:
9—Base—continuation of present dairy policy and assuming high debt.
10—Subsidize interest rate so that effective rate on all loans is 8 percent.
11—Restructure debt.
12—Base policy and new entrant.
13—New entrant and no price support for dairy.
14—New entrant and a 9-percent increase in feed costs.
NA=Not applicable.

SOURCE: Office of Technology Assessment.

Arizona Dairy—359 Cows

Given the high debt base (scenario 9), the 359-cow Arizona dairy had a 66-percent probability of survival and improved its financial position from the beginning to the end of the 10-year period (table F-11). The present value of ending net worth was $709,000, compared with a $471,000 beginning net worth. Total debt was reduced $244,000 over the 10-year period. The total equity-to-asset ratio increased from 0.45 at the beginning of the period to 0.63 at the end.

Restructuring debt (scenario 11) increased the probability of survival from 66 to 70 percent and improved the present value of ending net worth and equity-to-asset ratio compared with that of the high debt base (scenario 9).

A new entrant with high debt had a 52-percent probability to survive over the 10-year period (scenario 12). A new entrant under the no-dairy-price-support program (scenario 13) had only a 6-percent probability of survival. Increased feed costs (scenario 14) decreased a new entrant's probability of survival from 52 to 16 percent.

Florida Dairy—350 Cows

Given the high debt base (scenario 9), the 350-cow Florida dairy had only a 34-percent probability of survival (table F-12). Present value of ending net worth was $375,000; beginning net worth was $466,000. The total equity-to-asset ratio decreased from 0.44 at the beginning to 0.38 at the end of the 10-year period.

The interest subsidy (scenario 10) increased the probability of survival from 38 percent to 42 percent and increased the present value of ending net worth from $375,000 to $382,000, compared with the results of high debt base (scenario 9). The debt restructuring policy (scenario 11) improved present value of ending net worth for the 350-cow Florida dairy compared with the results of high debt base scenario (scenario 9). However, the improvement was considerably less than under the interest subsidy scenario.

A new entrant with high debt had only a 22-percent probability of survival (scenario 12). The probability of survival declined to 2 percent, given a 9-percent higher feed cost (scenario 13). The no-dairy-price-support program reduced the probabilities of survival of a new entrant and a high debt dairy from 22 to 2 percent (scenario 14).

Table F-12.—Comparison of Selected Policy Scenarios on a 350-Cow Florida Dairy, Assuming High Debt and New Entrant Conditions

	High debt				New entrants			
	Initial	Scenarios			Initial	Scenarios		
Criteria	situation	9	10	11	situation	12	13	14
Probability of survival (percent)	NA	34	38	36	NA	22	2	2
Probability of positive net present value (percent)	NA	34	38	36	NA	22	2	2
After-tax net present value mean ($1,000)	NA	6	26	23	NA	−75	−297	−263
Average present value of ending net worth ($1,000)	466	375	382	392	527	326	116	181
Total debts after 10 years ($1,000)	594	609	609	602	773	741	897	897
Average ending equity ratio (fraction)	0.44	0.38	0.39	0.39	0.41	0.31	0.10	0.17
Average internal rate of return (fraction)	NA	−0.05	−0.04	−0.04	NA	−0.05	−0.22	−0.03
Average annual net farm income ($1,000)	NA	−88	−76	−85	NA	−155	−194	−211

[a]The scenarios are:
9—Base—continuation of present dairy policy and assuming high debt.
10—Subsidize interest rate so that effective rate on all loans is 8 percent.
11—Restructure debt.
12—Base policy and new entrant.
13—New entrant and no price support for dairy.
14—New entrant and a 9-percent increase in feed costs.
NA=Not applicable.

SOURCE: Office of Technology Assessment.

Appendix G

Workgroups, Background Papers, and Acknowledgments

Workgroups

Seven technical workgroups comprised of preeminent physical, biological, and social scientists from the public and private sectors provided and/or reviewed much of the technical information for this study. OTA greatly appreciates the time and effort of each member of these workgroups.

Animal Technology Workgroup

Eugene C. Allen
Professor and Associate Dean
College of Agriculture
University of Minnesota
St. Paul, MN

Howard L. Bachrach
Consultant
ARS-USDA
Plum Island Animal Disease Center
Greenport, NY

John B. Campbell
Professor
North Platte Agricultural Experiment Station
North Platte, NE

Stanley E. Curtis
Professor
Department of Animal Science
University of Illinois
Urbana, IL

James R. Fischer
Associate Director
Agricultural Experiment Station
Michigan State University
East Lansing, MI

William Hansel
Liberty Hyde Bailey Professor of Animal Physiology
Department of Physiology
New York State College of Veterinary Medicine
Cornell University
Ithaca, NY

Bennie I. Osburn
Professor of Pathology
Associate Dean of Research
School of Veterinary Medicine
University of California
Davis, CA

Wilson G. Pond
Research Leader
ARS-USDA
Roman L. Hruska U.S. Meat Animal Research Center
Clay Center, NE

Plant Technology Workgroup

Joseph A. Berry
Staff Member
Plant Biology Department
Carnegie Institution of Washington
Stanford, CA

J.A. Browning
Chairman
Department of Plant Sciences
Texas A&M University
College Station, TX

Robert T. Fraley
Manager
Plant Molecular Biology Group
Monsanto Agriculture Products Co.
St. Louis, MO

George G. Kennedy
Associate Professor
Department of Entomology
North Carolina State University
Raleigh, NC

Chester G. McWhorter
Chief, Southern Weed Science Laboratory
Agricultural Research Service
Delta State Research Center
U.S. Department of Agriculture
Stoneville, MS

Louis G. Nickell
Vice President of Research and Development
Velsicol Chemical Co.
Chicago, IL

Soil and Water Technology Workgroup

Martin Alexander
Professor of Soil Microbiology
Department of Agronomy
Cornell University
Ithaca, NY

Larry L. Boersma
Professor of Soil Physics
Department of Soil Science
Oregon State University
Corvallis, OR

Charles H. Davis
Director of Chemical Division
National Fertilizer Development Center
Tennessee Valley Authority
Muscle Shoals, AL

George R. Foster
Hydraulic Engineer
National Soil Erosion Lab
ARS-USDA
Purdue University
West Lafayette, IN

Charles A. Francis
Professor
Department of Agronomy
University of Nebraska
Lincoln, NE

Richard R. Harwood
Program Officer
Winrock International
Arlington, VA

Economics and Policy Workgroup

Boyd M. Buxton
Research Leader
ERS-USDA
Department of Agriculture and
Applied Economics
University of Minnesota
St. Paul, MN

Stephen C. Cooke
Economist
Department of Agriculture and
Applied Economics
University of Minnesota
St. Paul, MN

B.R. Eddleman
Professor
Department of Agricultural Economics
Mississippi State University
Mississippi State, MS

S.R. Johnson
Professor of Economics and Administrator
Center for Agricultural Rural Development
Iowa State University
Ames, IA

Ronald D. Knutson
Professor
Department of Agricultural Economics
Texas A&M University
College Station, TX

James W. Richardson
Associate Professor
Department of Agricultural Economics
Texas A&M University
College Station, TX

Wesley B. Sundquist
Professor
Department of Agriculture and
Applied Economics
University of Minnesota
St. Paul, MN

Agricultural Finance and Credit Policy Workgroup

Peter J. Barry
Professor
Department of Agricultural Economics
University of Illinois
Urbana, IL

John R. Brake
Professor
Department of Agricultural Economics
Cornell University
Ithaca, NY

David A. Lins
Associate Professor
Department of Agricultural Economics
University of Illinois
Urbana, IL

Ronald Meekhof
Agricultural Economist
Economic Research Service
U.S. Department of Agriculture
Washington, DC

John Penson
Professor
Department of Agricultural Economics
Texas A&M University
College Station, TX

Rural Community Workshop

Frederick H. Buttel
Associate Professor
Department of Rural Sociology
Cornell University
Ithaca, NY

Cornelia Butler Flora
Professor
Department of Sociology, Anthropology, and Social Work
Kansas State University
Manhattan, KS

Dean MacCannell
Professor
Department of Applied Behavioral Sciences
University of California
Davis, CA

Louis E. Swanson
Assistant Professor
Department of Sociology
University of Kentucky
Lexington, KY

J.C. van Es
Professor
Department of Agricultural Economics
University of Illinois
Urbana, IL

Agricultural Research and Extension Policy Workgroup

George Hyatt
Past Director
Cooperative Extension Service
North Carolina State University
Raleigh, NC

Ronald D. Knutson
Professor
Department of Agricultural Economics
Texas A&M University
College Station, TX

Roy L. Lovvorn
Retired
North Carolina State University and U.S. Department of Agriculture
Raleigh, NC

Fred C. White
Professor
Department of Agricultural Economics
University of Georgia
Athens, GA

Background Papers

This report was possible because of the valuable information and analyses contained in the background reports commissioned by OTA. These papers were reviewed and critiqued by the workgroups, advisory panel, and outside reviewers. The papers are available in a separate five-part volume through the National Technical Information Service.*

Part A: The Emerging Technologies

Alexander, Martin, "Biological Nitrogen Fixation," paper prepared for the Office of Technology Assessment, 1985.
Allen, Eugene C., "Regulation of Growth and Development," paper prepared for the Office of Technology Assessment, 1985.
Bachrach, Howard L., "Genetic Engineering in Animals," paper prepared for the Office of Technology Assessment, 1985.
Battelle Columbus Laboratories, "Communication and Information Management," paper prepared for the Office of Technology Assessment, 1985.
Battelle Columbus Laboratories, "Crop Separation, Cleaning, and Processing," paper prepared for the Office of Technology Assessment, 1985.
Battelle Columbus Laboratories, "Engine and Fuels," paper prepared for the Office of Technology Assessment, 1985.
Battelle Columbus Laboratories, "Labor Saving Technology," paper prepared for the Office of Technology Assessment, 1985.
Battelle Columbus Laboratories, "Land Management," paper prepared for the Office of Technology Assessment, 1985.
Battelle Columbus Laboratories, "Monitoring and Control—Animal Agriculture," paper prepared for the Office of Technology Assessment, 1985.

*Request for "Technology, Public Policy, and the Changing Structure of American Agriculture, Volume II: Background Papers" should be directed to: National Technical Information Service, U.S. Department of Commerce, 5285 Port Royal Rd., Springfield, VA 22161 (703) 487-4650.

Battelle Columbus Laboratories, "Monitoring and Control—Plant Agriculture," paper prepared for the Office of Technology Assessment, 1985.

Battelle Columbus Laboratories, "Telecommunication," paper prepared for the Office of Technology Assessment, 1985.

Berry, Joseph A., "Enhancement of Photosynthetic Efficiency," paper prepared for the Office of Technology Assessment, 1985.

Boersma, Larry, "Water and Soil-Water-Plant Relations," paper prepared for the Office of Technology Assessment, 1985.

Browning, J.A., "Plant Disease and Nematode Control," paper prepared for the Office of Technology Assessment, 1985.

Campbell, J.B., "Livestock Pest Control," paper prepared for the Office of Technology Assessment, 1985.

Curtis, Stanley E., "Environment and Animal Behavior," paper prepared for the Office of Technology Assessment, 1985.

Davis, Charles H., "Chemical Fertilizers," paper prepared for the Office of Technology Assessment, 1985.

Fischer, James R., "Crop Residue and Animal Waste Use," paper prepared for the Office of Technology Assessment, 1985.

Foster, George R., "Soil Erosion, Productivity, and Tillage," paper prepared for the Office of Technology Assessment, 1985.

Fraley, Robert T., "Genetic Engineering in Plants," paper prepared for the Office of Technology Assessment, 1985.

Francis, Charles A., "Multiple Cropping," paper prepared for the Office of Technology Assessment, 1985.

Hansel, William, "Animal Reproduction," paper prepared for the Office of Technology Assessment, 1985.

Kennedy, George G., "Management of Insects and Mites," paper prepared for the Office of Technology Assessment, 1985.

Liebhardt, William, and Harwood, Richard, "Organic Farming," paper prepared for the Office of Technology Assessment, 1985.

McWhorter, C.G., and Shaw, W.C., "Present Status and Future Needs in Weed Control," paper prepared for the Office of Technology Assessment, 1985.

Nickell, Louis G., "Plant Growth Regulators," paper prepared for the Office of Technology Assessment, 1985.

Osburn, Bennie I., "Animal Disease Control," paper prepared for the Office of Technology Assessment, 1985.

Pond, Wilson G., "Animal Nutrition," paper prepared for the Office of Technology Assessment, 1985.

Part B: Economics and Policy Analysis

Buxton, Boyd M., "Economic, Policy, and Technology Factors Affecting Herd Size and Regional Location of U.S. Milk Production," paper prepared for the Office of Technology Assessment, 1985.

Cochrane, Willard W., "Farm Commodity Policy, Technological Advance and Structural Change," paper prepared for the Policy, Technology, and Structure of Agriculture Workshop, Office of Technology Assessment, 1983.

Cooke, Stephen C., "Size Economies and Comparative Advantage in the Production of Corn, Soybeans, Wheat, Rice, and Cotton in Various Areas of the United States," paper prepared for the Office of Technology Assessment, 1985.

Eddleman, B.R., "Economic Effects of Selected Policies and Technology on the Economic Viability of General Crops Farms in the Delta Region of Mississippi," paper prepared for the Office of Technology Assessment, 1985.

Richardson, James W., "Economic Impacts of Selected Policies and Technology on the Economic Viability of Three Representative Cotton Farms in the Texas Southern High Plains," paper prepared for the Office of Technology Assessment, 1985.

Richardson, James W., "Economic Impacts of Selected Policies and Technology on the Economic Viability of Three Representative Wheat Farms in the High Plains," paper prepared for the Office of Technology Assessment, 1985.

Sundquist, W.B., "Economic Impacts of Selected Farm Policies, Income Tax Provisions, and Production Technology on the Economic Viability of Corn-Soybean Farms in East Central Illinois and Irrigated Row Crop Farms in South Central Nebraska," paper prepared for the Office of Technology Assessment, 1985.

Part C: Agricultural Finance and Credit

Barry, Peter J., "Credit Policy and Structural Issues in Agriculture," paper prepared for the Policy, Technology, and Structure of Agriculture Workshop, Office of Technology Assessment, 1983.

Barry, Peter J., "Regulatory and Performance Issues for Financial Institutions: Their Effects on Technology Adoption and Structural Change in Agriculture," paper prepared for the Office of Technology Assessment, 1985.

Brake, John R., "Short-Term Credit Policies for Dealing With Farm Financial Stress and Their Impacts on Farm Structure and Adoption of New Technology," paper prepared for the Office of Technology Assessment, 1985.

Carman, Hoy F., "Tax Policy, Technology, and the Structure of Agriculture," paper prepared for the Policy, Technology, and Structure of Agriculture Workshop, Office of Technology Assessment, 1983.

Lins, David A., "Overview of Capital and Credit Markets Serving Agriculture: Their Impact on Technology Adoption and Structural Change," paper prepared for the Office of Technology Assessment, 1985.

Trechter, David, and Meekhof, Ronald, "The Role of Federal Credit Assistance Programs in the Process of Technological Change," paper prepared for the Office of Technology Assessment, 1985.

Part D: Rural Communities

Buttel, Frederick H., and Lancelle, Mark, "Emerging Agricultural Technologies, Farm Structural Change, Public Policy, and Rural Communities in the Northeast," paper prepared for the Office of Technology Assessment, 1985.

Flora, Jan L., and Flora, Cornelia Butler, "Emerging Agricultural Technologies, Farm Size, Public Policy, and Rural Communities: The Great Plains and the West," paper prepared for the Office of Technology Assessment, 1985.

MacCannell, Dean, and Dolber-Smith, Edward, "Report on the Structure of Agriculture and Impacts of New Technologies on Rural Communities in Arizona, California, Florida, and Texas," paper prepared for the Office of Technology Assessment, 1985.

McCannell, Dean, "Agribusiness and the Small Community," paper prepared for the Policy, Technology, and Structure of Agriculture Workshop, Office of Technology Assessment, 1983.

Skees, Jerry R., and Swanson, Louis E., "Examining Policy and Emerging Technologies Affecting Farm Structure in the South and the Interaction Between Farm Structure and Well-being of Rural Communities," paper prepared for the Office of Technology Assessment, 1985.

van Es, J.C., Chicoine, David L., and Flotow, Mark A., "Agricultural Technologies, Farm Structure, and Rural Communities in the Cornbelt: Policy Choices and Implications for 2000," paper prepared for the Office of Technology Assessment, 1985.

Part E: Agricultural Research and Extension, Agricultural Structure, and Environment and Natural Resources

Hite, James, "Environmental and Natural Resource Impacts of Emerging Technologies in American Agriculture," paper prepared for the Office of Technology Assessment, 1985, based on the Environment and Natural Resources Workshop, Nov. 15-16, 1984.

Hyatt, George, "Agricultural Extension Policy: Its Evolution and Current Issues," paper prepared for the Office of Technology Assessment, 1984.

Knutson, Ronald D., "Agricultural Research and Extension Policy: Contemporary Issues," paper prepared for the Office of Technology Assessment, 1985.

Lovvorn, Roy L., "Agricultural Research Policy: Its Evolution and Current Issues" paper prepared for the Office of Technology Assessment, 1984.

Sporleder, Thomas L., "Agricultural Structure Impacts of Emerging Technologies," paper prepared for the Office of Technology Assessment, 1985, based on the Agricultural Structure Workshop, Nov. 15-16, 1984.

White, Fred C., "Economic Impact of Agricultural Research and Extension," paper prepared for the Office of Technology Assessment, 1984.

Acknowledgments

OTA would like to thank the members of the advisory panel who helped guide the direction of this study and who reviewed drafts of this report. In addition, OTA acknowledges the following individuals for reviewing background papers, reviewing individual report chapters, or providing information to the project staff:

Walt Armbruster
Farm Foundation

Thomas Army
U.S. Department of Agriculture

Lloyd Bender
Bozeman, MT

Raymond Beneke
Iowa State University

David Bezdicek
Washington State University

Larry Bierman
J.R. Simplot Co.

J.L. Bittle
Scripps Clinic and Research Foundation

Robert Bjork
U.S. Department of Agriculture

Michael Boehlje
University of Minnesota

Elise Boulding
Dartmouth College

Dave Brown
U.S. Department of Agriculture

Orrin Burnside
University of Nebraska

Ward Butcher
Washington State University

G.S. Campbell
Washington State University

Hoy Carman
University of California—Davis

John Chapman
Valmont Industries

David Chicoine
University of Illinois

Estel Cobb
U.S. Department of Agriculture

R. James Cook
Washington State University

Nicholas Cross
Cattle Code America

Dave Dame
University of Florida

Machi Dilworth
U.S. Department of Agriculture

Paul Doak
Iowa State University

Mark Drabenstott
Kansas City Federal Reserve Bank

Roger Elmore
University of Nebraska

Robert Emerson
University of Florida

A. Farsaie
University of Maryland

Norman Faultner
Deere and Co.

Jan Flora
Kansas State University

Louise Fortman
University of California-Berkeley

Lloyd Frederick
Agency for International Development

William Fry
Cornell University

Robert Giaquinter
E.I. DuPont Experimental Station

Robert Goldberg
University of California-Los Angeles

Walter Goldschmidt
University of California-Los Angeles

J. Gorham
U.S. Department of Agriculture

Albert Grable
U.S. Department of Agriculture

Mary Nell Greenwood
U.S. Department of Agriculture

W.R. Guthrie
Butler Manufacturing Co.

William Harland
Bureau of the Census

Craig Harris
Michigan State University

Steve Hastings
University of Delaware

Virgil Hays
University of Kentucky

Paul Hedin
U.S. Department of Agriculture

William Heffernan
University of Missouri

Jim Hildreth
Farm Foundation

Fred Hines
U.S. Department of Agriculture

John Hopkin
Texas A&M University

Dixon Hubbard
U.S. Department of Agriculture

Wallace Huffman
Iowa State University

George Irwin
Farm Credit Administration

Ray Jackson
U.S. Water Conservation Laboratory

Richard Jensen
University of Arizona

Richard Just
University of Maryland

Terry Kinney
U.S. Department of Agriculture

Waldemar Klassen
U.S. Department of Agriculture

Terry Klopfenstein
University of Nebraska

John Knepler
DICKEY-john Corp.

Mark Lancelle
Cornell University

B.A. Larkins
Purdue University

Warren Lee
Ohio State University

Jerry Mannering
Purdue University

Tom McClure
Battelle Columbus Laboratories

Robert Miller
North Carolina State University

Page Morgan
Texas A&M University

A.J. Muehling
University of Illinois

Ralph Nave
U.S. Department of Agriculture

Paul Nixon
U.S. Department of Agriculture

John Nye
Purdue University

P.A. Oltenacu
Cornell University

Olga Owens
U.S. Department of Agriculture

Robert Papendick
U.S. Department of Agriculture

Kenneth Parks
Agrico Chemical Co.

Neal Peterson
U.S. Department of Agriculture

Hoyle Puckett
U.S. Department of Agriculture

Graham Purchase
U.S. Department of Agriculture

Philip Raup
University of Minnesota

Steve Rawlins
U.S. Department of Agriculture

Robert Riley
U.S. Department of Agriculture

Charles Rumburg
U.S. Department of Agriculture

Robert Schafer
U.S. Department of Agriculture

Paul Siegel
Virginia Polytechnic Institute

Jerry Skees
University of Kentucky

F.W. Slife
University of Illinois

Charles Smith
U.S. Department of Agriculture

Steve Sonka
University of Illinois

Vernon Sorenson
Michigan State University

Lauren Soth
Des Moines, IA

S.L. Spahr
University of Illinois

Thomas Sporleder
Texas A&M University

Richard Storm
Bureau of the Census

B. Stout
Texas A&M University

Harlee Strauss
American Chemical Society

Howard Teague
U.S. Department of Agriculture

E.D. Threadgill
University of Georgia

Daniel Tompkins
U.S. Department of Agriculture

David Trechter
U.S. Department of Agriculture

F.T. Turpin
Purdue University

David van Doren
Ohio State University

Frederick Weingarten
Office of Technology Assessment

Fred Westbrook
U.S. Department of Agriculture

Holly Wheeler
Applied Futures, Inc.

P.J. Wierenga
New Mexico State University

Kenneth Wilkinson
Pennsylvania State University

Ronald Wimberley
North Carolina State University

W. Fred Woods
U.S. Department of Agriculture

W.R. Woods
Purdue University

Index

Index